GCSE 9–1
COMBINED
SCIENCE
AQA REVISION GUIDE

Alessio Bernadelli

Kayan Parker

Mike Wooster

Authors Alessio Bernadelli (Physics), Kayan Parker (Biology) and Mike Wooster (Chemistry)
Editorial team Haremi Ltd
Series designers emc design ltd
Typesetting York Publishing Solutions Pvt. Ltd., INDIA
Illustrations York Publishing Solutions Pvt. Ltd., INDIA and Newgen KnowledgeWorks (P) Ltd, Chennai, India
App development Hannah Barnett, Phil Crothers and Haremi Ltd

Designed using Adobe InDesign
Published by Scholastic Education, an imprint of Scholastic Ltd, Book End, Range Road, Witney,
Oxfordshire, OX29 0YD
Registered office: Westfield Road, Southam, Warwickshire CV47 0RA
www.scholastic.co.uk

Printed by Bell & Bain Ltd, Glasgow
© 2017 Scholastic Ltd
1 2 3 4 5 6 7 8 9   7 8 9 0 1 2 3 4 5 6

British Library Cataloguing-in-Publication Data
A catalogue record for this book is available from the British Library.
ISBN 978-1407-17681-9

Acknowledgements

The publishers gratefully acknowledge permission to reproduce the following copyright material:

p10 Blamb/Shutterstock; p11 top photoiconix/Shutterstock; p11 bottom Lebendkulturen.de/Shutterstock; p14 Pressmaster/Shutterstock; p15 Tefi/Shutterstock; p17 top somersault1824/Shutterstock; p17 bottom ellepigrafica/Shutterstock; p18 BlueRingMedia/Shutterstock; p19 KYTan/Shutterstock; p20 Wikimedia/Quasar Jarosz; p21 Greg Amptman/Shutterstock; p22 top Dreamy Girl/Shutterstock; p22 bottom left and right Designua/Shutterstock; p23 luchunyu/Shutterstock; p26 Jose Luis Calvo/Shutterstock; p27 Christos Georghiou/Shutterstock; p30 joshya/Shutterstock; p33 BlueRingMedia/Shutterstock; p34 Designua/Shutterstock; p35 left and middle Designua/Shutterstock; p35 right sciencepics/Shutterstock; p36 top Alila Medical Media/Shutterstock; p36 bottomtoeytoey/Shutterstock; p37 BlueRingMedia/Shutterstock; p37 ellepigrafica/Shutterstock; p39 Health Survey for England 2012, Health and Social Care Information Centre, Open Government Licence v3.0; p43 BlueRingMedia/Shutterstock; p45 Sofiaworld/Shutterstock; p48 corlaffra/Shutterstock; p52 Timonina/Shutterstock; p56 OurWorldInData.org/malaria Creative Commons Share-Alike 4.0 International licence (data source WHO); p62 kirill_makarov/Shutterstock; p68 marina_ua/Shutterstock; p58 © 2008 Steph Sooksalee, Jack Tho, Nigel Britton Creative Commons Share-Alike licence; p69 ducu59us/Shutterstock; p70 Praisaeng/Shutterstock; p72 Alila Medical Media/Shutterstock; p73 Designua/Shutterstock; p74 Syda Productions/Shutterstock; p75 reproduced with permission of themedicalbiochemistrypage, LLC; p75 Sherry Yates Young/Shutterstock; p76 Tefi/Shutterstock; p82 Sutichak/Shutterstock; p83 snapgalleria/Shutterstock; p84 Designua/Shutterstock; p86 left Pakhnyushchy/Shutterstock; p86 right Michielde Wit/Shutterstock; p97 left Rob Hainer/Shutterstock; p97 right MusiggachartSMY/Shutterstock; p103 Peristenusdigoneutis Loan (Hymenoptera: Braconidae), by William H. Day, USDA ARS, from Cornell University: https://biocontrol.entomology.cornell.edu/parasitoids/peristenus.php; p104 Patrik Dietrich/Shutterstock; p105 alinabel/Shutterstock; p106 David Lee/Shutterstock; p109 photoiconix/Shutterstock; p110 top stockshoppe/Shutterstock; p110 bottom Brendan Somerville/Shutterstock; p112 Adapted from http://alltaskstraducoes.com.br/vdisk/27/diagram-of-the-greenhouseeffect-for-kids-i16.jpg; p116 Hurst Photo/Shutterstock; p118 bszef/Shutterstock; p124 Djordje Konstantinovic/Shutterstock; p127 Daxiao Productions/Shutterstock; p130 Andraž Cerar/Shutterstock; p137 nikkytok/Shutterstock; p141 chromatos/Shutterstock; p142 108MotionBG/Shutterstock; p145 top Billion Photos/Shutterstock; p145 bottom Fablok/Shutterstock; p146 Andrey Kucheruk/Shutterstock; p153 alice-photo/Shutterstock; p155 honglouwawa/Shutterstock; p160 Smith1972/Shutterstock; p162 Natalia Evstigneeva/Shutterstock; p164 left My name is boy/Shutterstock; p164 right Kucher Sergey/Shutterstock; p174 Stocksnapper/Shutterstock; p179 joker1991/Shutterstock; p190 Alexey Stiop/Shutterstock; p200 Perry Harmon/Shutterstock; p205 Pix One/Shutterstock; p209 Aun Photographer/Shutterstock; p212 PR Image Factory/Shutterstock; p217 IPCC, from Carbon Dioxide and Global Warming Case Study, Purdue University, http://iclimate.org/ccc/Files/carbondioxide.pdf; p224 steveball/Shutterstock; p225 kunmanop/Shutterstock; p227 PhotoStock10/Shutterstock; p239 top styleuneed.de/Shutterstock; p239 bottom left DSBfoto/Shutterstock; p239 bottom right ersin ergin/Shutterstock; p241 Gearstd/Shutterstock; p242 top left Ian Duffield/Shutterstock; p242 bottom left Maxisport/Shutterstock; p244 right Nicola Bertolini/Shutterstock; p245 Aleks Kend/Shutterstock; p250 garagestock/Shutterstock; p251 first image Viktorija Reuta/Shutterstock; p251 second image johavel/Shutterstock; p251 third image Aleksandr Petrunovskyi/Shutterstock; p251 fourth image Serz_72/Shutterstock; p252 first image BigMouse/Shutterstock; p252 second image monicaodo/Shutterstock; p252 third image all_is_magic/Shutterstock; p252 fourth image KittyVector/Shutterstock; p252 fifth image VectorBand/Shutterstock; p253 anweber/Shutterstock; p258 Icon Craft Studio/Shutterstock; p259 Artur Synenko/Shutterstock; p260 your/Shutterstock; p263 photosync/Shutterstock; p266 Guschenkova/Shutterstock; p267 Edgewater Media/Shutterstock; p272 M. Unal Ozmen/Shutterstock; p275 MIGUEL GARCIA SAAVEDRA/Shutterstock; p278 ducu59us/Shutterstock; p281 Joe White/Shutterstock; p283 Panimoni/Shutterstock; p286 top wellphoto/Shutterstock; p286 bottom Andreas Prott/Shutterstock; p289 Vadim Sadovski/Shutterstock; p292 Thipawan kongkamsri/Shutterstock; p294 Igor Bukhlin/Shutterstock; p297 MASTER PHOTO 2017/Shutterstock; p288 Alexandru Nika/Shutterstock; p300 Tatiana Popova/Shutterstock; p301 Pete Niesen/Shutterstock; p301 PHOTOMDP/Shutterstock; p304 Bruce Ellis/Shutterstock; p317 Fouad A. Saad/Shutterstock.

Note from the publisher:

Please use this product in conjunction with the official specification and sample assessment materials. Ask your teacher if you are unsure where to find them.

How to use this book

This Revision Guide has been produced to help you revise for your 9–1 GCSE in AQA Combined Science. Broken down into subjects, topics and subtopics it presents the information in a manageable format. Written by subject experts to match the new specification, it revises all the content you need to know before you sit your exams.

The best way to retain information is to take an active approach to revision. Don't just read the information you need to remember – do something with it! Transforming information from one form into another and applying your knowledge through lots of practice will ensure that it really sinks in. Throughout this book you'll find lots of features that will make your revision an active, successful process.

SNAP IT!

Use the Snap It! feature in the revision app to take a picture, film a video or record audio of key concepts to help them stick. Great for revision on the go!

DO IT!

Activities that get you to turn information from one form into another so that it really embeds in your memory.

Callouts Step-by-step guidance to build understanding.

WORK IT!

Worked examples with model solutions to help you see how to answer a tricky question.

NAIL IT!

Tips written by subject experts to help you in the revision process.

MATHS SKILLS

To help you with those tricky bits of maths that you need to know and remember.

Practical Skills

Revisit the key practicals in the specification.

STRETCH IT!

Questions or concepts that stretch you further and challenge you with the most difficult content.

CHECK IT!

Check your knowledge at the end of a subtopic with the Check It! questions.

Consolidate your revision with the Review It! questions at the end of every topic.

Review It!

H
Higher Tier only content is highlighted helping you to target your revision.
H

Use the AQA Combined Science Exam Practice Book alongside the Revision Guide for a complete revision and practice solution. Packed full of exam-style questions for each subject and subtopic, along with practice papers, the Exam Practice Book will get you exam ready!

Contents

Biology

Topic 1 Biology

Topic 2 Biology

Topic 3 Biology

Topic 4 Biology

Topic 5 Biology

4

Chemistry

Topic 6 Biology
Topic 7 Biology
Topic 1 Chemistry
Topic 2 Chemistry

Contents

Physics

Topic 1 Physics

Topic 2 Physics

Topic 3 Physics

Topic 4 Physics

Topic 5 Physics

Topic 6 Physics

Topic 7 Physics

HOW TO REVISE!

PLAN YOUR REVISION

Get ahead by planning your revision!

Work out the **time** you have available for revising.

Think about when you work at your best. Are you a morning or an evening person?

Allocate **MORE TIME** for the topics you struggle with.

Revision works best in **SMALL BURSTS**, so keep sessions **SHORT AND SWEET**!

Remember to allow time to **PRACTISE** applying what you have revised.

Use your **revision app** to put together a revision timetable.

LOOK AFTER YOURSELF

Help your brain by looking after your whole body!

Take regular **breaks** from revising – your brain needs time to digest information in order to retain it.

HOTEL

Keep **hydrated** by drinking plenty of water – dehydration stops your brain from working at its full capacity.

Regular **exercise** helps stimulate the brain and will help you relax.

Get plenty of **sleep**, especially the night before an exam.

EAT WELL and limit unhealthy snacks – your brain needs fuel for memory and concentration.

Find methods of **relaxation** that work for you throughout the revision period.

BE PREPARED!

Limit potential stress on the day of an exam by getting everything you need ready the night before.

30

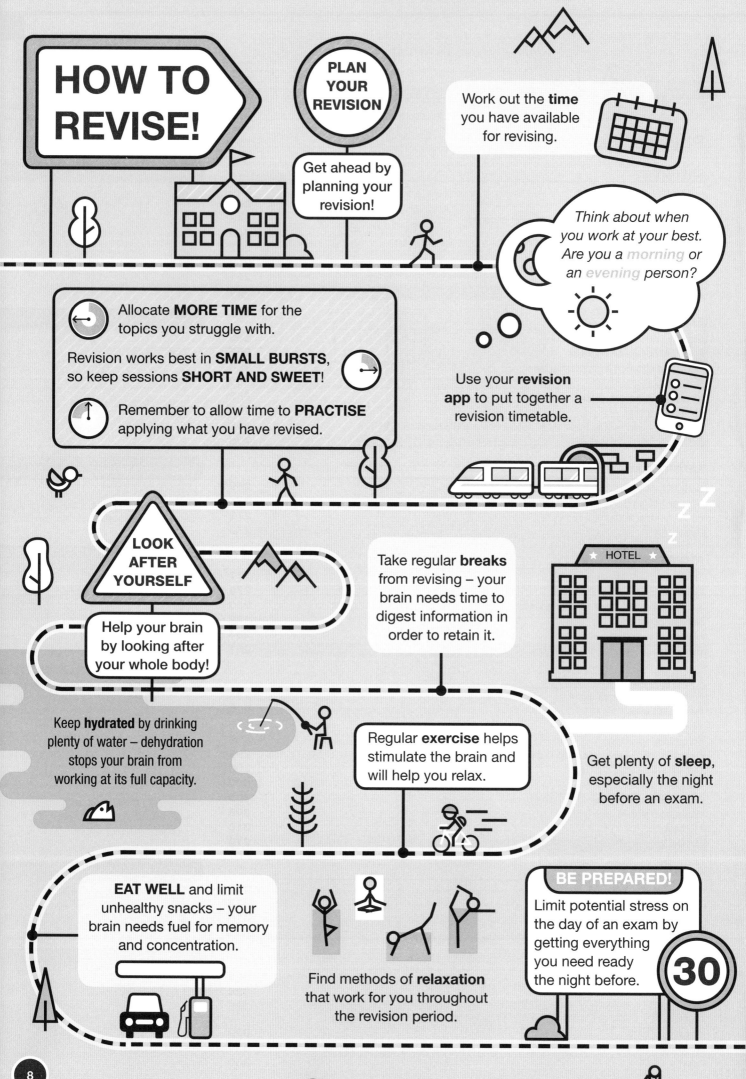

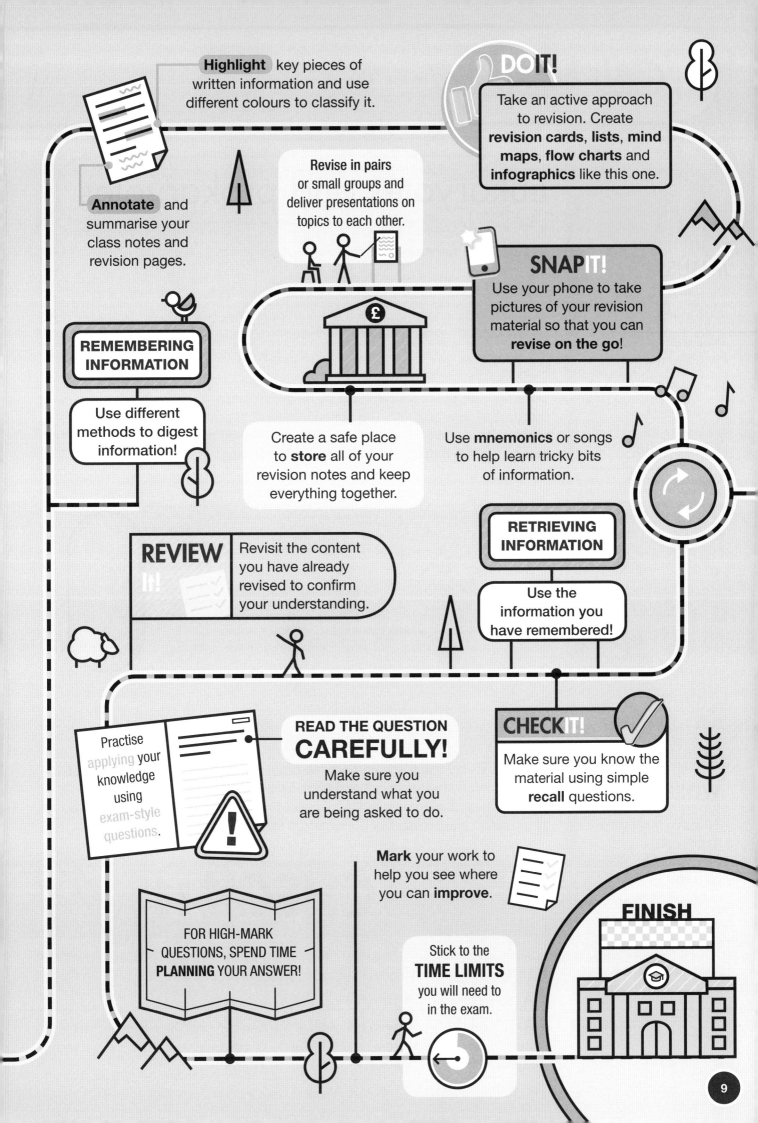

Highlight key pieces of written information and use different colours to classify it.

DO IT!
Take an active approach to revision. Create **revision cards, lists, mind maps, flow charts** and **infographics** like this one.

Annotate and summarise your class notes and revision pages.

Revise in pairs or small groups and deliver presentations on topics to each other.

SNAP IT!
Use your phone to take pictures of your revision material so that you can **revise on the go!**

REMEMBERING INFORMATION

Use different methods to digest information!

Create a safe place to **store** all of your revision notes and keep everything together.

Use **mnemonics** or songs to help learn tricky bits of information.

RETRIEVING INFORMATION

Use the information you have remembered!

REVIEW It!
Revisit the content you have already revised to confirm your understanding.

CHECK IT!
Make sure you know the material using simple **recall** questions.

Practise applying your knowledge using exam-style questions.

READ THE QUESTION CAREFULLY!
Make sure you understand what you are being asked to do.

Mark your work to help you see where you can **improve**.

FOR HIGH-MARK QUESTIONS, SPEND TIME **PLANNING** YOUR ANSWER!

Stick to the **TIME LIMITS** you will need to in the exam.

FINISH

Cell biology

Eukaryotes and prokaryotes

All cells are either eukaryotic or prokaryotic.

Eukaryotic cells

All animal cells and plant cells are eukaryotic. They have a cell membrane and cytoplasm with genetic material enclosed in a nucleus.

Prokaryotic cells

Bacterial cells are prokaryotic. They have cytoplasm and a cell membrane surrounded by a cell wall. The genetic material in prokaryotic cells is not enclosed in a nucleus. It is a single DNA loop in the cytoplasm. There may also be smaller rings of DNA called plasmids in the cytoplasm.

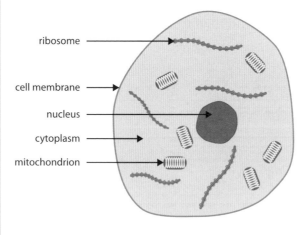

Eukaryotic cell

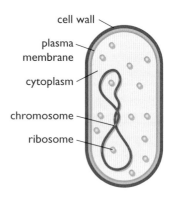

Prokaryotic cell

Scale and size of cells

All cells are very small and can only be seen with a microscope. Prokaryotic cells are much smaller (about one tenth smaller) than eukaryotic cells. Eukaryotic cells are measured in micrometres (μm). Prokaryotic cells can be measured in micrometers or nanometres (nm).

MATHS SKILLS

1 centimetre (cm)
= 10 mm

1 millimetre (mm)
= 1000 μm

1 μm = 1000 nm

In standard form,
1×10^4 is the same as writing 10 000.

WORKIT!

An *E.coli* bacterium measures 2×10^3 nm in diameter. How many μm is this?

2×10^3 nm is the same as 2000 nm.

There are 1000 nm in 1 μm

2000 nm / 1000 = 2 μm

CHECKIT! ✓

1 How is the genetic material stored in a prokaryotic cell?

2 A cell measures 5×10^3 nm. How many μm is this?

3 A bacterial cell is one-tenth the width of a eukaryotic cell. The eukaryotic cell is 2 μm wide. Calculate the width of the bacterial cell. Give your answer in nm using standard form.

Animal and plant cells

Animal and plant cells are both eukaryotic, but have some differences in their sub-cellular structure.

Animal cells

Animal cells have a nucleus, mitochondria, ribosomes, cytoplasm and a cell membrane, as well as some other sub-cellular structures.

Plant cells

Plant cells have the same sub-cellular structures as animal cells, but with the addition of chloroplasts, a permanent vacuole filled with cell sap, and a cell wall made out of cellulose. Algae also have a cellulose cell wall.

Function of sub-cellular structures

Sub-cellular structure	Function
Nucleus	Contains the genetic material
Mitochondria	Provides energy by carrying out respiration
Ribosomes	Carries out protein synthesis
Cytoplasm	Where most of the chemical reactions happen
Cell membrane	Controls the movement of substances in and out of the cell
Chloroplast	Absorbs light for photosynthesis
Permanent vacuole	Filled with cell sap to help keep the cell turgid
Cellulose cell wall	Gives strength to the cell and supports the plant
Plasmids	Additional genetic material

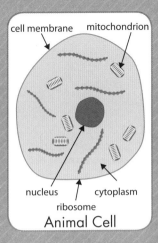

cell membrane   mitochondrion

nucleus   cytoplasm
ribosome
Animal Cell

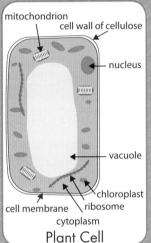

mitochondrion
cell wall of cellulose

nucleus

vacuole

chloroplast
cell membrane   ribosome
cytoplasm
Plant Cell

STRETCH IT!

Find out more about the functions of the sub-cellular structures. What happens inside the nucleus? How do the mitochondria carry out respiration?

CHECK IT!

1 Make a table to show which sub-cellular structures are found in animal cells, plant cells and prokaryotic cells.

2 Explain why active cells, such as muscle cells, contain more mitochondria than less active cells.

3 This one-celled organism lives in a pond and can carry out photosynthesis. Use your knowledge of cells to identify whether this organism is a plant. Justify your answer.

DO IT!

Look at pictures of cells in your textbook or online, and practise drawing them. Identify as many sub-cellular structures as you can.

11

Microscopy

NAILIT!

Magnification means how much larger the image is than the specimen.

Resolution (or **resolving power**) means how easily two points on the specimen can be distinguished from one another. The higher the resolution, the sharper the image will be.

Most cells are too small to be seen with the naked eye, so to get an understanding of what is happening inside cells we need microscopes. Microscopes have developed over the years to give higher magnifications and greater resolution.

Light microscopy

Light microscopes use light in order to view specimens. These have a low resolving power and a useful magnification of only about 1000 times (×1000), which means that the details within sub-cellular structures cannot be easily seen.

Electron microscopy

Electron microscopes use electrons to see the surface of cells, or inside the cells. These have a much higher resolving power than a light microscope with a magnification of one million times (×1 000 000). The development of the electron microscope has increased our understanding of sub-cellular structures as the sub-cellular structures within cells can now be seen in detail.

MATHS SKILLS

The magnification of the cell can be worked out using the formula:

$$\text{magnification} = \frac{\text{size of image}}{\text{size of real object}}$$

This can also be shown as a magnification triangle:

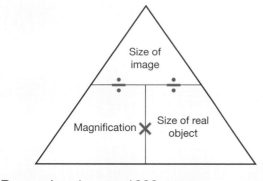

Remember 1 mm = 1000 µm

WORKIT!

A cell that is 17 micrometers (µm) in diameter appears to be 3.4 cm in diameter when viewed through a microscope. Calculate the magnification. (3 marks)

First write the formula:

$$\text{Magnification} = \frac{\text{size of image}}{\text{size of real object}} = \frac{3.4\,\text{cm}}{17\,\mu m} \quad (1)$$

Then make sure that both measurements are in the same units. In this case, it will be easier to put them both into µm.

$$\text{Magnification} = \frac{34\,000\,\mu m}{17\,\mu m} \quad (1)$$

Then do the division:

$$\text{Magnification} = \times 2\,000 \quad (1)$$

NAILIT!

You should be able to write your answer in standard form.

Standard form is a way of writing very large numbers. For example:

$15\,000\,000 = 1.5 \times 10^7$

CHECKIT!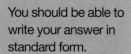

1 Give two advantages of using an electron microscope to view cells.

2 Calculate the magnification of a cell that is 12 µm wide and appears 3 cm wide under the microscope.

3 A cell 4 µm wide was magnified 12 000 times. Rearrange the magnification formula to work out the size of the image. Write your answer in µm using standard form.

Using a light microscope

For this particular practical you need to produce labelled scientific drawings using a light microscope. A magnification scale must be used. Make sure you know how to calculate the magnification of a specimen observed through a light microscope.

Observing plant and animal cells

How to set up a light microscope

1 Place the specimen on the stage.

2 Switch on the microscope so that the light passes through the specimen.

3 Make sure that the ×4 objective lens is clicked into place above the specimen.

4 Bring the specimen into focus by looking down the eyepiece lens and moving the coarse focus.

5 When the specimen is in focus, move the objective lenses so that the ×10 objective lens is clicked into place above the specimen.

6 If the specimen is out of focus, bring it into focus using small movements of the fine focus.

7 Repeat steps 5 and 6 with the ×40 objective lens.

8 You should now be able to observe your specimen.

Practical Skills

The magnification of the specimen will be the eyepiece lens multiplied by the objective lens.

In this case, it will be ×4 multiplied by ×10 = ×40 magnification.

Practical Skills

- Always use a sharp pencil.
- Draw a smooth line (no shading or sketching).
- Make drawings as big as the space allows.
- Label lines should be drawn with a ruler.
- Include a magnification scale.

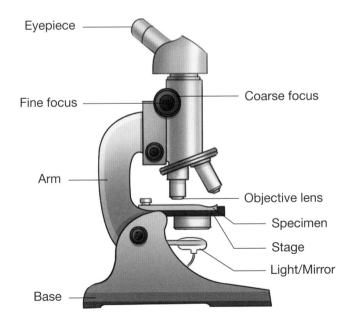

- Eyepiece
- Fine focus
- Coarse focus
- Arm
- Objective lens
- Specimen
- Stage
- Light/Mirror
- Base

A light microscope

MATHS SKILLS

Learn the magnification triangle below by heart.

You may be asked to rearrange the formula in the exam.

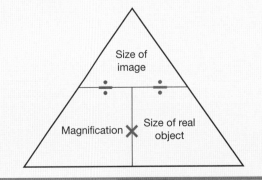

Size of image

Magnification ✕ Size of real object

Drawing and labelling plant and animal cells

You will need to be able to draw and label the plant and animal cells that you see.

Practise drawing and labelling plant and animal cells by drawing photos of cells found in your textbook or online. Try to identify the nucleus, cytoplasm, cell membrane and any other organelles you can see.

WORKIT!

Assume that this animal cell is 1 mm in width as seen through the microscope, at a magnification of ×100. Work out the actual size of the cell.

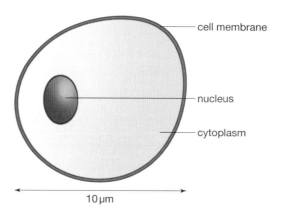

- cell membrane
- nucleus
- cytoplasm

10 μm

1mm divided by 100 = 0.01mm or 10μm. The width of the cell is 10μm.

NAILIT!

Microscope slides are often **stained** so that you can see the cells or tissues more clearly. Different stains can be used to identify sections of tissue, or specific organelles within a cell.

CHECKIT!

1 What is the magnification if the eyepiece lens is ×10 and the objective lens is ×40?

2 A cell is observed using a microscope at a magnification of ×400. The cell image is 2 mm in diameter. Calculate the actual size of the cell.

3 Outline how a student could observe some blood cells using a light microscope.

Cell specialisation

Most cells are specialised in order to carry out a particular function. The cell's structure and composition are modified so that it can carry out a particular role.

Cell organisation

Cells are the basic building blocks of all organisms. Groups of similar cells come together to form a tissue, such as muscle tissue. In tissue, all the cells work together to carry out a particular function. Different tissues can come together to form organs, such as the heart. Organs are organised into organ systems, such as the circulatory system. All of the organ systems make up the whole organism.

SNAPIT!

Create your own version of this table, or take a photo and learn the features of each specialised cell on the go.

Specialised cells

Specialised cell	Function	Specialised cell	Function
Sperm cells	Swim to the ovum (egg) for fertilisation. Have a tail for swimming. Packed full of mitochondria to provide energy. Sperm head (acrosome) contains enzymes to help break into the ovum.	Root hair cells	Take up water and mineral ions for the plant. Long thin hair to increase the surface area over which water can be taken up.
Nerve cells	Carry nerve impulses to and from the brain. Long thin axon allows nerve impulses to travel along the cell. Has many dendrites to pass nerve impulses to nearby nerve cells.	Xylem cells	Transport water from the roots to the leaves, as part of a tissue. Cells have no ends and are hollow to make a tube for water to move through. Lignin in the cell wall to waterproof the cells.
Muscle cells	Contract and relax as part of a muscle tissue, for movement. Packed full of mitochondria to provide energy.	Phloem cells	Transport sugars around the plant, as part of a tissue. Small holes in the end plates allow sugars to move through the cells.

CHECKIT!

NAILIT!

Q3 and 4 are asking you to **explain**, so give a reason for each feature, in as much detail as you can.

1 What is a specialised cell?

2 Describe how a nerve cell is specialised for carrying nerve impulses.

3 Using your knowledge of cell organisation, explain whether sperm cells can be regarded as a tissue.

4 Xylem is a specialised cell that transports water up the plant. Explain how the structure of the cell helps it to carry out this function.

Cell differentiation

In a developing embryo, none of the cells are specialised. The cells have the potential to become any type of cell. As the embryo develops, the cells begin to differentiate. They begin to become specialised to carry out a particular function.

The importance of cell differentiation

In a single-celled organism, such as a bacterium, the cell has to carry out all of the functions to keep the cell alive. In multicellular organisms, such as animals and plants, each type of cell becomes specialised to carry out one function. That way, the work is divided up by the cells. As the cells differentiate, they change their shape and acquire different sub-cellular structures.

How cells differentiate

In an embryo, the cells are not yet differentiated. These cells are called stem cells. As the cells divide, each cell is exposed to a different chemical or hormone, which makes it start to differentiate. The type of cell that the stem cell becomes depends on the hormone that it is exposed to. When a baby is born, most of the cells that make up its body are differentiated.

Stem cells

In a mature animal, cells divide only in order to replace or repair damaged tissue, and cannot differentiate into new cell types. In plants, many of the cells are able to differentiate into new cell types. This means that we are able to take cuttings from a mature plant to make many new plants.

STRETCHIT!

Find out about stem cells and how they are being used in medicine to grow new organs.

NAILIT!

This question is asking you to **explain**, which means that you should use your knowledge of the subject area to give a detailed account.

NAILIT!

If a question asks you to **outline**, give brief steps on how to do something.

WORKIT!

Explain why differentiation is important in a multicellular organism. (3 marks)

Each type of cell can be specialised to carry out one particular function. (1)

Cells can work together to carry out functions. (1)

This division of labour is more efficient for the organism. (1)

CHECKIT!

1 Name a cell that can become any type of specialised cell.

2 Where could you find cells that can differentiate into other cell types?

3 Outline how a cell becomes differentiated.

Mitosis and the cell cycle

Most of the cells in the body go through the cell cycle. The final stage of the cell cycle is mitosis, a type of cell division.

Chromosomes

The nucleus of a cell contains chromosomes made out of DNA. Each chromosome contains many genes. In the body cells, the chromosomes are found in pairs. In human body cells, there are 46 chromosomes.

The cell cycle

The cell cycle is tightly controlled and is divided into four stages:

G1 phase – First gap phase in which the organelles of the cell (except the chromosomes) are doubled.

S phase – During the synthesis phase, the DNA replicates to form two copies of each chromosome.

G2 phase – The second gap phase. Here, the chromosomes are checked for errors, so that they are not passed onto the daughter cells.

M phase – In the mitosis phase, the cell divides into two identical daughter cells.

Changes, such as mutations, in the cell can cause uncontrolled cell division. This can lead to cancer (see page 42).

Mitosis

During mitosis, the doubled chromosomes line up in the centre of the cell. One set of chromosomes is pulled to each end of the cell, and the nucleus divides to form two new nuclei. The cytoplasm and cell membranes then divide to form two identical cells.

Why does mitosis occur?

Mitosis is the process for:

* growth of multicellular organisms
* repair of damaged tissues
* replacement of cells
* asexual reproduction (single-celled organisms and some plants).

WORKIT!

Why is it important that the chromosomes are checked during the G2 phase? (2 marks)

The chromosomes have to be checked for errors. (1)

If the errors are passed onto the daughter cells, it could lead to mutations. (1)

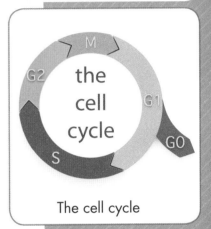

The cell cycle

① Interphase   ② Prophase   ③ Prometaphase   ④ Metaphase

⑤ Anaphase   ⑥ Telophase

Mitosis completed (two new daughter cells)

Mitosis

NAILIT!

You do not need to know the name of each stage of mitosis, but you do need to know what is happening to the chromosomes and the nucleus.

CHECK IT!

1 Give three reasons for mitosis in a plant.

2 Describe what happens during mitosis.

3 Calculate how many cells there would be if a single cell divided by mitosis 24 times.

Stem cells

Stem cells are undifferentiated cells that can become any type of specialised cell.

Stem cells in plants and animals

Stem cells are found in developing embryos. As the cells divide, they are exposed to different chemicals and hormones and begin to differentiate into specialised cells (see page 15). Some stem cells can be found in adults, but these have a limited number of specialised cells that they can become. Plants contain stem cells called meristem tissue in the rapidly dividing root and shoot tips.

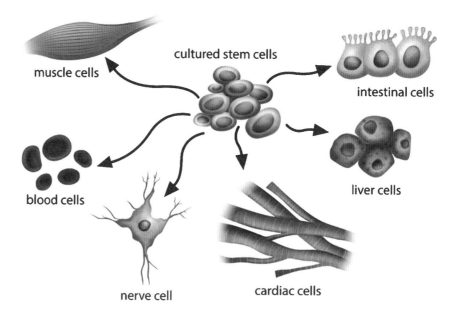

muscle cells

cultured stem cells

intestinal cells

blood cells

nerve cell

cardiac cells

liver cells

Functions of stem cells

- Stem cells in embryos become all the specialised cells in the body.

- Adult stem cells are found in different organs around the body. They replace cells in that particular organ. For example, stem cells in the skin can differentiate into different types of skin cell.

- Stem cells from adult bone marrow can form many types of cell including blood cells.

- The plant root tips and shoot tips contain meristem tissue that can become specialised cells.

Therapeutic cloning

Stem cells can be used to grow organs for transplant. An embryo is produced that has the same genes as the patient. Stem cells from this embryo can be used to grow an organ that will not be rejected from the patient's body. Stem cells can also be injected for other medical treatments. For example, treatment with stem cells may be able to help conditions such as diabetes and paralysis.

DOIT!

Make revision cards with the functions of stem cells/meristem tissue on one side and how this links to one of their uses on the other side.

Advantages	Disadvantages
No rejection of cells/organs by patient	Transfer of viral infection
No waiting time for transplants	Ethical/religious objections

Uses of stem cells in plants

- Cuttings taken from the meristem tissue in plants can be used to produce clones of plants quickly and economically.

- Meristem tissue is found in plants where new growth takes place, for example, root and shoot tips and flowering parts.

- Rare species of plants can be cloned to protect them from extinction.

- Crop plants with special features, such as disease resistance, can be cloned to produce large numbers of identical plants for farmers.

WORKIT!

Evaluate the practical risks and benefits, as well as social and ethical issues, of the use of stem cells in medical research and treatments. (4 marks)

Stem cells from embryos can be used to research many diseases that affect humans. (1) However, there is an ethical objection to using embryos as they could potentially grow into humans/animals. (1) Using stem cells in medical treatments means that the body will not reject the cells, (1) but there is a risk of transfer of viral infection from putting the stem cells into the body. (1)

STRETCHIT!

Find out how cuttings from plants can be used to produce many cloned plants.

NAILIT!

This question is asking you to **evaluate**, which means that you should make a judgement about the use of stem cells using the available evidence.

✓ CHECKIT!

1. Where is meristem tissue found in a plant?

2. Give two uses of stem cells.

3. Biologists are trying to conserve a rare species of orchid. Describe how they could use the properties of meristem tissue to do this.

Diffusion

Substances may move in and out of cells by diffusion.

What is diffusion?

Diffusion is the spreading out of the particles of any substance in solution, or particles of a gas, resulting in a net movement from an area of higher concentration to an area of lower concentration. In cells, diffusion happens across the cell membranes, to allow substances such as oxygen and carbon dioxide to get in and out of the cell. Diffusion also allows the waste product, urea, to get out of the cell. Diffusion is a passive process because it does not require any energy.

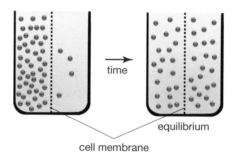

time

equilibrium

cell membrane

The diagram above shows the movement of particles across the cell membrane over time. The particles move from the more concentrated side to the less concentrated side, until both sides have the same number of particles (equilibrium).

Factors that affect the rate of diffusion

The rate of diffusion will increase if:

- The difference in the concentrations (the concentration gradient) increases.

- The temperature increases. Increasing the temperature increases the kinetic energy of the particles, which makes them move faster.

- The surface area of the membrane increases. Areas of the body with a large surface area, such the alveoli of the lungs, have a faster rate of diffusion.

SNAPIT!

Draw out the diagrams on this page, page 22 and page 25. Compare them to learn the differences between diffusion, osmosis and active transport. Take a photo to help with your revision.

MATHS SKILLS

Surface area to volume ratio is best understood if you imagine the organisms to be shaped into cubes.

WORKIT!

What is the surface area to volume ratio of a cube that has sides that are 2 cm in length? (3 marks)

The surface area of each side is
2 cm x 2 cm = 4 cm².

However, the cube has six sides, so the surface area is 4 cm² × 6 = 24 cm². (1)

The volume is 2 cm x 2 cm x 2 cm = 8 cm³. (1)

The surface area to volume ratio is 24:8.

This can be simplified to 3:1. (1)

Here are some other examples of working out the surface area to volume ratio:

Cube side (cm)	Surface area (cm²)	Volume (cm³)	Surface area to volume ratio
1	6	1	6:1
3	54	27	2:1

The need for exchange surfaces and transport systems

The smaller the cube (organism), the larger the surface area to volume ratio. For example, single-cell organisms, such as bacteria, have a relatively large surface area, compared to their volume. This allows sufficient transport of particles into and out of the cell to meet their needs.

Large, multicellular animals have much smaller surface area to volume ratios. This means that they cannot get all of the substances they need by diffusion alone. They need specialised exchange surfaces in order to get all of the substances that they need.

The effectiveness of the specialised exchange surface is increased when:

- There is a large surface area.
- The membrane is thin, to provide a short diffusion pathway.
- There is a good blood supply (in animals) to remove the diffused particles, and help to maintain a high concentration gradient.
- It is ventilated (in animals for gas exchange).

Specialised exchange surfaces

Mammalian lungs – many small alveoli give the lungs a large surface area for the diffusion of oxygen and carbon dioxide. A good blood supply helps to maintain a large concentration gradient.

Mammalian small intestines – villi and microvilli give the inner surface of the small intestine a large surface area for the diffusion of the products of digestion. A good blood supply helps to maintain a large concentration gradient.

Fish gills – have a large surface area for the diffusion of oxygen and carbon dioxide.

Plant roots – root hair cells have long hairs that increase the surface area for the movement of water (by osmosis – see page 22).

Plant leaves – have large air spaces inside the leaf and many holes (stomata) on the underside of the leaf, for the diffusion of carbon dioxide and oxygen.

You can see the gill slits clearly on this shark.

NAILIT!

Visking tubing is a selectively permeable membrane that can be used as a model cell.

Solutions inside and outside the Visking tubing can be tested at different times to observe diffusion across the membrane.

STRETCHIT!

Look at a diagram of villi and microvilli to understand how the surface area of mammalian intestines is increased.

CHECKIT!

1 What is diffusion?

2 Name two ways in which the rate of diffusion can be increased.

3 A cube has sides 4 cm in length. What is the surface area to volume ratio?

Osmosis

Water may move across cell membranes by osmosis.

What is osmosis?

Osmosis is the movement of water from a region of high water (low solute) concentration to a region of low water (high solute) concentration through a selectively permeable membrane.

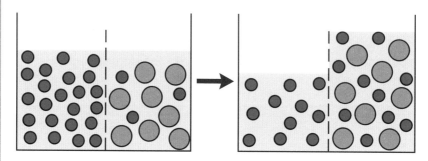

If cells are in a dilute solution, water moves into the cells. This causes plant cells to swell and become turgid, and animal cells to swell and eventually burst. If cells are in a concentrated solution, water moves out of the cell. This causes plant cells to become flaccid and plasmolysed, and animal cells to shrivel and crenate.

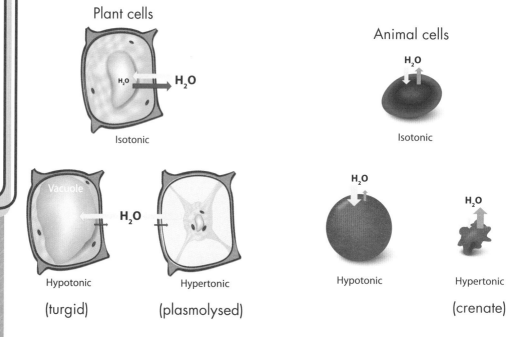

Plant cells

Isotonic

Vacuole

H_2O

Hypotonic
(turgid)

Hypertonic
(plasmolysed)

Animal cells

H_2O

Isotonic

H_2O

Hypotonic

H_2O

Hypertonic
(crenate)

Measuring the rate of water uptake

The amount of water taken up by osmosis can be measured using a Visking osmometer. A piece of Visking tubing containing a concentrated solution, is placed into a beaker of water. A glass tube is placed into the top of the Visking tubing. As the water from the concentrated solution moves into the Visking tubing, the liquid level in the glass tubing rises. This allows the volume of water that moved into the Visking tubing to be measured.

SNAP IT!

Take a photo of the tonicity diagrams on this page. Make sure you can describe the three types of tonicity: hypertonic, hypotonic and isotonic.

NAIL IT!

You may be asked to observe osmosis in plant cells using a light microscope.

In a hypertonic solution, the plasma membranes will come away from the cell walls (plasmolysed).

Measuring the change of mass in plants

It is possible to measure the rate of osmosis by placing plant tissue in solutions of different concentrations, and measuring the mass before and after using a weighing balance. If the plant tissue is placed into a dilute solution, water moves into the tissue, so the mass increases. If the plant tissue is placed into a concentrated solution, water moves out of the tissue, and the mass decreases.

WORKIT!

Calculate the percentage change in mass of plant tissue placed into 0%, 5% and 10% salt solutions. (3 marks)

Concentration (%)	Mass before (g)	Mass after 4 hours (g)	% change in mass
0	10	15	+50
5	10	12	+20
10	10	8	-20

$$\text{Percentage change} = \frac{15-10}{10} \times 100\% \quad (1)$$
$$= \frac{5}{10} \times 100\%$$
$$= 0.5 \times 100\% \quad (1)$$
$$= 50\% \quad (1)$$

If it is a percentage increase, add a plus sign, if it is a percentage decrease, add a minus sign.

To work out the percentage change, subtract the mass before from the mass after and divide by the mass before. Then multiply by 100%.

NAILIT!

When plotting a graph, make sure the independent variable (in this case, the salt solution concentration) is on the *x*-axis, and the dependent variable (in this case, the percentage change in mass) is on the *y*-axis.

CHECKIT!

1 What will happen to an animal cell if it is placed into a concentrated solution?

2 A disc of plant tissue with a mass of 8 g is placed into a dilute solution. After four hours, the mass of the disc has increased to 12 g. Calculate the percentage increase in mass.

3 Using the data from the table in the worked example, draw a graph to show the relationship between percentage change in mass and concentration.

NAIL IT!

You may also observe the action of fresh yeast being mixed with sugar (creaming yeast).

The two solids become a liquid by the osmosis of water out of the yeast cells.

NAIL IT!

If the mass has decreased, then water has moved out of the tissue.

MATHS SKILLS

You will be asked to work out the percentage change in mass for each piece of plant tissue. You should draw a graph of the percentage change in mass.

Investigating the effect of a range of concentrations of salt or sugar solutions on the mass of plant tissue

In this practical you will use your knowledge of osmosis (see page 22) to create hypotheses about plant tissue and plan experiments to test these hypotheses. You will need to recall how to work out the percentage change in mass.

Practical Skills

To investigate the effect of different concentrations of sugar/salt solutions on the mass of plant tissue (usually potato). You should:

- Measure the mass in grams before and after the tissues had been in the solutions.
- Calculate the percentage change in mass.

WORKIT!

A cube of potato had a mass of 5.0g, which decreased to 3.8g after being in sugar solution for 4 hours. What conclusions can you make about the sugar solution? (3 marks)

Water has moved out of the potato by osmosis (1) from a dilute solution to a concentrated solution. (1) The sugar solution is a more concentrated solution than the contents of the potato cells. (1)

CHECK IT! ✓

1 The graph shows the percentage change in mass of potato cubes placed into sugar solutions of 0, 1, 2, 3, and 4% concentration for 4 hours. Each potato cube was initially 5.0g in mass.

 a Calculate the percentage change in mass at 2% concentration.

 b Calculate the mass of the potato cube at 2% concentration after it had been in the sugar solution for four hours.

 c Give two variables that must be kept constant during this investigation.

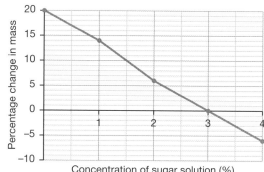

Active transport

Active transport is the movement of particles against the concentration gradient.

Active transport needs energy

Sometimes, cells need to take in, or release substances against the concentration gradient. This requires energy from respiration. The substances are transported from a more dilute solution to a more concentrated solution across the cell membrane using carrier proteins.

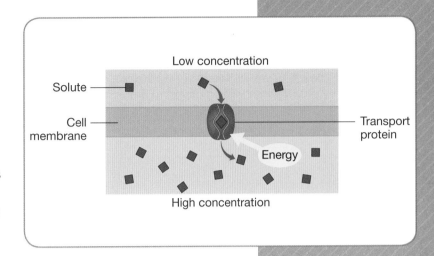

Low concentration

Solute

Cell membrane

Transport protein

Energy

High concentration

Active transport in plants

Root hair cells in plants need to take up mineral ions from the soil for healthy growth. There is already a high concentration of mineral ions inside the cell, so mineral ions need to move from a dilute solution in the soil to a concentrated solution inside the cell. The root hair cell uses energy to transport the mineral ions through protein carriers in the cell membrane.

Active transport in humans

In our bodies, sugar molecules need to be transported from relatively low concentrations in the small intestine to higher concentrations in the blood. Sugar is needed for respiration, so sugar molecules are actively transported across the wall of the small intestine into the blood.

WORKIT!

Explain the differences between diffusion, osmosis and active transport. (3 marks)

Many substances can move by diffusion or active transport, but only water can move by osmosis. (1)

Diffusion is the movement of particles from a concentrated solution to a dilute concentration, but active transport and osmosis involve the movement of particles or water from a dilute concentration to a concentrated solution. (1)

Diffusion and osmosis do not require energy, but active transport does require energy. (1)

NAILIT!

This question is asking you to **explain**, so you should compare the properties of each type of transport.

CHECKIT!

1 What is meant by the term concentration gradient?

2 Where does the energy for active transport come from?

3 Describe how mineral ions are taken into the plant root hair cell.

For additional questions, visit:
www.scholastic.co.uk/gcse

1 A plant cell is a eukaryotic cell. List three ways in which a prokaryotic cell is different to a eukaryotic cell.

2 a Describe how to use a light microscope to observe a specimen.

   b A magnified image of a cell is 30 000 µm in diameter and the actual diameter of the cell is 10µm. What is the magnification?

3 a Describe how a root hair cell is specialised and explain how these adaptations help the cell to carryout its function.

   b How does the root hair cell become specialised?

4 Some students were investigating the effect of placing cucumber slices into sugar solutions of different concentrations.

   a Write out a suitable method for how you could investigate this.

   b What are the independent and dependent variables in this investigation?

The results are shown below:

Concentration of sugar solution (%)	Mass of cucumber before (g)	Mass of cucumber after (g)	Percentage change in mass (%)
0.0	3.0	4.2	
0.5	3.1	3.9	
1.0	2.9	3.3	
2.0	3.2	2.8	

   c Using the results above, calculate the percentage change in mass for each cucumber slice.

   d Draw a graph of the percentage change in mass at each sugar concentration.

   e Use evidence from the graph to work out the percentage change in mass at 1.5% sugar concentration.

   f How could the students have made sure that their investigation was valid and reliable?

5 a Put the stages of mitosis in the correct order.
      Prophase   Anaphase
      Interphase   Telophase
      Prometaphase   Metaphase

The photograph below shows onion root tip cells in the process of mitosis.

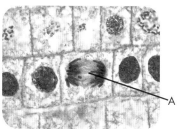

   b What stage of mitosis is cell A in?

   c Explain why onion root tip cells are undergoing mitosis.

6 Stem cells are being used in research to make organs for transplant.

   a What are stem cells?

   b Describe where stem cells can be found.

   c Discuss the advantages and disadvantages of using stem cells to make organs for transplant.

7 a Define diffusion.

   b The diagram below shows the concentration of salt ions on either side of a partially permeable membrane. In which direction will the salt ions move?

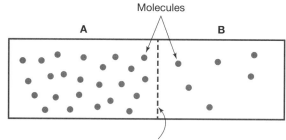

Molecules

A   B

Selectively Permeable Membrane

   c Explain your answer to (b).

8 a What type of diffusion requires protein channels in order to cross the cell membrane?

   b Compare diffusion and active transport.

Tissues, organs and organ systems

The human digestive system

The human digestive system breaks down all of the food that you eat; mechanically, by chewing, and chemically, using enzymes.

Which organs are parts of the human digestive system?

The human digestive system is made up of the stomach, small intestine, large intestine, pancreas, liver and gall bladder.

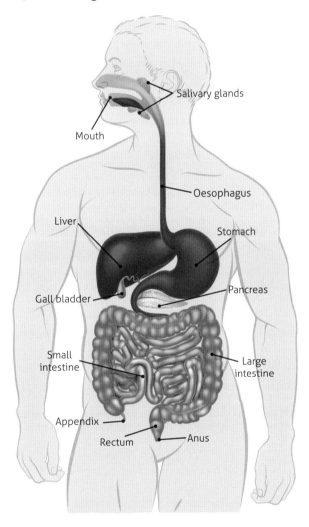

SNAPIT!

Find a picture of the digestive system, take a photo and learn it. You may be asked to label a similar diagram in the exam.

Chewed food mixes with saliva from the salivary glands, and then travels down the oesophagus to the stomach. Large lumps of food are broken down into smaller lumps in the stomach, before moving into the small intestine. Bile from the liver, and pancreatic juices from the pancreas are added to the food in the small intestine. These help with digestion.

Food needs to be broken down into small, soluble molecules so it can be absorbed into the bloodstream. Any food that is not digested (such as fibre) moves into the large intestine. Here, water is removed and a solid mass known as faeces passes out of the body through the anus.

Enzymes used in digestion

Enzymes	Site of production	Substrate	Products	Uses
Carbohydrase (Amylase)	Salivary glands (carried in saliva) Pancreas (carried in pancreatic juices)	Carbohydrate (starch)	Sugars	Respiration Builds new carbohydrates
Proteases	Stomach Pancreas (carried in pancreatic juices)	Protein	Amino acids	Builds new proteins
Lipases	Pancreas (carried in pancreatic juices)	Lipids (fats)	Fatty acids Glycerol	Builds new lipids

Bile is not an enzyme, but it is made by the liver and secreted by the gall bladder, into the small intestine, to help with lipid digestion. Bile does not digest lipids, but breaks them into smaller droplets (emulsifies) so that the lipase enzymes have a greater surface area to work on. Bile is alkaline, so it also neutralises hydrochloric acid from the stomach. This is important because the enzymes in the small intestine work best at a slightly alkaline pH.

Products of digestion

The products of digestion can be written as word equations:

$$\text{starch} \xrightarrow{\text{amylase}} \text{sugars}$$

$$\text{protein} \xrightarrow{\text{protease}} \text{amino acids}$$

$$\text{fat} \xrightarrow{\text{lipase}} \text{fatty acids + glycerol}$$

DOIT!

Learn these word equations by heart. You may be asked to write them out in an exam.

NAILIT!

The amount of energy in food is measured in kilojoules (kJ) using a calorimeter.

WORKIT!

Explain how new proteins are made in the body. (3 marks)

Proteases are made in the stomach/pancreas. (1)

Proteases break down protein from foods into amino acids. (1)

Amino acids are combined to make new proteins in the body. (1)

CHECKIT!

1 Where in the body are lipases made?

2 What is the role of amylase?

3 Explain why bile is needed in the digestion of lipids.

Enzymes

Enzymes increase the rate of chemical reactions in living organisms.

What is an enzyme?

An enzyme is a protein that acts as a biological catalyst. Enzymes allow chemical reactions to take place faster at lower temperatures than non-biological catalysts. All of the metabolic reactions in the body (such as breaking down or synthesising molecules) require enzymes.

The effect of temperature and pH on enzymes

Enzymes have specific (optimum) temperatures and pH at which they work best. Enzymes in the human body have an optimum temperature of 37°C. The optimum pH of the enzyme depends on where it acts.

- Amylase in saliva – pH7
- Pepsin in the stomach – pH2
- Lipase in the small intestine – pH8

The effects of temperature and pH on enzymes are shown in the table:

	Below optimum	Above optimum
Temperature	Less kinetic energy, so enzymes and their substrate move more slowly and fewer collisions occur. The rate of reaction decreases.	High temperatures denature the enzyme by breaking hydrogen bonds in its structure. The enzyme's active site changes shape and is no longer able to function.
pH	Loss of activity for the enzyme. The enzyme's active site changes shape and the rate of reaction decreases.	Loss of activity for the enzyme. The enzyme's active site changes shape and the rate of reaction decreases.

MATHS SKILLS

Carrying out rate calculations

The rate of reaction is the speed at which a reaction is taking place. To work out the rate of reaction, divide the amount of reactant used, or product made, by time.

WORKIT!

Hydrogen peroxide is converted into oxygen and water by the enzyme catalase. The speed of the reaction can be worked out by measuring the amount of oxygen collected in 5 minutes. Calculate the rate of reaction if 15 cm³ of oxygen is collected in 5 minutes. (2 marks)

$$\text{Rate of reaction} = \frac{\text{Amount of product made}}{\text{Time}} = \frac{15\,cm^3}{5\,min}\,(1) = 3\,cm^3/min\,(1)$$

You can also work out the rate of reaction using the gradient of a graph.

WORKIT!

This graph shows the amount of oxygen being released by catalase against time. Work out the initial rate of reaction. (2 marks)

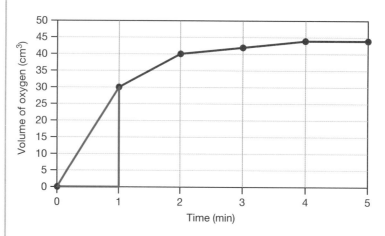

Draw a line down towards the x-axis, and a line across away from the y-axis.

Read off the vertical line and the horizontal line and calculate the gradient.

Here, the vertical line is 30 cm³, and the horizontal line is 1 min.

Draw a tangent to the curve (here shown in red) on the first (initial) part of the line. (1)

Initial rate of reaction is $\frac{30}{1}$ = 30 cm³/min (1)

DOIT!

Model the 'lock and key theory' using modelling clay. Use one colour to represent the enzyme, and one colour to represent the substrate.

The lock and key theory

Enzymes have an active site which is a specific shape. This shape is complementary to the substrate that the enzyme works on. This means that the substrate fits into the active site on the enzyme, like a key fits into a lock. Only the correct substrate can fit into the active site.

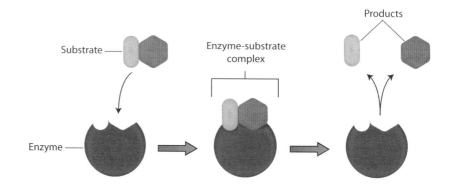

CHECKIT!

1 What is the effect of placing amylase enzyme into a solution of pH3?

2 20 cm³ of a product is digested by an enzyme in 4 minutes. Calculate the rate of reaction.

3 Explain why an enzyme is less effective if it is denatured.

Using qualitative reagents to test for a range of carbohydrates, lipids and proteins

For this practical you will need to recall the colours for each of the positive and negative food tests in the table below.

	Benedict's test	Iodine test	Biuret test	Emulsion test
Method	Add Benedict's reagent and heat for several minutes.	Add iodine to food sample.	Add Biuret reagent to food sample.	Add ethanol to food sample, then add distilled water.
Tests for…?	Sugars	Starch	Protein	Lipids
Positive result	Brick red Orange Yellow Green	Blue-black	Lilac	Emulsion formed
Negative result	Blue	Orange	Blue	No emulsion

SNAP IT!

Take a picture of this table and learn the food tests.

WORKIT!

A sample of milk was tested using Benedict's reagent, iodine and Biuret reagent. Describe and explain the results. (3 marks)

The Benedict's reagent turned the milk green, as there is a small amount of sugar in the milk. (1)

The iodine stayed orange, as there is no starch in milk. (1)

The Biuret reagent turned the milk lilac, as there is protein present in milk. (1)

NAIL IT!

Describe the results for each test and then explain why these results happened.

CHECK IT!

1 What could you conclude about a food sample that turned lilac with Biuret reagent?

2 Describe how to carry out the emulsion test.

3 What results would you expect to see if you carried out the iodine test and the Benedict's test on a sample of pasta? Explain your answer.

The effect of pH on amylase

NAILIT!

The rate of reaction can be worked out by measuring how long it takes for a reactant to be used up, or how long it takes for a product to be made. In this investigation, iodine is used to indicate when the reactant (starch) has been used up.

MATHS SKILLS

Calculating the rate of reaction

To work out the rate of reaction, divide the amount of reactant used, or product made, by time.

NAILIT!

Remember to always include the units in your answer.

Amylase is an enzyme found in saliva in the mouth. It digests starch (amylose) into sugars (maltose) and has an optimum pH of 7. In this practical you will need to use your knowledge of enzymes to hypothesize on how pH affects amylase activity. Remember to consider what your control variables will be.

Practical Skills

Investigating the effect of pH

- Use a measuring cylinder or syringe to measure out a volume of amylase and place this into a test tube.
- Use a measuring cylinder or syringe to measure out a volume of buffer of a particular pH and add this to the same test tube.
- Use a measuring cylinder or syringe to measure out a volume of starch solution, add this to the same test tube and set a timer.
- Gradually add iodine to work out when the starch has been digested into sugar. The iodine will be blue-black when starch is present in the test tube, and orange when the starch is no longer present.
- Record results every 30 seconds to make sure your samples are representative.
- Record the time taken for the amylase to digest the starch in buffers of different pH.
- Draw a graph showing the time taken for the amylase to digest the starch against pH.

WORKIT!

At pH4, 2 cm³ of starch solution was used up in 140 seconds. Calculate the rate of reaction. (2 marks)

$$\text{Rate of reaction} = \frac{\text{Amount of reactant used}}{\text{Time}} = \frac{2}{140} \text{ (1)} = 0.014 \text{ cm}^3/\text{s (1)}$$

CHECKIT! ✓

1 What does amylase do?

2 Explain what would happen to the rate of reaction of amylase at a pH above pH7.

3 2 cm³ of starch was used up in 60 seconds. Calculate the rate of reaction.

The heart

The human heart is located in the thorax (chest cavity) and is made of cardiac muscle.

The structure and function of the human heart

The heart is divided into the right side and the left side. The blood is pumped through both sides at the same time to pump the blood through a double circulatory system.

Deoxygenated blood from the body enters the right atrium of the heart through the superior and inferior vena cava. At the same time, oxygenated blood from the lungs enters the left atrium of the heart through the pulmonary vein. The muscles of the right and left atria contract at the same time, and push the blood into the right and left ventricles. A fraction of a second later, the muscles of the right and left ventricle contract at the same time.

The right ventricle pushes the blood into the pulmonary artery. This blood will now travel to the lungs where gas exchange takes place. The left ventricle pushes the blood into the aorta. This blood will now travel around the body. The blood vessels that supply the heart muscle with oxygen are called the coronary arteries.

Valves between the atria and the ventricles, and inside the pulmonary artery and the aorta, prevent blood from flowing backwards around the heart.

Pacemakers

A group of cells in the right atrium control the natural resting heart rate. They are the heart's natural pacemaker. If a person has an irregular heart rate, a small electrical device called an artificial pacemaker can be put in place to correct this.

You will not be expected to know the names of the valves.

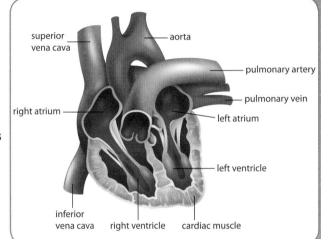

superior vena cava — aorta — pulmonary artery — pulmonary vein — right atrium — left atrium — left ventricle — inferior vena cava — right ventricle — cardiac muscle

SNAPIT!

Make your own basic diagram of a heart. Take a photo of it and learn the labels.

WORKIT!

Describe how oxygenated blood flows through the heart. (4 marks)

Oxygenated blood enters the left atrium (1) through the pulmonary vein. (1)
The muscles in the left atrium contract, pushing the blood into the left ventricle. (1)
When the left ventricle contracts, the blood leaves the heart through the aorta. (1)

✓ CHECKIT!

1 Which vein of the heart carries oxygenated blood?

2 What is the purpose of the valves in the heart?

3 Explain how the heart maintains a natural resting heart rate, and why this is important.

The lungs

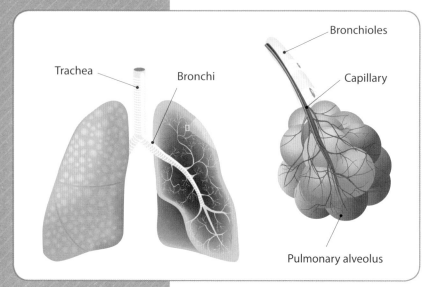

Trachea
Bronchi
Bronchioles
Capillary
Pulmonary alveolus

The lungs are located in the thorax (chest cavity).

The structure and function of human lungs

The lungs take in air from outside the body for gas exchange. Oxygen is taken in and carbon dioxide is released.

Air travels down the trachea and into the right and left bronchi. The air travels along smaller branches to the alveoli. Each alveolus is supplied with a blood capillary.

Gas exchange takes place in the alveoli. Oxygen diffuses from the alveoli into the blood capillary, and carbon dioxide diffuses from the blood capillary into the alveoli.

Adaptations of the lungs for gaseous exchange

- Many alveoli to increase the surface area.

- The alveoli wall and the capillary wall are only one cell thick to provide a short diffusion pathway.

- A good supply of blood capillaries moves oxygen away from the alveoli and maintains a high concentration gradient.

SNAPIT!

Make your own simple diagram of the lungs and label it. Take a photo and learn the labels.

WORKIT!

Describe how oxygen moves from the trachea and into the blood. (3 marks)

Oxygen moves from the trachea into the right and left bronchi. (1)

Oxygen then moves down smaller branches and into the alveoli. (1)

Oxygen diffuses out of the alveoli and into the blood capillaries. (1)

CHECKIT! ✓

1 Where in the lungs is the concentration of carbon dioxide gas greatest?

2 Why is a short diffusion pathway important in gas exchange?

3 Explain how having a good blood supply helps to increase the rate of diffusion.

Blood vessels

DOIT!

Make cue cards of arteries, veins and capillaries and match the blood vessels to the features and functions.

Blood vessels carry blood around the body. There are three types: arteries, veins, and capillaries.

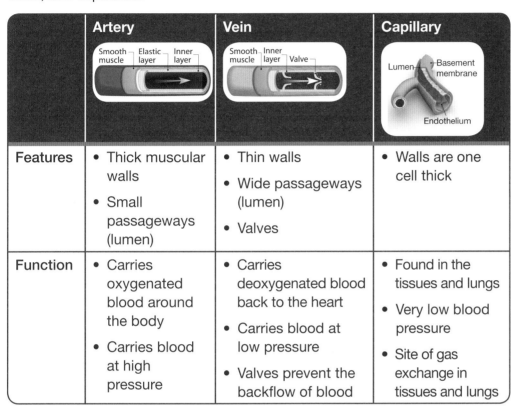

	Artery	Vein	Capillary
Features	• Thick muscular walls • Small passageways (lumen)	• Thin walls • Wide passageways (lumen) • Valves	• Walls are one cell thick
Function	• Carries oxygenated blood around the body • Carries blood at high pressure	• Carries deoxygenated blood back to the heart • Carries blood at low pressure • Valves prevent the backflow of blood	• Found in the tissues and lungs • Very low blood pressure • Site of gas exchange in tissues and lungs

Calculating blood flow

It is possible to work out the rate of blood flow, if you know the distance the blood travelled and the time that it took.

WORKIT!

Blood flows through a 12 mm artery in 0.2 seconds. What is the rate of blood flow? (3 marks)

$$\text{Rate of blood flow} = \frac{\text{Distance travelled by blood}}{\text{Time}} = \frac{12}{0.2}\ (1) = 60\,mm/s\ (2)$$

WORKIT!

Compare and contrast the structure and function of arteries and veins. (4 marks)

Arteries have thick walls, but veins have thin walls. (1)

Arteries have small lumens, but veins have wide lumens. (1)

Arteries carry blood at high pressure, but veins carry blood at low pressure. (1)

Most arteries carry oxygenated blood, but most veins carry deoxygenated blood. (1)

You could also have mentioned that veins have valves, but arteries do not.

✓ CHECKIT!

1 What type of blood vessel carries deoxygenated blood to the heart?

2 What is the name of the blood vessel that carries oxygenated blood away from the heart?

3 Explain the structure and function of capillaries.

Blood

Blood is a fluid which consists of: plasma – a yellow fluid; red blood cells – carry oxygen around the body; white blood cells – protect the body from pathogens; platelets – needed to form scabs if there is a cut; large, insoluble proteins; dissolved substances, such as hormones, oxygen, carbon dioxide.

Blood cells

There are two types of blood cell: red blood cells and white blood cells.

	Red blood cell	White blood cell
Function	Carry oxygen around the body	Protect the body from pathogens
Adaptation	Biconcave in shape – increased surface area for oxygen to diffuse in	Phagocytes can change shape – to engulf pathogens
	No nucleus – more room to pack in haemoglobin	Neutrophils have a lobed nucleus – to squeeze through small spaces
	Packed with haemoglobin – to carry more oxygen	Lymphocytes have a lot of rough endoplasmic reticulum (with attached ribosomes) – to produce antibodies

NAILIT!

Red blood cells are smaller than white blood cells, and are red. White blood cells are larger than red blood cells and are usually stained light purple with a pink-purple nucleus.

Practical Skills

Identifying blood cells

You need to be able to recognise blood cells from a drawing and from a photograph taken with a microscope.

WORKIT!

Identify cells A and B in the photograph. (2 marks)

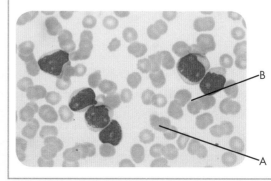

Cell B is a white blood cell. (1)

Cell A is a red blood cell. (1)

CHECKIT! ✓

1 Name two of the main components of blood.

2 How are red blood cells adapted to carry oxygen?

3 Explain why lymphocytes need a large amount of rough endoplasmic reticulum.

Coronary heart disease

Coronary heart disease (CHD) is a non-communicable disease that affects the heart and the coronary arteries.

What causes CHD?

In CHD, layers of fatty material and cholesterol build up inside the coronary arteries (the arteries that supply the heart muscle with oxygen). This makes the passageway (lumen) inside the artery narrower, and reduces the flow of blood.

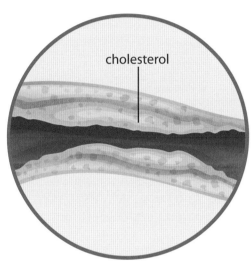

Coronary artery

CHD symptoms

There are three main symptoms associated with CHD:

- Angina – chest pains, often brought on by exercise, as the blood supply to the muscles of the heart is restricted.

- Heart failure – muscle weakness in the heart, or a faulty valve, means that the heart does not pump enough blood around the body at the right pressure.

- Heart attack – the blood supply to the muscles of the heart is suddenly blocked, usually by a clot.

Faulty valves

Heart valves are faulty if they:

- Do not open properly – blood cannot pass into the atrium or ventricle.

- Do not close properly – blood can flow backwards around the heart.

In both cases, the heart is put under extra strain and has to pump harder to get the blood through the heart.

STRETCH IT!

Find out about stem cell research into heart and heart valve transplants.

Treatments for CHD

Blocked or restricted coronary arteries can be treated with a stent. This is a small tube that holds the artery open. Medicines called statins may be given. Statins reduce blood cholesterol levels and slow down the rate of fatty material deposit.

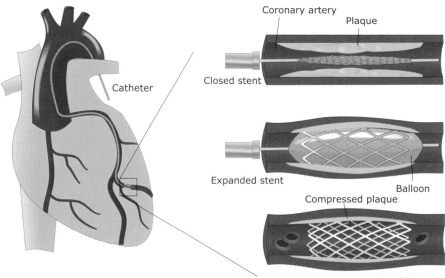

Faulty valves can be treated using a mechanical valve, or a transplanted valve from an animal (usually a pig) or a human donor.

Patients with severe heart failure may have a heart transplant from a human donor, or have an artificial heart while waiting for a transplant. Research is being carried out to make replacement hearts and heart valves from a patient's own stem cells.

NAILIT!

You need to compare the different methods of treatment of CHD and demonstrate the advantages and disadvantages of each method.

DOIT!

Make cue cards for each treatment for CHD and put the advantages and disadvantages of each one on the back. These can be technical, social or ethical reasons.

WORKIT!

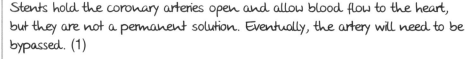

Evaluate the methods of treatment for CHD. (4 marks)

Stents hold the coronary arteries open and allow blood flow to the heart, but they are not a permanent solution. Eventually, the artery will need to be bypassed. (1)

Statins reduce blood cholesterol levels but only slow down the rate of fatty material deposit. The fatty material will eventually build up if the patient does not eat a healthy diet. (1)

Faulty valves can be replaced with mechanical or biological valves. Mechanical valves will only last for a while/biological valves may be rejected by the body. (1)

Heart transplants give the patient a new heart but there may be rejection/long waiting list for transplants. (1)

CHECKIT! ✔

1 Name two symptoms of CHD.

2 Describe the causes of CHD.

3 Explain what would happen if a heart valve did not close properly.

Health issues

Health is a state of physical and mental well-being. A person's physical or mental health may be affected by many factors such as diet, stress, genetic background and behavioural choices.

What is a disease?

Diseases stop parts of the body working properly and cause **symptoms**. There are two types, shown here in the table:

	Communicable	Non-communicable
Cause	Pathogen, such as virus, bacteria, fungi or protozoa	Genetic Lifestyle – diet, stress, exercise levels
Spread	Transmitted from one person, animal or plant to another	Non-transmissible

Diseases that interact

Defects in the immune system	→	person is more likely to suffer from infectious diseases
Viruses living in cells	→	could be a trigger for cancer
Immune reactions initially caused by pathogens	→	could trigger allergies, such as asthma or skin rashes
Severe physical ill health	→	could lead to depression or other mental health problems

Sampling

Sampling is used to find patterns in peoples' lifestyle choices and their incidence of disease. This is called epidemiological data. For example, epidemiological data has shown that people with diets high in saturated fat are more likely to suffer from coronary heart disease.

MATHS SKILLS

Scatter diagrams and correlations

You should be able to identify a correlation between two variables by looking at a scatter diagram.

These scatter diagrams show the three possible types of correlation:

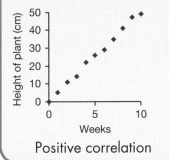

Positive correlation

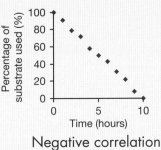

Negative correlation

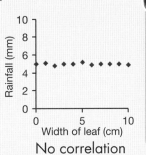

No correlation

SNAP IT!

Take a photo of the three types of correlation graph and learn them on the go.

NAIL IT!

Bar charts show data from distinct categories and the bars do not touch. Histograms show numerical data and the bars touch.

MATHS SKILLS

Translating disease incidence information

Make sure you can write numerical data using information from a graph. You should also be able to draw a graph from numerical data.

DO IT!

Look at health data graphs online and practise interpreting them.

WORKIT!

This graph shows the percentage of males and females diagnosed with diabetes at different ages. Describe the pattern using data from the graph. (3 marks)

In all age groups, except for age 16-34, more males than females were diagnosed with diabetes. (1)

The percentage of females diagnosed with diabetes increased from 1% at age 16-34, to 13% at age 75+. (1)

The percentage of males diagnosed with diabetes increased from 0.5% at age 16-34, to 17.5% at age 75+. (1)

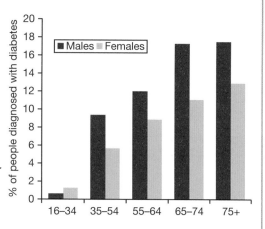

WORKIT!

In a class of 30 students, six ate five portions of fruit and vegetables per day, eight ate four portions, ten ate two portions, three ate one portion and two ate no portions. One student ate six portions.

Draw a frequency table to show this data. (2 marks)

Portions of fruit and vegetables	Frequency
0	II
1	III
2	IIII IIII
3	0
4	IIII III
5	IIII I
6	I

(2)

CHECK IT! ✓

1 In an experiment, more product was produced the longer the reaction was carried out. A graph was plotted of mass produced against time. Name the type of correlation shown.

2 Describe and explain how you would show the data in the frequency table above as a graph.

3 Suggest how defects in a person's immune system could make them more likely to suffer from infectious diseases.

Effect of lifestyle on health

Lifestyle and substances in the environment can lead to an increase in non-communicable diseases.

Risk factors for non-communicable diseases

Poor diet, smoking and lack of exercise	→	increased risk of cardiovascular disease
Obesity	→	increased risk for type 2 diabetes
Drinking alcohol	→	causes cirrhosis of the liver and affects brain function
Smoking	→	increased risk of lung disease such as emphysema and lung cancer
Smoking and alcohol	→	effect on unborn babies
Carcinogens, such as ionising radiation	→	increased risk of cancer including skin cancer

Many diseases are caused by the interaction of a number of factors.

WORKIT!

The pie chart shows the total number of deaths (in millions) from non-communicable diseases (NCD) in the world in 2008. What percentage of deaths were from cardiovascular disease? (2 marks)

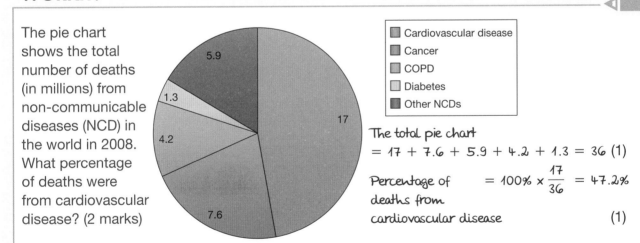

Legend:
- Cardiovascular disease
- Cancer
- COPD
- Diabetes
- Other NCDs

Values shown: 5.9, 1.3, 4.2, 7.6, 17

The total pie chart
$= 17 + 7.6 + 5.9 + 4.2 + 1.3 = 36$ (1)

Percentage of deaths from cardiovascular disease $= 100\% \times \dfrac{17}{36} = 47.2\%$ (1)

CHECKIT!

1   Name two risk factors associated with cardiovascular disease.

2   Give one financial cost of living with a non-communicable disease.

3   The number of deaths worldwide from cardiovascular disease in 2030 is predicted to be 23.6 million. Using data from the pie chart above, calculate the percentage increase in deaths this would be since 2008.

Cancer

Cancer is a non-communicable disease with many interacting risk factors.

What is cancer?

Mutations in cells are caused by a number of different factors. When normal cells divide, they are checked for mutations. If any mutations exist, the cell is not able to go through mitosis (a type of cell division, see page 17). Also when normal cells divide, they lie next to each other in a uniform way. Cancer cells are abnormal. They divide without being checked, and form a mass, lying over the top of one another. A mass of cancer cells is called a tumour.

Types of tumour

There are two types of tumour:

Benign – growths of abnormal cells, contained in one area, usually within a membrane. They do not invade other parts of the body.

Malignant – growths of abnormal cells that invade neighbouring tissues and spread through the blood to different parts of the body, where they form secondary tumours.

Causes of cancer

There are many interacting risk factors associated with cancer, including:

- genetic risk factors
- smoking
- drinking alcohol.
- poor diet
- ionising radiation

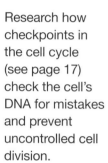

STRETCH IT!

Research how checkpoints in the cell cycle (see page 17) check the cell's DNA for mistakes and prevent uncontrolled cell division.

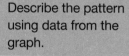

NAIL IT!

Describe the pattern using data from the graph.

WORKIT!

This graph shows the annual deaths from lung cancer in Australia per 1000 in 2015, and the number of cigarettes smoked per day. Describe the evidence from the graph that shows that smoking may cause lung cancer. (3 marks)

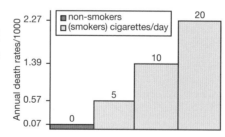

The more cigarettes smoked per day, the more deaths from lung cancer. (1)

0.07 non-smokers per 1000 of population died from lung cancer (1) compared with 2.27 smokers per 1000 of population who smoked 20 cigarettes per day. (1)

CHECK IT!

1 Name two risk factors for developing cancer.

2 Describe how a tumour is formed.

3 Suggest why malignant tumours are more difficult to treat than benign ones.

Plant tissues

Different parts of plants are made from different types of tissue.

The leaf

The leaf is a plant organ. It is made of several plant tissues: epidermal tissues, palisade mesophyll, spongy mesophyll, xylem and phloem. The lower epidermal tissues contain guard cells. These control the opening and closing of small holes called stomata. This is where carbon dioxide and oxygen enter and exit the leaf.

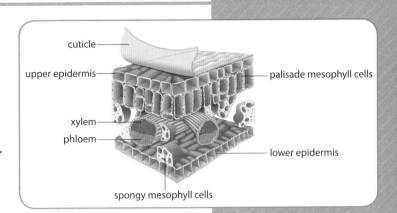

cuticle

upper epidermis

palisade mesophyll cells

xylem

phloem

lower epidermis

spongy mesophyll cells

Function of plant tissues

Plant tissue	Function
Epidermal tissues	Allow light to reach the palisade tissues – thin and transparent
Palisade tissues	Light absorption for photosynthesis – tightly packed, column shaped cells, packed with chlorophyll
Spongy mesophyll	Site of gas exchange – gases dissolve in thin layer of water on the surface of the loosely packed cells, and then diffuse into air spaces
Xylem	Transport of water to the leaf – narrow, hollow tubes with a waterproof layer in the cell walls
Phloem	Transport of sugar sap around the plant – narrow, almost hollow tubes
Meristem tissue	Divides rapidly at the root and shoot tips to provide new plant cells – can become any type of plant cell

NAILIT!

The diagram above shows a transverse section of a leaf. You may be asked to observe and draw this.

DOIT!

Look at a transverse section of a leaf using a microscope and identify the different tissues. Practise drawing the leaf.

WORKIT!

Explain how the spongy mesophyll tissue is adapted for its function. (3 marks)

The cells are loosely packed to allow space for the gases to diffuse. (1)

The cells have a thin layer of water on the surface for the gases to dissolve into. (1)

The cells are close to the lower epidermis/guard cells/stomata so that the gases can diffuse in and out of the leaf. (1)

✓ CHECKIT!

NAILIT!

In Q2, limit your drawing to 4–5 cells.

1 Which plant tissue transports water to the leaf?

2 Draw and label the palisade tissue.

3 Explain why it is important that the upper epidermis is thin and transparent.

Transpiration and translocation

Transpiration is the loss of water from the top part of the plant. Translocation is the movement of sugar sap up and down the plant.

What is transpiration?

The root, stem and leaves form a plant organ system for transport of substances around the plant. In transpiration, water moves into the root hair cells and into the xylem by osmosis (see page 22). Water travels up the xylem using the forces of adhesion and cohesion, until it reaches the leaf. In the leaf, water diffuses by osmosis into the cells, and then evaporates onto the surface of the spongy mesophyll cells. Water vapour then diffuses out of the leaf through the stomata.

What is translocation?

Photosynthesis in the leaves produces sugars. Some of these are needed in respiration, and to make molecules for the plant. The rest is stored in the roots as an energy store.

Sugar sap is made of dissolved sugars, and moves up and down the plant in the phloem tissue. Phloem tissue is made of many phloem cells joined end to end. The end plate of each phloem cell has many small pores in it and is called a sieve plate. The phloem cells join together to make a long narrow, almost hollow tube. Mature phloem cells are alive, but only have a limited number of sub-cellular structures.

Tissues in transpiration

1. Root hair cells are adapted to take in water by having a long hair to provide a large surface area (see page 15). The cytoplasm inside the root hair cell is a concentrated solution compared to the dilute solution in the soil, so water moves into the cells by osmosis. Mineral ions are also taken up by the root hair cells, by active transport (see page 25). Root hair cells have many mitochondria to provide the energy from respiration to do this.

2. Xylem tissue is made of many xylem cells joined end to end. The end plate of each xylem cell is removed, to make a long narrow, hollow tube for water and dissolved mineral ions to travel up. The movement of water through the xylem is called the transpiration stream.

   Mature xylem cells are dead. They do not contain a nucleus or any sub-cellular structures. The cell walls of the xylem contain lignin, which makes the cells waterproof. This helps the water to stay inside the xylem. Lignin also gives strength to the xylem and therefore the stem of the plant.

3. Guard cells form pairs on the lower epidermis of the leaf. When there is plenty of water, the cells become turgid and allow the stomata to open. When there is little water, the walls of the guard cells push together to close the stomata. This controls the loss of water from the leaf. Gases can also enter and exit the cell through the open stomata.

DO IT!

Make a mind map on transpiration and another on translocation, using all of the main points and then adding more detail.

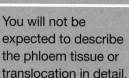

NAIL IT!

You will not be expected to describe the phloem tissue or translocation in detail.

DO IT!

Paint the underside of a leaf with clear nail varnish and allow to dry. Carefully peel the varnish off and mount on a slide to view the impression left by the epidermal cells with a microscope. You should be able to see the guard cells and stomata.

Measuring the rate of transpiration

The rate of transpiration can be measured by measuring the amount of water taken up by the plant. The simplest way to do this is to weigh the plant, and then weigh the plant again after a few hours. The change in mass is the mass of the water that has left the plant.

Increasing the rate of transpiration

The rate of transpiration can be increased by:

- increasing temperature
- increasing light levels
- increasing air movement
- decreasing humidity.

WORKIT!

The rate of transpiration of a plant was found by measuring the loss of water over four hours. The investigation was carried out at 20°C, 40°C, and 60°C. Which temperature do you think had the fastest rate of transpiration? Justify your answer. (4 marks)

60°C (1) because the rate of transpiration is higher with increased temperatures. (1)

The water molecules will have more kinetic energy at higher temperature so rate of diffusion is higher (1) and more water could diffuse out of the leaves. (1)

✓ CHECKIT!

1 Describe the process of transpiration.

2 Name two factors that increase the rate of transpiration.

3 In an investigation, 12 cm³ of water was lost by transpiration in 4 hours. Calculate the rate of transpiration.

For additional questions, visit:
www.scholastic.co.uk/gcse

1   a   Name two enzymes found in the human digestive system.

    b   What are carbohydrates digested into?

    c   Describe how an enzyme works.

    d   What would be the effect of placing an enzyme with an optimum pH of 7 into a solution with a pH of 3?

2   Describe the food tests for the following food groups:

    a   Sugars   b   Starch   c   Protein.

3   a   i   Which blood vessel carries deoxygenated blood to the heart?

        ii Which blood vessel carries oxygenated blood to the heart?

    b   Describe how the blood is kept flowing through the heart in the right direction.

    c   How is the heartbeat controlled?

4   a   Describe the pathway of the air from the mouth to the site of gaseous exchange in mammals.

    b   Describe how the gaseous exchange surface in mammals is adapted.

    c   Blood flows through a 10 mm section of capillary in the lungs in 0.4 seconds. Calculate the rate of blood flow.

5   A patient with CHD has previously had a heart attack.

    a   What causes a heart attack?

    b   The doctor advises the patient to either take statins or have a stent fitted in their coronary artery. Evaluate the advantages and disadvantages of each treatment.

6   a   In plant tissues, guard cells open and close to form small pores called stomata. What is the function of the stomata?

    b   An investigation into the number of open stomata at different temperatures was carried out. The results are shown in the table below:

Temperature	Number of open stomata			Mean number of open stomata
	1	2	3	
10	14	16	12	14
20	34		38	36
30	49	51	44	
40		78	81	82

    i   Complete the table.

    ii   Correct any mistakes in the table.

    iii Describe what the data shows.

    iv   Suggest a reason for these results.

    c   Increased humidity decreases the rate of transpiration. What factors will increase the rate of transpiration?

Infection and response

Communicable diseases

SNAPIT!

Take a photo of this table and learn one or two examples of each type of pathogen.

Communicable diseases are caused by pathogens that are spread from one person or animal to another.

Pathogens

Pathogens are microorganisms that cause disease. Not all microorganisms are pathogens.

There are four groups of pathogens:

Pathogen	Appearance	Transmission	Diseases
Bacteria	Single-celled Prokaryotic	Airborne – coughing and sneezing Direct contact with infected person/animal/sharp object In food and water	Tuberculosis Tetanus Cholera Salmonella food poisoning
Viruses	Much smaller than a bacterium Needs a host	Airborne – coughing and sneezing Direct contact with infected person	Common cold Influenza HIV
Fungi	Single-celled Eukaryotic Chitin cell wall	Airborne – carried on wind Direct contact with infected person/animal Indirect contact – towels and changing room floors	Ringworm Athlete's foot
Protists	Single-celled Eukaryotic No cell walls	Vector – carried by insect	Malaria Dysentery

Bacteria and viruses may reproduce rapidly inside the body. Bacteria may produce poisons (toxins) that damage tissues and make us feel ill. Viruses live and reproduce inside cells, causing cell damage.

Reducing the spread of diseases

The spread of diseases can be reduced by:

- regular handwashing
- using tissues
- vaccination programmes
- preparing food safely
- access to clean drinking water
- clean, well-ventilated homes
- practising safe sex
- reducing the population of some insects in infected areas, e.g. mosquitoes.

Outbreaks of disease

Sometimes a disease begins to affect many people in a short amount of time. This is called an outbreak.

From 2014–2016, there was an outbreak of a virus called Ebola in West Africa. Ebola is a severe, often fatal disease, that is easily spread from human to human by direct contact, contact with the bodily fluid of an infected person, e.g. blood.

The outbreak was controlled by preventing the movement of people in affected areas and treating infected people quickly.

WORKIT!

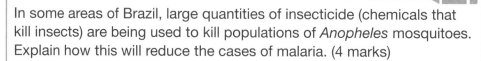

In some areas of Brazil, large quantities of insecticide (chemicals that kill insects) are being used to kill populations of *Anopheles* mosquitoes. Explain how this will reduce the cases of malaria. (4 marks)

Malaria is caused by a protist/protozoa/plasmodium (1) carried by the *Anopheles* mosquito. (1)

The protist/protozoa/plasmodium gets into the human body when the mosquito bites them. (1)

Fewer mosquitoes mean that people are less likely to get bitten. (1)

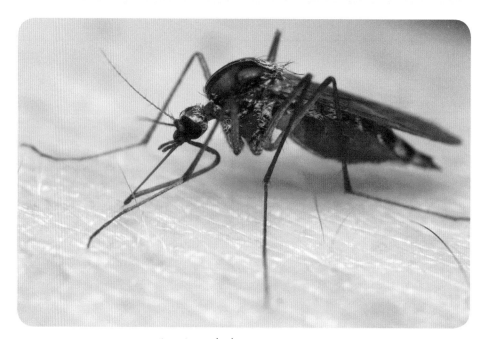

An *Anopheles* mosquito

NAILIT!

For Q3, think about pathogens and insects that thrive in damp conditions, and how most viruses and bacteria can be spread.

CHECKIT! ✓

1 Name two diseases caused by a virus.

2 Describe how bacteria may be transmitted.

3 Suggest why people living in damp, crowded housing may be more susceptible to communicable diseases.

Viral diseases

Viruses live and reproduce inside cells, causing cell damage. Some viruses can cause serious illnesses and even death if not treated quickly.

Measles

Measles is spread by breathing in droplets from sneezes and coughs. The symptoms of measles are a red skin rash and a fever. Measles is a serious illness and can result in death if complications arise. Most babies and young children in the UK are vaccinated against measles. This means that they are unlikely to catch the measles virus.

HIV

Human immunodeficiency virus (HIV) is transmitted through sexual contact, or contact with infected blood, such as when drug users share needles. The virus causes a flu-like illness in the first instance. Untreated, HIV attacks the cells of the immune system, developing into acquired immune deficiency syndrome (AIDS). This is where the immune system no longer functions, and the person may die from other infections or cancer. HIV can be treated with antiretroviral drugs, which control the progression of the disease.

Tobacco mosaic virus (TMV)

TMV is a widespread plant virus that affects many types of plant, including tobacco and tomato plants. The virus appears as a black 'mosaic' pattern on the leaves. This affects plant growth, as the dark areas cannot carry out photosynthesis. TMV spreads from plant to plant through direct contact, or in infected soil. TMV is treated by removing infected plants, and washing hands before planting new plants.

WORKIT!

Describe how the spread of HIV could be reduced. (3 marks)

Practise safe sex. (1)

Check blood products for HIV to prevent transmission through blood transfusions. (1)

Give pregnant women with HIV medicines to prevent transmission of HIV to the fetus. (1)

STRETCHIT!

Find out why measles has been on the rise in the UK for the past few years.

NAILIT!

HIV is caused by contact with infected body fluids. It can be passed from mother to child.

DOIT!

Put the diseases measles, HIV and TMV into a table with the headings: virus, causes, symptoms, treatment. This will be easier to remember than a page of text.

CHECKIT!

1 Describe how measles is spread.

2 Describe the symptoms of untreated HIV.

3 Describe the effect of removing plants infected with TMV from a field.

Bacterial diseases

Salmonella food poisoning and gonorrhoea are both caused by bacteria.

Salmonella food poisoning

Salmonella food poisoning is caused by the bacterium *Salmonella enterica*, in food prepared in an unhygienic way. The symptoms of this type of food poisoning are fever, abdominal cramps, vomiting and diarrhoea. The salmonella bacteria move into the digestive system and secrete toxins that cause the symptoms. Most people recover without treatment, but in severe cases, people may need to go into hospital to be treated for dehydration.

The way to prevent salmonella food poisoning is to prepare and store food in a hygienic way, and make sure that food is cooked thoroughly. Salmonella can be found in poultry (chickens and turkeys) and eggs, so in the UK, poultry are vaccinated against salmonella.

Gonorrhoea

Gonorrhoea is a sexually transmitted disease (STD) caused by the bacterium *Neisseria gonorrhoeae*. The symptoms are a thick yellow or green discharge from the vagina or penis, and pain when urinating. Untreated, gonorrhoea can lead to severe complications and in rare cases, infertility or septicaemia.

Gonorrhoea can be treated with antibiotics, but new strains of *Neisseria gonorrhoeae* have become resistant to some antibiotics, such as penicillin. In the UK, there are currently two antibiotics that can be used to treat this infection. The spread of gonorrhoea can be controlled by practising safe sex, for example, using a condom.

WORKIT!

A person wants to avoid getting salmonella food poisoning. What advice would you give to them? (4 marks)

Do not eat undercooked eggs or poultry, or drink unpasteurised milk. (1)

Wash your hands with soap and water before handling food. (1)

Keep food properly refrigerated. (1)

Clean cooking surfaces before preparing food. (1)

CHECKIT! ✓

1 Which type of medicine is used to treat bacterial infections?

2 Explain why poultry in the UK are vaccinated against salmonella.

3 Describe and explain the consequences of antibiotic resistance in treating gonorrhoea.

Fungal and protist diseases

Fungi and protists can cause a range of diseases, including rose black spot and malaria.

Rose black spot

Rose black spot is a fungal disease, spread by water or wind, that affects roses. It is caused by the fungus, *Diplocarpon rosea*, and infects the leaves of the rose with black or purple spots. This affects photosynthesis, as the areas of the leaf covered in spots cannot photosynthesise. This reduction in photosynthesis will affect the growth of the plant. Infected leaves turn yellow and drop to the ground.

Rose black spot can be treated with fungicides (chemicals that kill fungi), or removing and destroying the infected leaves.

Malaria

Malaria is caused by a type of protist. These protists live in the saliva of the female *Anopheles* mosquito and are spread to humans when the mosquito bites them. The mosquito is a vector. The protists spend some of their life cycle in the human body, and move back into the mosquito when the mosquito bites an infected person. In this way, the protists can be spread from person to person.

Malaria causes a severe fever, which can occur several times over a person's lifetime, and can be fatal. Malaria can be treated with antimalarial drugs. The main way to prevent getting malaria is to prevent yourself being bitten by a mosquito, using mosquito nets at night, and long-sleeved clothes or insect repellent during the day. Many people also take antimalarial drugs as a preventative. Mosquitoes like to breed in small pools of water. Another way of preventing the spread of malaria is to prevent the mosquitoes from breeding by removing pools where they breed, or killing mosquitoes using pesticides (chemicals that kill pests).

NAILIT!

You will not be expected to recall the full names of the fungi or protist.

STRETCHIT!

Find out about the life cycle of the malaria protist.

NAILIT!

Mosquitoes are part of a food chain. The food chain will be disrupted if the mosquitoes are removed.

WORKIT!

Compare the advantages and disadvantages of removing mosquitoes from an area. (4 marks)

Removing the mosquitoes will reduce the chances of getting malaria. (1)

Removing the mosquitoes will remove a food source in the food chain. (1)

Some animals may decrease in number due to the lack of mosquitoes (1) but it will save some human lives. (1)

✓ CHECKIT!

1 Which type of medicine is used to treat malaria?

2 How do mosquito nets help to prevent malaria?

3 Describe how to prevent the spread of rose black spot.

Human defence systems

DOIT!

To get a better idea of how phagocytosis happens, look at a video online of phagocytes engulfing pathogens.

The human body has many different types of defence against pathogens.

Non-specific defence systems

These non-specific defences try to stop all pathogens getting into our bodies.

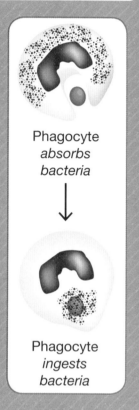

Phagocyte *absorbs bacteria*

Phagocyte *ingests bacteria*

	Defence
Skin	Physical barrier against pathogens
	Breaks in the skin form scabs
	Sweat glands produce sweat that inhibits pathogens
Nose	Small hairs and mucus trap airborne particles
Trachea and bronchi	Mucus traps pathogens and is moved up to the throat by small hairs called cilia
Stomach	Hydrochloric acid and protease enzymes in the stomach kill pathogens

Specific immune systems

If pathogens get past the non-specific defences, there are two types of white blood cell that can destroy them.

1 *Phagocytes*

These are a type of white blood cell that can engulf (take inside their body) any type of pathogen. They do this by extending their cell membrane around the pathogen until it is surrounded. The pathogen then enters the phagocyte and is digested by enzymes. This process is called phagocytosis.

2 *Lymphocytes*

These are a type of white blood cell that attack specific pathogens. They produce antibodies that have several functions:

- attach to pathogens and prevent them from entering cells
- attach to pathogens and target them for phagocytosis
- act as antitoxins by attaching to toxins.

WORKIT!

Describe and explain the body's defences against the influenza virus. (4 marks)

The influenza virus cannot pass through the skin, which is a non-specific barrier. (1)

Mucus traps any inhaled virus, which is then moved up to the throat by cilia. (1)

Phagocytes will engulf the virus by phagocytosis. (1)

Lymphocytes will produce antibodies against the virus. (1)

CHECKIT! ✓

1 Name two non-specific defences against pathogens.

2 Outline the process of phagocytosis.

3 Explain how the specific immune system stops bacterial toxins making a person feel ill.

Vaccination

Vaccinations are given to people to prevent them from developing illnesses from pathogens.

How does vaccination work?

Vaccination is the introduction into the body of small quantities of dead or inactive forms of a pathogen into the body. These dead or inactive pathogens stimulate the white blood cells to make antibodies. If the same pathogen re-enters the body, white blood cells respond quickly to make the correct antibodies, and prevent illness.

Vaccination programmes

In the UK, most babies and young children are vaccinated against a number of pathogens. The more people who are vaccinated against a pathogen, the less likely the pathogen is to spread through the population. This is called herd immunity. There will always be a few people who cannot be vaccinated. They will be protected by herd immunity and are unlikely to become ill from that pathogen.

Around the world, vaccination programmes are used to prevent the spread of disease. Vaccination programmes successfully rid the world of smallpox in 1979.

WORKIT!

What is herd immunity? (2 marks)

When most of a population is vaccinated against a pathogen. (1)
Non-vaccinated people are protected from infection with that pathogen. (1)

DOIT!

Write out a step-by-step of how vaccinations work.

NAILIT!

You do not need to know details of vaccination schedules or side effects associated with specific vaccines.

STRETCHIT!

Find out about global vaccination programmes to reduce the spread of tuberculosis (TB).

CHECKIT!

1 What is in a vaccine?

2 Suggest why the UK no longer routinely vaccinates against TB.

3 In 1998, a doctor falsely claimed that the MMR vaccine caused autism. Using data from the graph describe and explain the percentage of children in the UK vaccinated with the MMR vaccine and the incidence of measles.

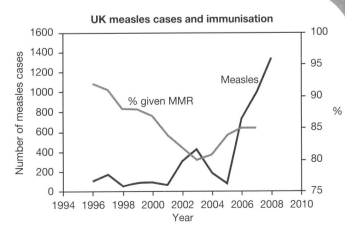

Antibiotics and painkillers

Antibiotics

Antibiotics, such as penicillin, are medicines that destroy or inhibit the growth of bacteria. They are used to treat bacterial infections, such as kidney infections, or diseases, such as pneumonia. Antibiotics can be given as pills, a cream, or an injection. Specific bacteria are killed by specific types of antibiotic, so it important to be prescribed the correct one. They cannot be used to treat any diseases caused by a virus, such as influenza.

Antibiotic resistance

Since antibiotics were first used in the 20th century, the number of deaths from infectious bacterial diseases has greatly reduced. However, many bacterial strains are resistant to some antibiotics. Some bacterial strains, such as MRSA, are resistant to most antibiotics. Antibiotic resistance is increasing and this is concerning as it may lead to increased deaths. Methods to reduce antibiotic resistance are:

- to be careful with how antibiotics are prescribed
- to take the full course of prescribed antibiotics
- to discover new antibiotics.

Painkillers and other medicines

Painkillers do not kill pathogens, but can be used to treat the symptoms of disease, such as headache and fever.

Antiviral drugs are medicines that are used to treat diseases caused by a virus. However, this is difficult because when viruses invade the body, they move inside the cells. Any medicine that kills the virus would also have to attack the body's own cells, and this could cause damage inside our bodies.

WORKIT!

The graph shows the percentage of *Acinibacter*, a type of bacteria that are resistant to the antibiotic, imipenem, in different years. What data from the graph shows that antibiotic resistance is increasing? (3 marks)

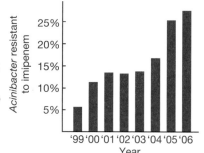

From 1999 to 2005, the percentage of *Acinbacter* that are resistant to imipenem increases. (1)

In 1999, 6% of *Acinbacter* are resistant to imipenem. (1)

In 2005, 27% of *Acinbacter* are resistant to imipenem. (1)

CHECKIT! ✓

1 Give an example of an antibiotic.
2 Explain why antibiotics should not be given to treat a viral disease.
3 Calculate the percentage increase in the antibiotic resistance of *Acinibacter* to imipenem from 1999 to 2005.

New drugs

New drugs are discovered every year. They may be extracted from plants or microorganisms, or synthesised in a laboratory.

Discovering new drugs

Traditionally, new drugs were discovered by extracting compounds from plants or microorganisms. For example:

- the painkiller aspirin comes from willow bark
- the heart drug digitalis comes from foxgloves
- penicillin comes from the *Penicillium* mould.

New drugs are still discovered in this way, but chemists in the pharmaceutical industry can also alter already existing compounds to make more effective medicines.

Testing new drugs

Before a new drug can be given to a patient, it must be tested for safety and effectiveness.

Step 1: preclinical testing – test new drug on cells, tissues and animals in a laboratory. The drug is tested for toxicity (is it poisonous?), efficacy (how well does it work?) and dose (how much of it do we need?). If the drug is toxic or not effective it does not pass to step 2.

Step 2: clinical trials – a very low dosage of the new drug is given to healthy volunteers and patients. If the drug is safe, it moves to step 3.

Step 3: clinical trials – different doses of the new drug are given to healthy volunteers and patients to find the optimum dose for the drug.

Step 4: wider clinical trials – the new drug is given to patients and the efficacy monitored. Often the drug is compared to another commonly used drug, or a placebo (something that looks like the new drug but contains no medicine). To test the efficacy without bias, the patients and the doctor are not told which drug they are taking. This is called a double blind trial.

The results of clinical trials are peer reviewed (judged by other scientists) and published in scientific or medical journals.

DO IT!

Read about Alexander Fleming and how he discovered penicillin from some dirty plates!

 STRETCH IT!

Sometimes clinical trials don't go to plan. Find out about an incidence when the clinical trial had to be stopped early.

WORKIT!

Explain why new drugs need to be tested before being given to patients. (3 marks)

To make sure that the drug is not toxic/ poisonous. (1)

To check the efficacy/how well the drug works. (1)

To find the correct dosage for the drug. (1)

✓ CHECKIT!

1 How are new drugs discovered?

2 Describe what happens during preclinical testing.

3 Why is it important to avoid bias during clinical trials?

Infection and response

1 a What is a pathogen?

   b Name two communicable diseases that are caused by bacteria.

   c Describe how bacterial diseases are transmitted.

   d What type of drugs can be used to treat bacterial infections?

2 a Compare the structure of bacteria and viruses.

   b Measles is a virus that is often caught by small children.

      i Describe the symptoms of measles.

      ii Identify some ways in which measles could be prevented.

   c TMV is a virus that affects plants.

      i What does TMV stand for?

      ii Explain how TMV affects the growth of plants.

3 The graph below shows the number of deaths from malaria from 2000 to 2015.

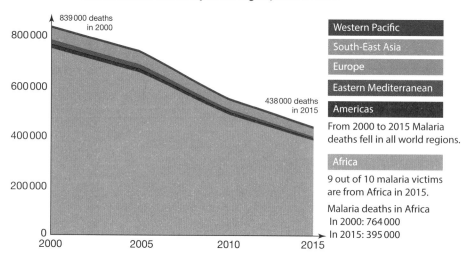

Global malaria deaths by world region, 2000 to 2015

Western Pacific

South-East Asia

Europe

Eastern Mediterranean

Americas

From 2000 to 2015 Malaria deaths fell in all world regions.

Africa

9 out of 10 malaria victims are from Africa in 2015.

Malaria deaths in Africa
In 2000: 764 000
In 2015: 395 000

839 000 deaths in 2000

438 000 deaths in 2015

   a What type of microorganism causes malaria?

   b i Describe the pattern of the graph.

      ii Suggest reasons for this pattern.

4 a Describe two non-specific defences the body has against pathogens.

   b Compare two types of white blood cell that defend the body against pathogens.

5 A new antibiotic (C) is being tested in a laboratory, along with two other antibiotics (A and B) that are in common use. The results are shown in the table below:

Antibiotic	Number of bacteria killed			Mean number of bacteria killed
	1	2	3	
A	34	35	37	35
B	44	39	42	42
C	140	126	132	133

   a i Describe the results of the table.

      ii Suggest a reason for these results.

   b Explain why it is important to keep making new antibiotics.

Photosynthesis

Photosynthesis is the process of using energy from sunlight to make sugars.

The process of photosynthesis

Photosynthesis takes place in the leaves. **Chloroplasts** in the cells of the leaf are packed with a pigment called chlorophyll. **Chlorophyll** absorbs sunlight and uses it to convert carbon dioxide and water into glucose (a type of sugar) and oxygen. The word equation for this is:

$$\text{carbon dioxide} + \text{water} \xrightarrow{\text{light}} \text{glucose} + \text{oxygen}$$

The chemical equation for photosynthesis is:

$$6CO_2 + 6H_2O \xrightarrow{\text{light}} C_6H_{12}O_6 + 6O_2$$

This is an **endothermic** reaction because more energy is taken in by the reaction than is given off. The **light** energy from the Sun is converted into **chemical** energy in the plant.

Reactants for photosynthesis

The carbon dioxide needed for photosynthesis enters the leaf through the stomata. The water gets to the leaf through the xylem (see page 43).

Products of photosynthesis

Product of photosynthesis	
Glucose	Used in respiration
	Makes other important molecules such as starch, fats and amino acids
Oxygen	Used in respiration
	Released from the plant through the stomata

DO IT!

Write out the word equation and the chemical equation until you have learned them by heart.

WORK IT!

Explain the effect on photosynthesis if a plant is left in a bright room but not watered. (3 marks)

Photosynthesis cannot occur. (1)

The plant would only have carbon dioxide and sunlight. (1)

The plant needs water to carry out photosynthesis. (1)

NAIL IT!

Don't forget to balance your equation – you need to count how many of each atom you have on each side of the equation.

CHECK IT!

1 What are the products of photosynthesis?

2 Explain where the concentration of chlorophyll would be highest in the leaf.

3 The chemical formula for glucose is $C_6H_{12}O_6$. Explain where the carbon atoms have come from.

Rate of photosynthesis

The rate of photosynthesis changes depending on a number of different, interacting factors.

Factors that affect the rate of photosynthesis

The rate of photosynthesis increases if:

- the temperature increases (up to an optimum temperature)
- light intensity increases
- carbon dioxide concentration increases
- the amount of chlorophyll increases.

Limiting factors

H

The rate of photosynthesis can be measured by measuring the volume of oxygen produced by a plant in a certain time, under certain conditions. The shape of the graph for each condition is shown below:

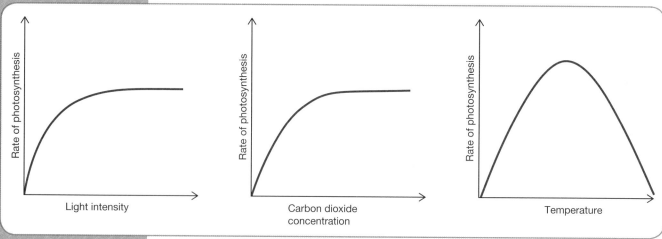

Copy out these graphs and take a photo to help you learn their shapes.

Effect of light and carbon dioxide

As the light intensity, or the carbon dioxide concentration increases, the rate of photosynthesis increases, until a certain point. After this point, the rate of photosynthesis remains constant. That is because one of the other factors is limiting the rate of photosynthesis. This is called the limiting factor. For example, in the first graph, as the light intensity increases and rate of photosynthesis increases, light intensity is the limiting factor. When the rate of photosynthesis no longer increases some other factor (such as the concentration of carbon dioxide) has become the limiting factor.

Effect of temperature

As the temperature increases, the rate of photosynthesis increases, until the optimum temperature for the enzymes and proteins that carry out photosynthesis is reached. Above this temperature, the enzymes and proteins begin to denature and can no longer carry out their function. Therefore the rate of photosynthesis decreases.

Measuring the rate of photosynthesis

In an experiment, the volume of oxygen given off by a plant is measured over time. The rate of photosynthesis can be calculated by dividing the volume of oxygen collected over time:

$$\text{Rate of photosynthesis} = \frac{\text{Volume of oxygen}}{\text{Time}}$$

WORKIT!

$12\,cm^3$ of oxygen is collected in 3 minutes. Calculate the rate of photosynthesis. (3 marks)

$\text{Rate of photosynthesis} = \frac{12\,m^3}{3\,min} = 4\,cm^3/min$ (3)

H

The inverse square law and light intensity in photosynthesis

In an investigation into the effect of light on the rate of photosynthesis, a lamp is moved away from the plant. As the distance from the plant increases, the light intensity decreases. The light intensity is inversely proportional to the square of the distance.

H

Increasing the rate of photosynthesis

Farmers need to increase the rate of photosynthesis in plants in order to increase the yield (the volume of plants, fruit or vegetables produced). They do this by:

- growing plants in a greenhouse to increase the temperature
- increasing the hours of light by switching lights on at night
- increasing the amount of carbon dioxide inside the greenhouse.

NAILIT!

Remember to include units when you answer a question.

STRETCHIT!

Read up on the greenhouses at Thanet Earth to get an idea of how photosynthesis is controlled to increase the yield of tomatoes.

CHECKIT!

H 1 Explain the term 'limiting factor'.

2 Describe and explain what would happen to the rate of photosynthesis if the carbon dioxide concentration were decreased.

3 Calculate the rate of photosynthesis if $28\,cm^3$ of oxygen is collected in 4 minutes.

Investigating the effect of light intensity on the rate of photosynthesis

In this practical you will need to use your knowledge of photosynthesis to hypothesize about how light intensity will affect the rate of photosynthesis.

The rate of photosynthesis at different intensities of light can be calculated by measuring the volume of oxygen given off by the plant in a certain time.

STRETCHIT!

There are many methods for carrying out this investigation and a diagram in the exam may not be the same as the one you used.

Find two alternative methods for investigating the rate of photosynthesis.

MATHS SKILLS

Calculating the rate of photosynthesis

The rate of photosynthesis can be calculated by dividing the volume of oxygen released by the time taken.

Practical Skills

Select suitable apparatus to collect the oxygen gas given off by a piece of pondweed. You will need to:

- make sure that the stem of the pondweed is cut underwater, to prevent any air bubbles getting into the stem
- make sure that the oxygen gas cannot escape from the glassware
- make sure there is only one source of light
- keep all variables, except for light intensity, constant.

WORKIT!

The data from the investigation is shown in the table below.
Calculate the rate of photosynthesis at each distance. (4 marks)

Distance of lamp from plant (cm)	Volume of oxygen released (cm³)	Time (min)	Rate of photosynthesis
0	200	5	200/5 = 40 cm³/min
10	200	5	200/5 = 40 cm³/min
20	100	5	100/5 = 20 cm³/min
30	60	5	60/5 = 12 cm³/min

(4)

NAILIT!

Think about any stage of your investigation that was not as accurate as it could have been. How could you improve the reliability and validity of your investigation?

CHECKIT!

1 Describe the pattern of the data in this investigation.

2 Give two variables that need to be controlled in this investigation.

3 Suggest how you could improve this investigation.

Uses of glucose

Glucose is a type of sugar. It is made by the process of photosynthesis, and has the chemical formula, $C_6H_{12}O_6$.

How is glucose used?

More glucose is made by photosynthesis than is needed, so it can be stored as starch or oil to be used later. Starch and oils can be broken down into glucose by enzyme action.

Use of glucose	Animals	Plants
Used in respiration to make energy	✓	✓
Converted into insoluble glycogen for storage	✓	
Converted into cellulose, which strengthens the cell wall		✓
Converted into insoluble starch for storage		✓
Used to produce fat for storage	✓	✓
Used to produce oil for storage		✓
Used to produce amino acids for protein synthesis		✓

To make proteins, plants also need to take up nitrates from the soil.

SNAPIT!

Take a photo of the table and learn how glucose is used.

WORKIT!

Compare the way in which glucose is used in animals and plants. (4 marks)

Glucose is used for respiration in both animals and plants. (1)

Glucose is used for making amino acids in plants. (1)

Glucose is stored as glycogen in animals, but as starch in plants. (1)

Glucose is used to produce fat for storage in animals, but usually used to produce oils for storage in plants. (1)

NAILIT!

Think back to digestion (page 27). Enzymes are used to digest starch into sugars, such as glucose.

✓ CHECKIT!

1 Where in a plant is glucose made?

2 What is the source of glucose for animals?

3 Explain the importance of storing glucose as glycogen or starch.

Respiration

Respiration happens in all living cells. It is the process of converting glucose into energy.

The process of respiration

Respiration is an exothermic reaction, because it releases more energy than is taken in. The chemical energy in glucose is converted into energy for the cell to:

- carry out chemical reactions to build larger molecules
- move
- keep warm.

Aerobic respiration

Aerobic respiration uses oxygen. It takes place in the cytoplasm of the cell, and in the mitochondria. The word equation for aerobic respiration is:

glucose + oxygen → carbon dioxide + water

The chemical equation for aerobic respiration is:

$$C_6H_{12}O_6 + 6O_2 \rightarrow 6CO_2 + 6H_2O$$

Aerobic respiration is carried out most of the time in animals and plants and releases a lot of energy, compared to anaerobic respiration.

Anaerobic respiration

Anaerobic respiration takes place without oxygen. It only takes place in the cytoplasm of the cell and does not involve the mitochondria. Muscle cells carry out this type of respiration when they are exercising quickly, for example, when a person is sprinting. It does not release very much energy, as the glucose is not completely oxidised, and can only be carried out in humans for a short time.

The word equation for anaerobic respiration in muscles is:

glucose → lactic acid

The lactic acid builds up in the muscle cells, causing muscle fatigue and cramp. The lactic acid is broken down using oxygen after the exercise is finished. This is called the oxygen debt.

STRETCH IT!

Find out how respiration occurs inside the mitochondria.

Plant and yeast cells can carry out anaerobic respiration for a much longer time, and produce different products. This process is called fermentation.

The word equation for anaerobic respiration in yeast and plants is:

$$glucose \rightarrow ethanol + carbon\ dioxide$$

Uses of fermentation

Product of fermentation	Use
Carbon dioxide	Causes bread and cakes to rise
Ethanol	Making beer/wine/spirits

WORKIT!

Compare the processes of aerobic and anaerobic respiration. (4 marks)

Aerobic respiration uses oxygen and anaerobic respiration does not use oxygen. (1)

Aerobic respiration releases more energy than anaerobic respiration. (1)

In humans, aerobic respiration produces carbon dioxide and water, whereas anaerobic respiration produces lactic acid. (1)

In yeast, aerobic respiration produces carbon dioxide and water, whereas anaerobic respiration produces carbon dioxide and ethanol. (1)

✓ CHECKIT!

1 Where in the cell does aerobic respiration take place?

2 Describe how yeast is used to make wine and bread.

3 The graph below shows yeast being grown anaerobically at different temperatures. Describe and explain the pattern.

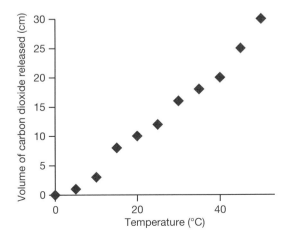

Response to exercise

The human body responds to the increased demand for energy during exercise.

The effect of exercise

During exercise, the muscle cells require more energy in order to keep moving. This means that the rate of cellular respiration must increase. In order to do this, the cells need more oxygen and glucose. These are supplied to the cell by:

- an increase in heart rate so that the blood flows to the cells more quickly
- an increase in the breathing rate to oxygenate the blood more quickly
- an increase in breath volume to take in more oxygen with each breath.

DO IT!

Test your pulse before and after one minute of vigorous exercise. Test your pulse again after five minutes rest. Has it returned to normal?

Practical Skills

Investigations into exercise

The heart rate can be measured by measuring the pulse rate. Count the number of pulses in 15 seconds and multiply the number by four. This gives the beats per minute.

Anaerobic exercise

When the body exercises vigorously, the cells may respire anaerobically. This means that no oxygen is used, and there is a build up of lactic acid in the muscles (see page 62) due to the incomplete oxidation of glucose. After a short time, the muscles will be fatigued and stop contracting efficiently.

H

The oxygen debt

Lactic acid is removed from the cells and transported through the blood to the liver. The liver uses oxygen to convert the lactic acid into useful products, such as glucose. This process requires extra oxygen and is called the oxygen debt.

H

WORKIT!

A person goes for a five minute jog. They measure their pulse rate before and after exercising, and find their pulse rate has increased. Explain why. (3 marks)

The muscle cells need to respire more quickly (1)
to provide more energy for the muscle cells to move. (1)
The heart beats faster to get oxygen and glucose to the muscle cells more quickly. (1)

CHECKIT! ✓

1 A person counts 12 pulses in 15 seconds. What is their heart rate?
2 Explain why the breathing rate increases during exercise.
3 Explain why sprinters need to breathe deeply after finishing a race.

Metabolism

Metabolism is all of the enzyme-controlled reactions in the body. The energy for these reactions comes from respiration in the cells.

Synthesis and breakdown of molecules

There are two types of metabolic reaction:

1 Complex molecules are made out of simpler ones. For example:

> **sugars** → complex carbohydrates
>
> **amino acids** → proteins
>
> **3 fatty acids + glycerol** → lipids

These reactions are important in the body for growth, repair and energy storage.

2 Complex molecules are broken down into simpler ones. For example:

> **complex carbohydrates** → sugars
>
> **proteins** → amino acids
>
> **lipids** → 3 fatty acids + glycerol

These reactions are important in digestion, to make the molecules small enough to be absorbed into the bloodstream in the small intestine.

Metabolism reactions

Here are some other important metabolic reactions:

- respiration
- in plants – formation of amino acids from glucose and nitrate ions
- in animals – breakdown of excess proteins to urea for excretion in urine.

DO IT!

Make revision cards of each metabolic reaction and learn some examples.

WORKIT!

Explain why metabolism needs energy. (3 marks)

Energy is needed to break the bonds in a large complex molecule. (1)

Enzymes need energy to carry out a reaction. (1)

Energy is needed to form new bonds between smaller molecules. (1)

CHECKIT!

1 Give an example of a metabolic reaction.

2 Explain why the production of complex molecules is important in the body.

3 Suggest why respiration is described as a metabolic reaction.

NAILIT!

Look back at the process of respiration on page 62.

1 a What is the chemical formula of glucose?

   b Name two uses of glucose in the human body.

   c Compare how glucose is stored in plants and animals.

2 a Explain why during exercise the breathing rate, breath volume, and heart rate increase.

   b Describe how you could measure the pulse rate.

   c Explain what is meant by an oxygen debt after anaerobic exercise.

3 Respiration is a metabolic reaction.

   a Give one further example of a metabolic reaction.

   b i What is the word equation for aerobic respiration?

      ii What is the balanced symbol equation for aerobic respiration?

4 a What are the products of photosynthesis?

   b Name two factors that increase the rate of photosynthesis.

   c What is meant by the term 'limiting factor'?

5 Some students investigated the rate of photosynthesis at different light intensities.

   a i Draw the expected shape of the graph.

      ii Label the axes correctly.

   b Explain the shape of the graph you have drawn.

6 An investigation into the rate of photosynthesis in an aquatic plant at different temperatures was been carried out. The results are shown in the table below:

Temperature (°C)	Volume of oxygen released (cm³)	Time (min)	Rate of photosynthesis (cm³/min)
20	100	5	20
30	200	5	
40	150	5	
50	50	5	
60	0	5	

   a Calculate the rate of photosynthesis for each temperature.

   b Identify the optimum temperature for photosynthesis in this plant.

   c i Explain why the volume of oxygen increased from 20°C to 30°C.

      ii Explain why the volume of oxygen decreased from 30°C to 60°C.

   d Suggest how the researchers could make their data more reliable.

Homeostasis and response

Homeostasis

Homeostasis keeps the internal conditions of the body constant, whatever the outside environment may be.

The role of homeostasis

The body works best under optimum conditions. Enzymes work best at a temperature of 37°C, and our cells work best with optimum blood glucose levels and water levels. The process that regulates these conditions is called homeostasis. If the internal or external environment changes, it is the role of homeostasis to bring the conditions back to the optimum again.

Examples of homeostasis

	Action if it increases	Action if it decreases
Temperature	Mechanisms to cool the body down	Mechanisms to warm the body up
Blood sugar levels	Release of **insulin** so that cells take up more glucose	Release of **glucagon** to break down glycogen into glucose
Water levels	Less **ADH** is released and more urine is produced	More **ADH** is released and less urine is produced

How homeostasis works

The optimal conditions are under automatic control. If conditions change, nervous or chemical responses bring them back to normal.

Control systems

Receptors ⟶ Coordination centre ⟶ Effector

Cells that detect stimuli

Receives and processes information from receptors

Responds to restore optimum levels

A stimulus can be any change in the internal or external environment. It is detected by a receptor, that sends a nerve impulse to the coordination centre, usually the brain (but could be the spinal cord or pancreas). The coordination centre sends a nerve impulse to an effector (muscle or gland) which brings about a response.

CHECKIT!

1 What is homeostasis?

2 Give an example of a condition that needs to be kept at an optimum level in the body.

3 Describe the role of the coordination centre in controlling homeostasis.

The human nervous system

Central Nervous System
brain
spinal cord

Peripheral N.S.

ganglion

nerve

The human nervous system is made of nerve cells, or neurones.

The structure of the human nervous system

The human nervous system is divided into two sections: the central nervous system (CNS) and the peripheral nervous system (PNS).

The central nervous system is made up of the brain and spinal cord and coordinates all of the nerve impulses. The peripheral nervous system (PNS) is made up all of the other nerves and sends and receives nerve impulses.

How the nervous system works

The nervous system allows the body to react to the surroundings and coordinate the body's behaviour. Receptors detect a stimulus and send electrical impulses along the nerves to the spinal cord. The spinal cord sends the electrical impulses to the brain. The brain is a coordinator and sends more electrical impulses back down the spinal cord, and along the nerves to muscles or glands (effectors). The effectors can then respond to the stimulus by either moving a muscle or causing a gland to secrete.

stimulus → receptor → coordinator → effector → response

DOIT!

Write each of the steps in the flow chart on cards and practise putting them into the right order.

NAILIT!

In Q3 think about the function of each part of the CNS and PNS.

WORKIT!

A car driver sees that the traffic light has turned to red. Describe the action of the nervous system. (4 marks)

Retinal cells in the eye are receptors to the red colour (the stimulus). (1)

An electrical impulse is sent to the brain (the coordinator). (1)

The brain sends an electrical impulse to the effectors, the muscles in the foot. (1)

The foot muscles move and put on the brake. This is the response. (1)

CHECK**IT!** ✓

1 What are the brain and spinal cord made from?

2 Describe the role of the coordinator.

3 Compare the structure and function of the central nervous system and the peripheral nervous system.

Reflexes

Reflexes are automatic, rapid responses to stimuli that do not involve the conscious part of the brain.

The structure of neurones

There are three types of neurone in a reflex arc; a sensory neurone, a relay neurone and a motor neurone.

The electrical impulses move along the long axon in the direction of the arrows. The dendrites at the end of the neurone pass the electrical impulse to the next neurone by causing a chemical to diffuse across the small gap (the synapse). When the chemical reaches the next neurone, it causes the electrical impulse to continue. The round cell bodies, which contain the nucleus, are in different positions on the different types of neurones.

The reflex arc

In a reflex arc, the receptor detects a stimulus and sends an electrical impulse along the sensory neurone to the spinal cord. It passes to a relay neurone in the spinal cord, which sends the electrical impulse straight back along a motor neurone. The motor neurone sends the electrical impulse to an effector which carries out a response. A reflex action does not involve the conscious part of the brain.

SNAP IT!

Sketch these neurones to learn their shape and take a photo for recalling later.

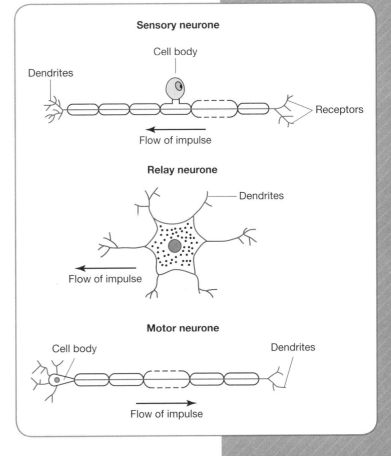

Sensory neurone

Cell body

Dendrites

Receptors

Flow of impulse

Relay neurone

Dendrites

Flow of impulse

Motor neurone

Cell body

Dendrites

Flow of impulse

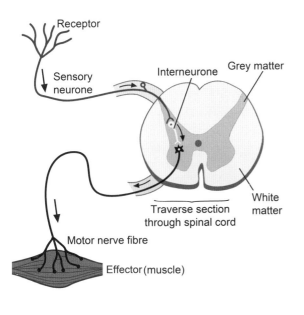

Receptor

Sensory neurone

Interneurone

Grey matter

Traverse section through spinal cord

White matter

Motor nerve fibre

Effector (muscle)

STRETCH IT!

Find out more about the structure of neurones.

The importance of the reflex arc

Reflex arcs are quicker than normal reactions and protect us from harm. They allow us to move away from danger or prevent something dangerous from happening.

Some examples of reflexes are:

- blinking when something is near to your eye
- the pupil in your eye contracting in bright light
- coughing reflex when something irritates the top of your windpipe (trachea)
- knee jerk when your lower leg is struck with a small hammer just below your knee.

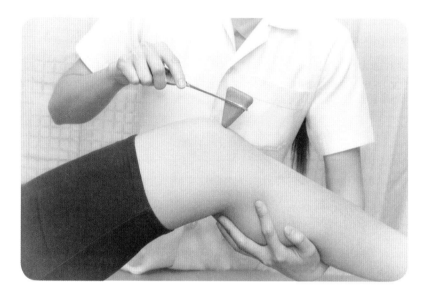

WORKIT!

Compare and contrast the three neurones used in a reflex arc. (4 marks)

Sensory neurones, motor neurones and relay neurones all have an axon and carry an electrical impulse. (1)

Sensory neurones carry an electrical impulse to the spinal cord, motor neurones carry the electrical impulse away from the spinal cord. (1)

Sensory neurones, motor neurones and relay neurones all have dendrites and pass the electrical impulse onto nearby neurones. (1)

Sensory neurones, motor neurones and relay neurones all have a cell body, but in different locations on the neurone. (1)

CHECKIT! ✓

1 What is a reflex?

2 Give two examples of a reflex.

3 Describe what happens when a light is shone into someone's eye.

Investigating the effect of a factor on human reaction time

In this experiment you will be expected to plan an investigation choosing appropriate apparatus and techniques to measure the process of reaction time.

Practical Skills

There are many factors affecting reaction time which you can investigate, for example, before and after drinking a caffeinated drink, right hand versus left hand, morning versus afternoon, and before and after eating.

Popular reaction experiments include timing how long it takes to catch a ruler or using a computer program where a button is pressed when an image comes up on the screen.

Make sure that you:

• carry out your reaction experiment before and after your condition/treatment

• repeat your results three times so you can calculate a mean

• show your results as a table and a graph.

DO IT!

Look at your investigation and note down wherever the data was not valid or reliable, and what you could do to improve it.

NAIL IT!

Look at the headings of the columns as well as the data in the table to help with part c of the Work It!

WORKIT!

A student wanted to test the effect of caffeine on reaction times. Their data is shown in this table.

Condition	Reaction time 1 (s)	Reaction time 2 (s)	Reaction time 3 (s)	Mean
Before caffeine	0.62	0.65	0.66	0.64
After caffeine	0.34	0.4	0.38	0.37

a Which variables would need to be kept constant during this investigation? (3 marks)

The amount of caffeine drunk. (1) The time of day when the drink was taken. (1) The person who drank the caffeine drink. (1)

b What conclusion can you make about the data? (1 mark)

The reaction time decreases after taking caffeine. (1)

c Correct any mistakes in the table. (2 marks)

The top of the mean column should read 'mean time (s)'. (1)
The 0.4 time for after caffeine, reaction time 2 should be 0.40. (1)

✓ CHECKIT!

1 In the Work It! investigation, what is the dependent variable?

2 Why should investigations be repeated?

3 Suggest why drinking caffeine decreases the reaction time.

Human endocrine system

The human endocrine system is made up of many glands that secrete hormones into the blood. Hormones act as chemical messengers.

How the endocrine system works

Each gland makes one or more hormones, which are then secreted directly into the bloodstream.

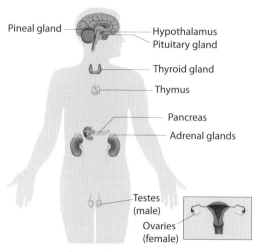

The Endocrine System

Gland	Hormone
Pituitary gland	Many hormones, including growth hormone, follicle-stimulating hormone (FSH) and luteinising hormone (LH)
Pancreas	Insulin and glucagon
Thyroid	Thyroxine
Adrenal gland	Adrenaline
Ovary	Oestrogen and progesterone
Testes	Testosterone

The blood carries the hormone to a target organ or cells, where the hormone has an effect.

The pituitary gland

The pituitary gland in the brain controls many of the other glands in the body by releasing hormones that affect them. For example, the pituitary releases thyroid-stimulating hormone (TSH), which targets the thyroid and causes it to release thyroxine. For this reason, the pituitary gland is often called the 'master gland'.

Endocrine system versus nervous system

The endocrine system acts more slowly than the nervous system, but the effects of the endocrine system last for a longer time.

CHECK IT! ✓

1 What is the name of the endocrine gland on top of the kidneys?

2 Give one example of an endocrine gland and the hormone that it produces.

3 Give one similarity and one difference between the endocrine and nervous systems.

Control of blood glucose concentration

STRETCH IT!

Find out how a lack of insulin affects people with diabetes.

Blood glucose concentration is monitored and controlled by the pancreas.

The role of the pancreas

The pancreas monitors the blood glucose concentration as the blood passes through it.

- If the blood glucose concentration is too high, for example, just after a meal, then the pancreas releases insulin into the blood.

- If the blood glucose concentration is too low, for example, several hours after eating, then the pancreas releases glucagon into the blood.

SNAP IT!

Make a copy of the negative feedback cycle below and take a photo.

Insulin

Insulin is a hormone that targets all cells in the body. It causes the cells to take up glucose (a type of sugar). This decreases the concentration of glucose in the blood back to the normal levels. In the liver and muscle cells, any excess glucose is converted into glycogen for energy storage.

Glucagon

Glucagon is a hormone that targets the liver and muscle cells. Glucagon causes the liver and muscle cells to break down glycogen into glucose and release it into the blood. This increases the concentration of blood glucose back to the normal levels.

Negative feedback cycle

The actions of insulin and glucagon can be described as a negative feedback cycle. In a negative feedback cycle, when something increases or decreases from the normal level, the effect works in the opposite direction to bring it back to the normal level. The diagram on the right shows the negative feedback cycle of insulin and glucagon.

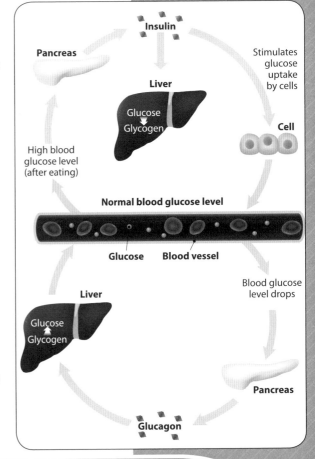

CHECK IT!

H 1 Where is glucagon made?

H 2 Describe the role of glycogen.

H 3 Explain how the blood glucose level is maintained using a negative feedback cycle.

Diabetes

In people with diabetes, either no insulin is produced, or the body stops responding to the insulin in the blood.

Types of diabetes

Not being able to control body sugar levels is dangerous for the body. The normal blood glucose level is 90 mg/dl, rising to 135 mg/dl immediately after a meal.

- When the blood glucose level is too low, it is called hypoglycaemia. This can cause shaking, dizziness, and even coma.

- When the blood glucose level is too high, it is called hyperglycaemia. This can cause kidney damage or damage to the eyes.

Type 1 diabetes

Type 1 diabetes is caused when the pancreas does not produce enough insulin. It is often identified in children, and is thought to have a genetic cause. People with type 1 diabetes have to inject insulin several times a day.

Type 2 diabetes

Type 2 diabetes is caused by lifestyle factors, such as obesity and poor diet. The pancreas still produces insulin but the body does not respond to it in the correct way. People with type 2 diabetes have to follow a carbohydrate-controlled diet and exercise regularly to control their condition.

Blood glucose levels in diabetes

After a meal, the blood glucose levels increase in people with and without diabetes. In those without diabetes, insulin is released and the blood glucose levels start to return to normal. For those with diabetes, the blood glucose levels remain high unless treated.

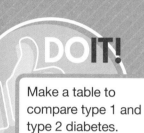

DO IT!

Make a table to compare type 1 and type 2 diabetes.

NAIL IT!

One of the risk factors for developing diabetes is a high BMI. BMI is measured using a person's height and weight:

$$BMI = \frac{weight\ (kg)}{height^2\ (m^2)}$$

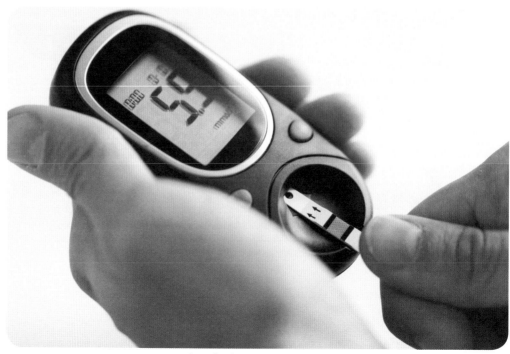

Blood glucose monitor

WORKIT!

This graph shows the blood glucose levels in a person with type 2 diabetes and a person who does not have diabetes. Describe and explain the line of the graph for the person with type 2 diabetes. (4 marks)

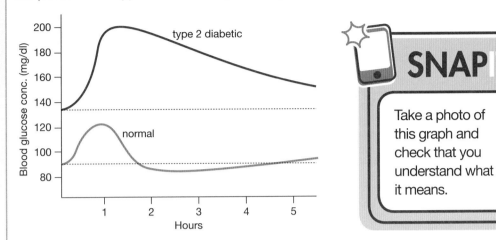

SNAPIT!

Take a photo of this graph and check that you understand what it means.

The blood glucose level starts at 135 mg/dl and increases to 200 mg/dl in 1.25 hours. (1)

From 1.5 to 5.5 hours the blood glucose levels decrease slowly to 150 mg/dl. (1)

This is because body does not respond in the correct way to the insulin produced by the pancreas, (1) and glucose in the blood is not taken up by the cells. (1)

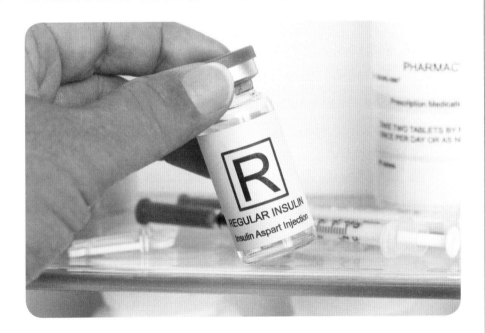

CHECKIT!

1 What treatment is given to people with type 1 diabetes?

2 Describe the symptoms of a person with diabetes when the blood glucose levels are too low.

3 What advice would you give to a person who is worried about developing type 2 diabetes?

Hormones in human reproduction

Human reproduction involves the interaction of many hormones.

Hormones during puberty

During puberty, the sex hormones are released in males and females. These hormones cause secondary sex characteristics to develop.

- In males testes secrete testosterone. Males develop deeper voices, chest and face hair, a more muscular body, and pubic hair. The genitals develop and sperm are produced.

- In females ovaries secrete oestrogen and progesterone. Females develop breasts, broader hips and pubic hair. Menstruation starts.

The menstrual cycle

Menstruation, or the menstrual cycle, lasts an average of 28 days, and is controlled by four hormones:

- Follicle stimulating hormone (FSH) – causes eggs (ova) in the ovaries to mature.

- Luteinising hormone (LH) – stimulates the release of an ovum (ovulation).

- Oestrogen – repairs the lining of the uterus after menstruation.

- Progesterone – maintains the uterus lining.

SNAP IT!

Make a sketch of the graph and check that you understand what it means.

DO IT!

Make a flow diagram of the steps involved in releasing the different hormones during the menstrual cycle.

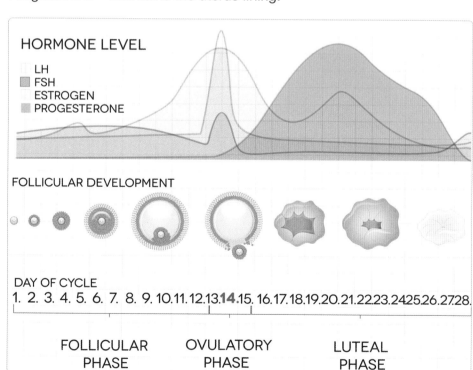

Interaction of hormones in the menstrual cycle

The pituitary gland releases FSH at the start of the menstrual cycle. FSH acts on the ovaries and causes many ova to mature. It also causes the ovaries to secrete oestrogen. Oestrogen slows down the production of FSH, so that usually only one ovum reaches full maturity. Oestrogen also stimulates the pituitary gland to release LH.

LH peaks around day 14 of the menstrual cycle, causing the mature ovum to be released from one of the ovaries. The release of the ovum causes large amounts of progesterone to be secreted by the ovaries. This causes a negative feedback loop, which stops FSH and LH from being produced by the pituitary gland. This in turn causes a decrease in oestrogen.

WORKIT!

Describe what happens to the hormone levels in a female body if an ovum is not fertilised. (3 marks)

The progesterone levels decrease. (1)

The oestrogen levels decrease. (1)

This means that FSH is no longer inhibited and the levels of FSH increase. (1)

NAILIT!

To answer this question, look at day 27 on the menstrual cycle graph on the previous page.

 STRETCHIT!

At Higher Tier you should be able to explain the interactions of FSH, oestrogen, LH and progesterone, and their role in the control of the menstrual cycle.

✓ CHECKIT!

1 Name the hormone involved in developing male secondary sex characteristics.

2 What is the role of LH?

3 Suggest what happens to the levels of hormone in the female body if the ovum is fertilised.

Contraception

Contraception is any method used to prevent pregnancy.

Non-hormonal methods of contraception

This type of contraception uses barrier methods to prevent sperm reaching the egg. Non-hormonal methods include:

- using condoms/diaphragms – trap sperm
- spermicidal agents – kill or disable sperm
- intrauterine device which prevents the implantation of an embryo (may also release hormones)
- surgical methods (sterilisation) – cut and tie sperm ducts in males to prevent sperm leaving the penis, or cut and tie oviducts (Fallopian tubes) in females to prevent eggs travelling from the ovaries to the uterus
- abstaining from intercourse around the time that an egg may be in the oviducts.

Hormonal methods of contraception

Some methods of contraception use hormones to prevent the release of an egg during the menstrual cycle. For example:

- oral contraceptives (the pill) – contain oestrogen and progesterone so that FSH is inhibited, and no eggs mature (see pages 76–77)
- injection, skin patch, or implant under the skin – contain progesterone to inhibit the maturation and release of eggs. These last several months.

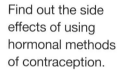

DO IT!

Write each method of contraception onto revision cards.

STRETCH IT!

Find out the side effects of using hormonal methods of contraception.

WORKIT!

Evaluate the hormonal and non-hormonal methods of contraception. (4 marks)

Condoms do not have any side effects, but are not 100% effective at preventing pregnancy. (1)

Oral contraceptives are effective at preventing pregnancy, but have some side effects on the female body. (1)

Condoms and oral contraceptives allow couples to choose the time to start a family. (1)

Surgical methods are very effective at preventing pregnancy but are difficult to reverse. (1)

CHECK IT!

1 Give an example of a non-hormonal method of contraception.

2 Describe how the hormones in oral contraceptives prevent eggs from maturing.

3 Suggest why some women may prefer to use a hormonal injection, rather than use oral contraceptives.

Using hormones to treat infertility

FSH and LH are used in in vitro fertilisation as fertility drugs.

In vitro fertilisation (IVF)

IVF is a type of fertilisation that happens outside of the body. Eggs are surgically removed from the mother and fertilised by the father's sperm in a laboratory. The fertilised eggs develop into embryos and are then placed surgically into the woman's uterus.

FSH and LH are given to the woman in order to produce enough eggs for IVF. These hormones cause several eggs to reach maturity so that several eggs can be fertilised by IVF at the same time.

Advantages and disadvantages of IVF

Advantages

- It allows the mother to give birth to her own baby.
- The baby will be the genetic offspring of the mother and father.

Disadvantages

- The success rates are not high – many couples need to have several rounds of IVF before a successful birth.
- It can lead to multiple births (two, three or more babies at once) as several embryos are implanted at once. This is less safe for the mother and the babies, and can lead to health problems.
- It is emotionally and physically stressful.

STRETCHIT!

Find out about mothers who have had multiple births through IVF.

DOIT!

Write each advantage and disadvantage of IVF onto revision cards.

NAILIT!

In the Work It! think about techniques and equipment that would be needed.

WORKIT!

IVF has been around since 1978. Suggest why IVF was not available before this date. (3 marks)

Developments in light microscopes that allowed the sperm and eggs to be viewed. (1) Developments in surgical techniques to remove eggs and implant embryos safely. (1) Developments in forms of FSH and LH that can be safely given to women undergoing IVF. (1)

✓ CHECKIT!

H 1 Which hormones are given to women as fertility drugs?

H 2 Describe what happens during IVF.

H 3 What social and ethical issues are associated with IVF treatments?

Negative feedback

Negative feedback is a mechanism that keeps the body functioning at set levels. If something goes above or below the set level, negative feedback brings it back again.

Adrenaline

Adrenaline is a hormone produced by the adrenal gland. It is secreted into the bloodstream when the body is under stress, for example, if a person is scared or excited. Adrenaline has several effects that help prepare the body for 'fight or flight':

- heart rate increases – to increase blood flow around the body
- breathing rate increases – to increase the amount of oxygen in the blood
- blood is redirected to the muscles – more blood flows to the muscles and less to the skin and intestines
- stimulates the liver – to breakdown glycogen into glucose to provide more glucose for muscles to use for respiration
- dilated pupils – more light enters the eyes.

Thyroxine

Thyroxine is a hormone secreted by the thyroid gland. Thyroxine stimulates the basal metabolic rate, which is important in growth and development.

If the thyroxine levels in the body are low, the thyroid gland is stimulated by the hypothalamus to secrete more; If the thyroxine levels in the body are high, the thyroid gland is stimulated by the hypothalamus to stop producing thyroxine.

This is a type of negative feedback.

DOIT!

Draw a diagram showing the negative feedback cycle for thyroxine.

WORKIT!

A person is feeling very tired and lacks in energy. Describe what will happen inside the body to return the person to their normal basal metabolic rate. (3 marks)

The hypothalamus stimulates the thyroid gland. (1) The thyroid gland secretes thyroxine. (1) Thyroxine stimulates the basal metabolic rate. (1)

CHECKIT! ✓

H1 Which organ system is the thyroid gland part of?

H2 Name two effects of adrenaline.

H3 Explain why the secretion of adrenaline is not controlled by negative feedback.

1 a Name two neurones in the nervous system.

  b Compare a voluntary nerve response to a reflex.

  c Students A and B have carried out an investigation into reaction times before and after drinking a drink containing 50 mg caffeine. Their results are shown below:

Condition	Student A	Student B
Before 50 mg caffeine	0.45	0.68
After 50 mg caffeine	0.40	0.59

  i Student A says this proves that caffeine increases reaction rates. Are they correct? Justify your answer.

  ii How could they improve their investigation?

2 a Define Homeostasis.

  b Insulin is a hormone that is made by the pancreas. Explain how insulin helps to control the concentration of glucose in the blood.

H c Explain the role of glucagon in controlling blood glucose concentration.

  d Diabetes is a condition that affects many people in the UK.

  i Give two differences between type 1 and type 2 diabetes.

  ii Describe how the blood glucose level after a meal would be different in a person with diabetes.

  iii People with diabetes have difficulty controlling their blood glucose concentration. Explain why.

3 a FSH is a hormone produced by the pituitary gland. Explain its role in the menstrual cycle.

  b FSH is given to women undergoing in vitro fertilisation (IVF).

  i Explain the role of FSH in IVF.

  ii What are the advantages and disadvantages of IVF treatment.

4 a Adrenaline and thyroxine are hormones produced by the endocrine system. When might adrenaline be released?

  b What effect does thyroxine have on the body?

  c Describe how the release of thyroxine is a type of negative feedback.

Inheritance, variation and evolution

DO IT!

Make a table comparing sexual and asexual reproduction.

WORKIT!

Describe how plants reproduce sexually. (3 marks)

Plants produce gametes/pollen/ovules. (1)

Gametes contain half of the genetic information needed to make a new plant. (1)

The gametes combine to produce seeds and grow a new plant. (1)

Sexual and asexual reproduction

Sexual reproduction requires two parents, whereas **asexual** reproduction only requires one parent.

Sexual reproduction

Sexual reproduction requires a male and a female. Both the male and the female produce gametes (sperm or pollen in males, eggs or egg cells in females.) Gametes are produced by meiosis (see page 83) and have half of the genetic information of a body cell. During meiosis, the genetic information is mixed, so that each of the gametes has different genetic material. This leads to variety in the offspring.

When the male gamete and the female gamete meet, the nuclei from each fuse together. This makes the first cell of the new organism – the zygote. In plants, the zygote is a seed. In animals, the zygote divides by mitosis into a ball of cells called an embryo. In mitosis, all of the cells are genetically identical.

Asexual reproduction

Some organisms can reproduce without another parent, using mitosis. All of the offspring are genetically identical to the parent. These are natural clones. There are several different mechanisms of asexual reproduction:

- Binary fission – bacteria divide their genetic material and elongate to double their normal size. They then break into two identical daughter cells.

- Budding – yeast double their genetic material and sub-cellular structures and form small buds on their surface. The buds develop and then break off from the parent yeast.

- Runners, bulbs and tubers – plants use these methods to grow new plants. For example, strawberry plants send out runners. Where they find soil, the runner develops roots and forms a new plant.

CHECKIT! ✓

1 Give two examples of asexual reproduction.
2 What is a zygote?

Meiosis

Meiosis is a type of cell division that produces gametes. The gametes are sperm and eggs in animals, and pollen and egg cells in plants.

The process of meiosis

When a cell in the reproductive organs divides by meiosis, it goes through two rounds of cell division to produce four daughter cells. Each daughter cell has half of the number of chromosomes as the original body cell. The chromosomes are randomly assorted into the four daughter cells so that they are genetically different from each other. This is the cause of variation.

Fertilisation

Fertilisation happens when a male and a female gamete join together and the two nuclei fuse. When this happens, the new cell (the zygote) has a full set of chromosomes. The zygote divides by a type of cell division called mitosis (see page 17). The zygote divides many times to form an embryo. All of the cells in the embryo are genetically identical. As the embryo develops, the cells begin to differentiate into specialised cells.

Meiosis

| Replication

Cell Division

Cell Division

SNAPIT!

Sketch the diagram of meiosis and recall the number of chromosomes in the parent and daughter cells. Remember to take a photo!

NAILIT!

You will not be expected to know the stages of meiosis.

WORKIT!

Describe what happens during meiosis. (3 marks)

A cell in a reproductive organ goes through two rounds of division. (1)

The number of chromosomes in the daughter cells is halved. (1)

Chromosomes are randomly assorted into four daughter cells to cause variation. (1)

✓ CHECKIT!

1 An animal's body cells contain 12 chromosomes. How many chromosomes will the gametes contain?

2 What is mitosis?

3 Explain how zygotes contain the same number of chromosomes as the body cells of that organism.

DNA and the genome

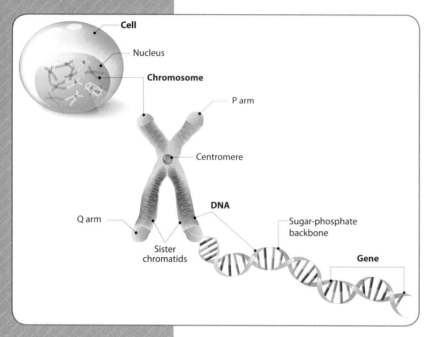

DNA (deoxyribonucleic acid) is the genetic material of the cell.

The structure of DNA

DNA is found in the nucleus of a cell. It is made of two strands, wound around each other to form a double helix. DNA is the material that makes up genes. Each gene contains a code to make a specific protein. Many genes are folded into large structures called chromosomes.

The genome

The genome of an organism is all of the genetic material of that organism. The genome contains all of the information needed to make that organism.

In humans, the genome is all 46 chromosomes in the nucleus. The human genome project has sequenced the whole of human genome and this is now being studied.

Understanding the genome

The human genome is being studied to find out:

- which genes are linked to diseases
- how to treat inherited disorders
- how to trace human migration patterns from the past.

> You will only need to know these reasons to study the human genome.

SNAPIT!

Sketch the diagram and labels to learn the structure of the DNA molecule. Take a photo to revise from later.

WORKIT!

Describe the genetic material of a cell. (4 marks)

The genetic material of a cell is made out of DNA. (1)

A small section of DNA is called a gene. (1)

Each gene codes for a particular protein. (1)

Many genes fold into chromosomes. (1)

CHECKIT!

1 What is a genome?

2 Describe the shape of the DNA molecule.

3 Explain the importance of studying the human genome.

Genetic inheritance

You inherit some genes from your mother and some from your father. Many characteristics are controlled by many genes, but where the characteristic is controlled by a single gene, the pattern of inheritance can be worked out using genetic diagrams.

Genetics keywords

Gamete – a sex cell, for example, sperm or an egg, containing half of the genetic material.

Chromosome – a large structure in the nucleus, made of DNA.

Gene – a short section of DNA that codes for a specific protein.

Allele – a version of a gene.

Dominant – an allele that is always expressed.

Recessive – an allele that is not expressed when a dominant allele is present.

Homozygous – when two copies of the same allele are present.

Heterozygous – when two different alleles are present.

Genotype – the alleles that are present in the genome.

Phenotype – the characteristics that are expressed by those alleles.

Single gene inheritance

Some characteristics are controlled by a single gene. The genotype is shown by two letters, a capital letter to represent the dominant allele, and a lower case letter to represent the recessive allele. The phenotype depends on the alleles of the gene that are present.

DO IT!

Put each keyword on a revision card with the definition on the back and test yourself.

Fur colour in mice

The fur colour gene in mice controls whether the fur colour is black or brown. There are two alleles, a dominant one, which is labelled B, and a recessive one that is labelled b.

- Mice with two dominant alleles, BB – black fur.
- Mice with one dominant and one recessive allele, Bb – black fur.
- Mice with two recessive alleles, bb – brown fur.

Red-green colour blindness in humans

The gene for red-green colour blindness is found on the X sex chromosome. If a female has two recessive alleles, or a male has one recessive allele, then that person will have red-green colour blindness.

- Male with the dominant allele, XY – normal vision.
- Male with the recessive allele, X^cY – red-green colour blindness.
- Female with one dominant and one recessive allele, XX^c – normal vision.
- Female with two recessive alleles, X^cX^c – red-green colour blindness.

Predicting inheritance

If the genotypes of the parents are known, it is possible to work out the possible genotypes and phenotypes of the offspring. It is also possible to calculate the probability of each genotype and phenotype.

For example, if two black mice that have the genotype Bb breed together, the possible genotypes of the offspring are BB, Bb or bb. The probabilities of each genotype are 1:2:1.

Therefore, out of four potential offspring, three would have black fur, and one would have brown fur, in a ratio of 3:1.

STRETCHIT!

Practise working out genotypes and phenotypes and the probabilities of each for the offspring, with parents of different genotypes.

WORKIT!

Describe and explain how the phenotype of a characteristic is controlled. (4 marks)

Each gene has two alleles that are present. (1)

Dominant alleles are always expressed if at least one dominant allele is present. (1)

Recessive alleles are only expressed if both of the alleles are recessive. (1)

The expressed alleles give the phenotype. (1)

NAILIT!

For Q2a work out the genotypes of the parents first.
The woman will have the genotype, X^{c}X and the man will have the genotype, XY.

CHECKIT! ✓

NAILIT!

Remember, probability is represented as a number between 0 and 1, **not** a percentage (%).

1 What does 'homozygous dominant' mean?

2 **a** A woman who is heterozygous for red-green colour blindness has a child with a man with normal vision. What are the possible genotypes for the child?

H b What is the probability that they will have a boy with red-green colour blindness?

Punnett squares

A Punnett square is a type of genetic diagram. You can use these to predict the genotypes of offspring.

How to use a Punnett square

Punnett squares can be used to work out the genotypes of offspring, if you know the genotypes of the parents. A small grid, 2 x 2 is drawn with the genotype of one parent on top of the grid, and the genotype of the other parent on the side of the grid.

Working out ratios

A ratio shows how many of the offspring are expected to have each phenotype. For example, in the Punnett square to the right, the ratio is 3:1. Three offspring would be expected to have the dominant characteristic and one offspring would be expected to have the recessive characteristic.

The letter at the top of each column is written in the boxes below. The letter at the side is written in each row. This gives four possible genotypes for the offspring:

	A	**a**
A	AA	Aa
a	Aa	aa

WORKIT!

In some flowers, petal colour is controlled by a single gene. The dominant allele, R, gives the petals a red colour. The recessive gene, r, gives petals a white colour.

a   What are the possible genotypes of the offspring if one parent has the genotype Rr, and the other parent has the genotype rr? (2 marks)

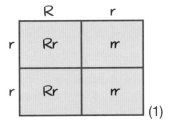

	R	r
r	Rr	rr
r	Rr	rr

(1)

The possible genotypes are Rr and rr. (1)

b   What percentage of the offspring will have white petals? (1 mark)

50% (1)

c   What is the ratio of phenotypes? (1 mark)

1:1 (1)

NAILIT!

All students need to be able to complete a Punnett square diagram. Only Higher Tier students need to be able to create their own Punnett square diagrams to make predictions about genetic crosses.

DOIT!

Practise using Punnett squares to work out the possible genotypes with parents of different genotypes. Try using RR and Rr, then Rr with rr. Are there any other combinations?

Family trees

It is possible to work out the genotypes of parents and offspring by looking at family trees.

Genetic pedigree A

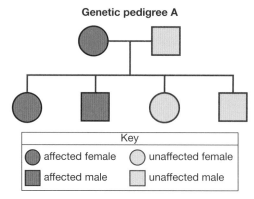

Key	
● affected female	○ unaffected female
■ affected male	☐ unaffected male

This example shows a disease caused by a single dominant allele. The mother and two of her children have one dominant allele and therefore have the disease. The father and the other two children have two recessive alleles and do not have the disease.

STRETCH IT!

Look at the pedigree for the British royal family and look at the inheritance of the gene for haemophilia.

WORK IT!

This family tree shows the inheritance of cystic fibrosis. People with two recessive alleles, cc, will have cystic fibrosis.
What are the genotypes of the parent of the man with cystic fibrosis? Explain how you know this. (3 marks)

Both parents must have the genotype Cc. (1)

The parents need to have at least one recessive allele to pass on to their son. (1)

If they had two recessive alleles, they would have cystic fibrosis. (1)

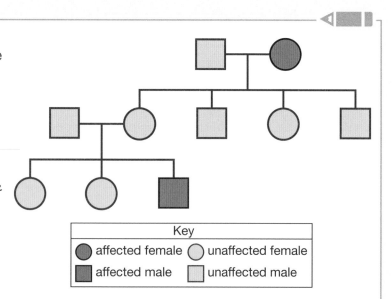

Key	
● affected female	○ unaffected female
■ affected male	☐ unaffected male

CHECK IT! ✓

NAIL IT!

Use a Punnett Square to work out the answer to Q3.

1 What is a dominant allele?

2 If the results of a Punnett square show that one offspring is homozygous dominant, two offspring are heterozygous, and one offspring is homozygous recessive. What is the ratio of their phenotypes?

H 3 In the above worked example, the woman with cystic fibrosis has four children. The father does not have a recessive allele for cystic fibrosis. What are the possible genotypes of the children?

Inherited disorders

Some disorders are caused by a single gene and can be inherited.

Disorders caused by a single gene

Inherited disorders can be caused by a dominant or a recessive allele. For example:

- Polydactyl – having an extra finger or toe. Caused by a dominant allele.
- Cystic fibrosis – a disorder that affects the cell membranes of the lungs and pancreas. Caused by a recessive allele.

People who know that they carry an allele for a disorder may wish to go through genetic counselling. This is when they discuss the likelihood of passing on their allele. Some people wishing to have children may go through embryo screening.

Embryo screening

Embryos can be screened to see if they contain the allele for the disorder. Only embryos without the allele will be implanted into the mother's uterus.

Advantages and disadvantages

Advantages	Disadvantages
Can select embryos without the allele for a disorder	Embryos with the undesired allele are discarded Procedure is expensive Could be used for non-disease alleles

Sex determination

Human body cells have 23 pairs of chromosomes. 22 pairs of chromosomes are autosomal chromosomes and two are sex chromosomes (allosomal). Human females have two X chromosomes. Human males have an X and a Y chromosome.

The possible sex chromosomes of the offspring can be worked out using a Punnett square (see punnett square to the right). ◄——

The probability of the offspring having male or female sex chromosomes is 0.5.

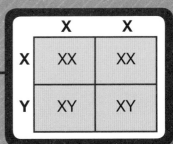

	X	X
X	XX	XX
Y	XY	XY

CHECKIT!

1 Name a disorder caused by a dominant allele.

2 Which sex chromosomes are present in human males?

3 A couple who are both heterozygous for cystic fibrosis want to have a child together. What is the chance they will have a child with cystic fibrosis?

DOIT!

Write out the advantages and disadvantages of embryo screening onto revision cards and learn them.

Variation

Organisms are different from one another. This is called variation.

Causes of variation

There is extensive genetic variation within a population of a species.

The two causes of variation are:

1. Genetic – the genes you inherit from your parents control many aspects of your phenotype. For example, your hair and eye colour.

2. Environmental – the world around you can have an influence of some aspects of your phenotype. For example, if you live in a hot climate, you may develop more pigmentation in your skin.

Many phenotypes are influenced by both genetic and environmental factors, for example, height.

Mutations

Mutations are a change in the base sequence of the DNA and happen continuously. Most mutations are repaired by the cell, but some remain in the genome. All variants arise from mutations. Most of the time, the new variant, for example, a change in a protein, will have no effect on the phenotype of the organism. Some variants will influence the phenotype. Rarely, a variant will lead to a new phenotype.

If the new phenotype is advantageous to the organism in its environment, the mutation will spread throughout the population by natural selection.

If the new phenotype is not advantageous to the organism, then that organism is less likely to have offspring and the mutation will not spread through the population.

NAILIT!

The **Human genome project** was completed in 2003, and mapped the entire human genetic code. This is being used to find out which proteins each gene codes for, research inheritable diseases, and discover new drug targets for medicine.

DOIT!

Make a table of phenotypes such as hair colour, eye colour, hair length and height, and decide whether they are caused by genetics only, environment only, or both.

WORKIT!

Explain why variation in organisms of the same species is important. (3 marks)

If all organisms are genetically identical, they will all be susceptible to the same diseases. (1) Variation allows some individuals to survive better than others if the environment changes, (1) through the process of natural selection. (1)

CHECKIT! ✓

1 Give an example of a phenotype that is influenced by genes and the environment.

2 What is a mutation?

3 Describe what happens to a mutation that gives an individual an advantage in its environment.

Evolution

Evolution is the gradual change in inherited characteristics of organisms over time through natural selection.

Natural selection

Within any population there is variation. Some individuals have characteristics that are better suited to the environment and are more likely to survive. Therefore, those individuals are more likely to have offspring and pass on those characteristics. This is natural selection.

Evolution

Life started on the Earth as very simple life forms more than three billion years ago. Since then, those simple life forms have gradually evolved into more complex life forms, through the process of natural selection.

Speciation

A species is a group of similar individuals that can breed together to produce fertile offspring. If the phenotypes of two populations of a species become so different that they can no longer interbreed, then they are no longer the same species, but two different species. This process of forming new species is called speciation.

SNAPIT!

Make your own version of the diagram and learn the process of natural selection. Take a photo to revise from later.

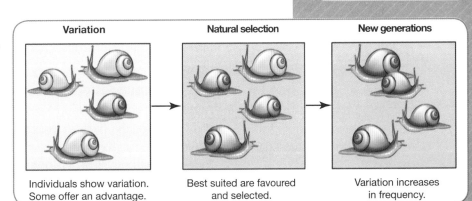

Variation	Natural selection	New generations
Individuals show variation. Some offer an advantage.	Best suited are favoured and selected.	Variation increases in frequency.

WORKIT!

A population of rabbits that lived in the Arctic had individuals with brown or white fur. Arctic wolves were eating the rabbits. The same population today has mostly individuals with white fur. Explain the mechanism that brought about these changes. (4 marks)

The rabbits with white fur were camouflaged in the snow. (1) They were less likely to be eaten by the arctic wolves. (1) They passed on the white fur characteristic to their offspring. (1) This is natural selection. (1)

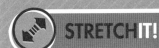

STRETCHIT!

Find out about Darwin's finches and how they became separate species.

CHECKIT!

NAILIT!

For Q3 remember to include natural selection in your answer.

1 What is a species?

2 Describe what would happen to individuals in a population who do not have advantageous characteristics.

3 Suggest how speciation would occur in two populations of a species that have been isolated from one another.

Selective breeding

NAIL IT!

Selective breeding is also called artificial selection because plants or animals that do not have the desired characteristics are not allowed to breed.

Food crops and domesticated animals have been selectively bred for thousands of years.

The process of selective breeding

Selective breeding is the breeding together of organisms with desired characteristics to produce offspring with the same desired characteristics. Selective breeding can lead to inbreeding where some breeds are particularly prone to disease or inherited defects. This can be seen in some breeds of dog.

Food plants

Two plants with the desired characteristics are selected. The pollen from one plant is transferred to the other plant for sexual reproduction. The resulting seeds are planted and grown into new plants. The best of these plants are bred together to make new plants. This continues for many generations until all of the plants produce offspring with the desired characteristics. For example, cauliflower plants that produce large cauliflower heads.

STRETCH IT!

Research selective breeding in wheat and see how modern wheat looks compared with the ancestral type.

WORKIT!

A gardener wants to produce tomato plants with larger tomatoes. Describe how the gardener could achieve this. (3 marks)

Choose two tomato plants that produce the largest tomatoes. (1) Breed these two plants together. (1) Choose the offspring that produce the largest tomatoes and breed them together. (1)

Domesticated animals

A good example of selective breeding in domesticated animals is milk production in cows. A cow that produces a high yield of milk is bred with a bull that produces daughters with a high yield of milk. The daughters that produce the highest yield of milk are bred with other bulls that have produced high yield daughters. This process continues for many generations until all female offspring produce high yields of milk.

Desirable characteristics include:

- disease resistance in food crops, e.g. blight-free potato plants
- large/unusual flowers
- animals that produce more meat/milk
- domestic dogs with a gentle nature.

The benefits and risks of selective breeding

Benefits	Risks
All offspring have desired characteristics Can eliminate diseases Increases the productivity/yield	Takes a long time to produce organisms with desired characteristics Loss of variety in species Animals may suffer discomfort, for example, cows with heavy udders Risk of inbreeding if two animals are too closely related

CHECK IT! ✓

1 Give an example of selective breeding.

2 Name two benefits of the selective breeding of plants.

3 Explain why selective breeding is also known as artificial selection.

Genetic engineering

What is genetic engineering?

Genetic engineering is the addition of a gene from an organism into another organism's genome.

Plants and animals have been selectively bred over many generations in order to produce organisms with desired characteristics. Genetic engineering aims to achieve the same outcome over one generation by adding a gene to that organism from another organism, usually from another species. The desired gene is cut out from the genome of the original organism and inserted into the organism to be modified using a number of different techniques. Organisms that have been modified in this way are called genetically modified organisms, or GMO.

DO IT!

Write the benefits and risks of selective breeding and genetic engineering onto revision cards.

Examples of genetic engineering:

- disease-resistant plants
- plants that produce bigger fruits
- bacteria that produce human insulin (for people with type 1 diabetes)
- gene therapy to overcome some inherited disorders, e.g. severe combined immunodeficiency (SCID).

Benefits and risks of genetic engineering

Benefits	Risks
Can produce organisms with the desired characteristics more quickly GMOs increase yield/productivity Food plants can be more nutritious, e.g. golden rice Crops can be grown in harsher environments Crops can be made resistant to pesticides Medicines can be made more cheaply, e.g. human insulin	Eating GMOs could potentially cause harm to people Can spread inserted gene to other species GM crops could cause harm to insect species Difficult to get the inserted gene into the right place in the genome Vegans/vegetarians/religious groups may object to xenotransplantations.

GM crops

Crops that have been genetically modified are called GM crops. Some examples of GM crops are:

- drought-resistant crops that can be grown in very dry conditions
- pest-resistant crops that contain genes that kill the pest
- herbicide-resistant crops
- crops that have had vitamins added, e.g. beta carotene (the precursor to vitamin A) can be added to rice
- crops that have had genes added that make the fruit last longer.

GM crops have a greater yield and therefore can feed more people. However, some people are concerned that eating GM crops will have unforeseen health consequences. Others also fear that genes such as pesticide-resistance genes will pass to weeds, causing them to also be resistant to pesticides. Pest-resistant plants may also cause harm to other insects, besides the pest, that eat the plant.

NAIL IT!

Pest-resistant crops can be made using the bacterium, *Bacillus thuringiensis*. These bacteria contain a gene that produces an insect-killing toxin. This gene is added the plant, so that the plant produces the toxin and is lethal to any insect that eats it.

Gene therapy

Gene therapy is the insertion of a gene into the genome of a person who has a disease or disorder, in order to remove the disease. For example, severe combined immunodeficiency (SCID) is a disorder that causes the immune system to no longer function. It is caused by a faulty allele. Gene therapy has successfully replaced the faulty allele in children with SCID with a normal allele, eliminating the disease. Trials to replace the faulty alleles of other diseases, such as cystic fibrosis, have proved difficult because it is difficult to get the gene into the correct place in the genome.

The process of genetic engineering

A common method of inserting a gene into a genome is to use vectors. These are usually a plasmid or a virus. Plasmids are small circles of DNA, originally found in bacteria.

1 The desired gene is cut out of the original genome using enzymes called restriction enzymes.

2 The same enzymes are used to cut open a vector.

3 The gene is inserted into the vector.

4 The vector is inserted into the cells of animals, plants or microorganisms at an early stage in their development. The organisms will develop with the desired characteristic.

This diagram shows the desired gene being inserted in bacteria.

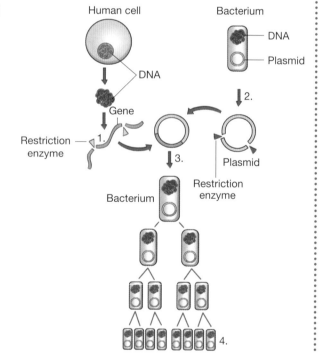

STRETCHIT!

Find out about the research into gene therapy for cystic fibrosis.

SNAPIT!

Sketch a copy of the genetic engineering diagram and add your notes to describe each stage. Take a photo to revise from later.

WORKIT!

Compare and contrast the benefits and risks of genetic engineering in plants. (4 marks)

Food plants can be made more nutritious, but it can be difficult to get the gene into the correct place in the genome. (1)

Can make plants resistant to herbicides, but there is the possibility of passing on this resistance gene to weeds. (1)

Can make plants resistant to pests, but could kill other insects not just the pests. (1)

GM crops produce a greater yield, but could potentially cause harm in people. (1)

CHECK IT!

1 Give an example of genetic engineering.

H 2 Describe simply how an organism is genetically modified.

3 Explain how gene therapy could be used to treat people with a disorder caused by a faulty allele.

Evidence for evolution

The theory of evolution has a lot of evidence to support it, such as the fossil record, antibiotic resistance in bacteria, and knowledge of genetic inheritance. The theory of evolution is now widely accepted.

The fossil record

Evolution is the gradual change of simple life forms into more complex ones. This can be seen in the fossil record. Fossils are the remains of organisms from millions of years ago and can be found in rocks. The oldest, simplest fossils are found in the oldest rocks. More complex fossils are found in the newer rocks.

How fossils are made

Fossils are formed if:

- parts of the organism's body, e.g. shell or bones, have not decayed because the organism died in a place where one or more of the conditions needed for decay are absent

- parts of the organism, e.g. shell or bark, are replaced by minerals as they decay

- traces of the organism, e.g. footprints or leaf prints, are preserved.

The fossil record is incomplete. This is because the soft part of an organism's body often decays, which means that there are few traces of many early life forms. Geological activity, such as earthquakes and volcanic eruptions have destroyed many fossils as well.

Resistant bacteria

Bacteria developing antibiotic resistance shows the mechanism of natural selection in a short time frame. Bacteria divide rapidly and can mutate rapidly as well. Any mutations can be quickly passed on to offspring, and to nearby bacteria by a process called conjugation.

There is variation in any population of bacteria. When a population of bacteria is exposed to an antibiotic, most of the population will die. If there is a bacterium with a mutation that gives it resistance to the antibiotic then that bacterium will survive, and pass its mutation to its offspring. As the mutation is spread through the population, the population will gradually become resistant to that antibiotic. We call this is a resistant strain. One strain of bacteria called MRSA (methicillin-resistant *Staphylococcus aureus*) is resistant to many types of antibiotic.

To reduce the rate of antibiotic-resistant strains of bacteria developing we should:

- make sure that antibiotics are not prescribed inappropriately, e.g. to treat a virus

- make sure that a course of antibiotics is finished, even if the person is feeling better. This kills all of the bacteria so none survive and form resistant strains

- restrict the use of antibiotics in agriculture.

NAILIT!

Evidence for human evolution can be found by looking at the fossil skeletons of human ancestors, such as Lucy (*Australopithecus afarensis*), and finding stone tools.

STRETCHIT!

Find out about whole preserved organisms such as woolly mammoths found in the ice.

NAILIT!

Natural selection can also be seen in the resistance of rats to the rat poison, **warfarin.**

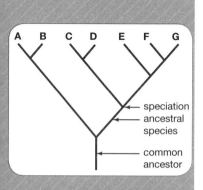

Scientists can develop new antibiotics that there is no resistance to. However, this takes a long time and costs a lot of money. It is difficult to make new antibiotics at the speed at which bacteria are becoming resistant to them.

Evolutionary trees

Organisms can be organised into evolutionary trees based on their DNA sequences, physical characteristics and behaviour. The more similar the species are, the more closely related the species are.

Where branches occur, speciation has taken place. Organisms that are close together on a branch are more closely related. In this evolutionary tree, species A and B are very closely related, and E and G are less closely related.

Extinction

When there are no longer any members of a species left on Earth, that species is extinct. Some examples of extinct species are the dinosaurs and woolly mammoths. We know that these species once existed because we can see their bodies in the fossil record. There are many reasons why a species may become extinct: loss of habitat, loss of food sources, hunting of the species.

DO IT!

Look at evolutionary trees in your textbook and online and work out how the species are related.

WORKIT!

This evolutionary tree shows the evolution of five species.

a   Which two species are the most closely related? (1 mark)

   Armadillo and human (1)

b   Which two species are the least closely related? (1 mark)

   Shark and human (1)

c   Describe how the evolutionary tree is worked out. (2 marks)

   The DNA sequences, physical characteristics, and behaviour of each species is compared. (1)
   The more similar two species are, the more closely related they are. (1)

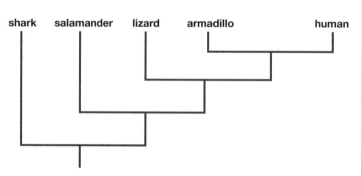

CHECKIT!

1 What is a fossil?

2 Give two reasons why a species might become extinct.

3 Explain how antibiotic resistance in bacteria shows the mechanism of natural selection.

Classification

All organisms are classified into groups based on their physical characteristics.

The Linnaean system

In the 18th century, Carl Linnaeus proposed the organisation of all species into groups:

Group	Example
Kingdom →	Animal
Phylum →	Vertebrate
Class →	Mammal
Order →	Carnivore
Family →	Felidae
Genus →	Panthera
Species →	Leo

Linnaeus put the species into groups depending on their structure and characteristics.

The binomial system

Linnaeus also proposed a system of giving all organisms two names to describe their genus and species. For example, the big cats all belong to the genus, *Panthera*, and so all have the same first name. However, different species of big cat have a different second name. So lions are called *Panthera leo*, and tigers are called *Panthera tigris*. This two-name system is called the binomial system, and is written in italics.

Lions and tigers are in the same genus but are different species, so they have many characteristics in common but cannot breed to produce fertile offspring

DO IT!

Use a mnemonic to learn the order of the groups, e.g. King Phillip Came Over From Great Spain.

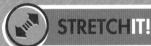

STRETCH IT!

Have a look at similar species and see how they are classified using the Linnaean system.

New models of classification

Technological advances in the 20th century mean that scientists have developed a better understanding of:

- the internal structures of the cell using microscopes
- biochemical processes, such as the amino acid sequences of key proteins
- genetics, including the sequencing of genomes.

This means that people can compare cell structures, amino acid sequences in proteins, and DNA sequences in different species and see how closely related they are.

Three-domain system

In the late 20th century, Carl Woese proposed the three-domain system of classification. In this system all organisms are organised into one of three domains:

Archaea – primitive bacteria that live in extreme environments such as volcanoes

Bacteria – true bacteria

Eukaryota – all animals, plant, fungi and protists.

Using biochemical analysis, Woese found that archaea bacteria were evolutionarily more closely related to eukaryotes than bacteria, and should therefore have their own domain.

WORKIT!

Describe how new technologies have changed the way that we classify organisms. (3 marks)

We can observe the internal structures of cells, and work out the amino acid sequence in proteins, and DNA sequences. (1)

We can use this information to see how closely related species are. (1)

New three-domain system proposed. (1)

CHECKIT!

1 Fill in the missing groups.

   kingdom _____ class _____ _____ genus species.

2 What is the binomial system?

3 Explain why archaea and bacteria are in two different domains.

1 a Define asexual reproduction.

   b Compare daughter cells made by mitosis and meiosis.

   c Describe fertilisation.

2 a What is a gene?

   b What is a chromosome?

   c What is a dominant allele?

   d Explain the difference between the terms homozygous and heterozygous.

3 a Name the two causes of variation.

   b Explain how variation leads to natural selection.

   c A farmer wanted to selectively breed crops with a high yield. What characteristics might a farmer select for?

   d The field where the crops were grown contained an insect pest. Suggest how the farmer could use genetic engineering to overcome the pest problem.

4 a What is meant by binomial classification?

   b Copy and complete the table to show the classification of the Wolf, *Canis lupis.*

Kingdom	Phylum		Order		Genus	Species
	Vertebrate	Mammal	Carnivore	Canidae		lupis

   c Name two ways in which fossils can form.

   d Explain how antibiotic resistance in bacteria shows the mechanism of natural selection.

5 a The colour of fur in mice is controlled by a single gene. The dominant allele, B, produces black fur. The recessive allele, b, produces brown fur.

     i   Use a Punnett square to work out the genotypes of the offspring of two black mice with the genotype, Bb.

     ii What is the ratio of the genotypes?

   b Red-green colour-blindness is a recessive disorder inherited on the X chromosome. If a colour-blind woman and a man who is not colour-blind had a child, what is the probability that their child will be colour-blind?

H 6 a What is genetic engineering?

   b Describe how human genes can be inserted into bacteria.

   c Evaluate the advantages of genetic engineering of human genes into other organisms.

Ecology

Communities

Organisms within an ecosystem are organised into communities.

Organisation in an ecosystem

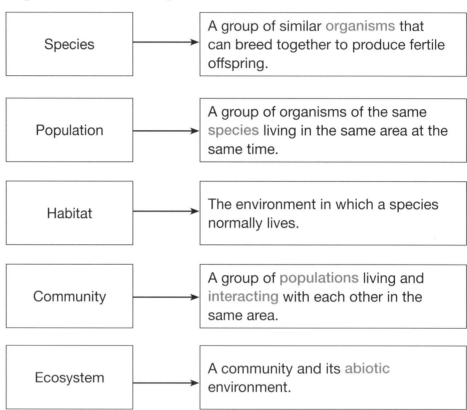

Species	A group of similar organisms that can breed together to produce fertile offspring.
Population	A group of organisms of the same species living in the same area at the same time.
Habitat	The environment in which a species normally lives.
Community	A group of populations living and interacting with each other in the same area.
Ecosystem	A community and its abiotic environment.

Each habitat has its own community, made up of populations of different species. These species interact with each other to compete for resources. All of the habitats make up the ecosystem.

Levels of organisation

Population → Community → Ecosystem

Competition

Individuals in a community compete with others of the same species (intraspecific competition) and with individuals from other species (interspecific competition). Individuals compete for:

- food – animals
- water – animals and plants
- territory/space – animals and plants
- light – plants
- mineral ions – plants
- mates – animals.

DO IT!

Make a table of factors that species compete for, and whether the competition for these is intraspecific, interspecific or both.

Interdependence

Species depend on each other for survival. They depend on each other to provide food, shelter, pollination and seed dispersal. If one species is removed from a habitat, it can affect the whole community. This is called interdependence. For example, bees pollinate many flowering species. If the bee population decreases, the plants will not be pollinated and the plant species will also decrease.

Predator–prey relationships

In a stable community, the numbers of predators and prey rise and fall in cycles. An increase in the number of prey in a population provides more food for predators, so the number of predators in a population rises, after a slight delay. This in turn causes a decrease in the number of prey in a population and then a decrease in the number of predators in a population. This change in population sizes can be recorded over many years and displayed as a predator–prey cycle.

WORKIT!

This graph shows the interaction of leopards and gazelles. Describe and explain the shape of the graph. (5 marks)

As the population of gazelles increases, the population of leopards increases. (1) This is because there is an increase in the food supply for the leopards. (1)

As the population of gazelles decreases, the population of leopards decreases. (1) This is because there is a decrease in the food supply for the leopards. (1)

There is a delay between the increase or decrease of the gazelle population and the increase or decrease of the leopard population. (1)

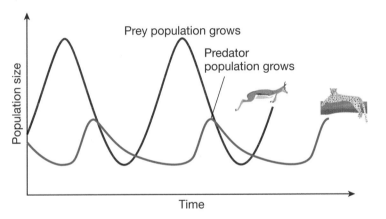

CHECKIT!

1 What is a community?

2 Two species of caterpillar are eating leaves in the same tree. What type of competition is this?

3 Mayfly nymphs are eaten by a species of frog that is in turn eaten by grass snakes. Suggest what would happen to the population of frogs and mayfly nymph if the grass snake population suddenly decreases.

Abiotic factors

An abiotic factor is a non-living condition that can affect where organisms live, for example, temperature.

Examples of abiotic factors

There are many abiotic factors that can affect a community. These include:

- temperature
- light intensity
- moisture levels in the soil and air
- pH of the soil.

- wind intensity and direction
- carbon dioxide levels
- oxygen levels of the water

How abiotic factors affect a community

Most organisms cannot live in extreme environments where it is very hot or cold, or there is no access to water. There are very few microscopic organisms that can survive in these conditions (see extremophiles on page 104). Even in stable communities, changes in the abiotic factors can affect the growth and number of organisms in a population.

DOIT!

Make a table to show which of these abiotic factors would affect animals, plants or both.

WORKIT!

The light intensity in a woodland area was measured as well as the height of the flowers in that area. The results are shown in this graph. Describe and explain the heights of the flowers in different light intensities. (3 marks)

As the light intensity decreases, the height of the flowers increases. (1)

The height of the flowers increases from 11 cm to 13 cm between 5 to 3 lux, and from 13 cm to 18 cm between 3 and 1 lux. (1)

This is because the flowers are trying to get to the light in order to photosynthesise. (1)

Mention photosynthesis in your answer.

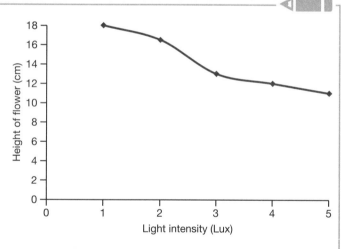

CHECKIT!

1 Give two examples of an abiotic factor.

2 Explain how the pH of the soil may affect organisms in a community.

3 Some trees live on a very windy hillside. Suggest how the wind may affect their growth.

Biotic factors

A biotic factor is any living component that affects the population of another organism or the environment.

Examples of biotic factors

Biotic factors affect the community. Some examples of biotic factors are: food availability; predators; microorganisms in the soil; pathogens; competition within and between species; parasites; symbiosis; pollination.

How biotic factors affect the community

All organisms belong to a food chain or food web. The increase or decrease of one species will therefore directly or indirectly impact the population numbers of another species. Disease, predation, and parasites can also decrease a population's numbers, but symbiosis is beneficial to both species involved.

DOIT!

Write each biotic factor on a revision card and write a short description of how it can affect another organism or the environment.

 STRETCHIT!

Research an example of symbiosis.

WORKIT!

Tarnished plant bugs (TPBs) are a pest that eats alfalfa plants in the USA. Parasites that kill TPBs were introduced to alfalfa fields in 1982. Use data from the graph to explain the effect of the parasites on the TPBs. (4 marks)

At first, as the percentage of parasitism increased, the number of TPBs remained constant. (1)

Around 1987-89, as the percentage of parasitism increased, the number of TPBs decreased. (1)

This is because the parasites killed some of the TPBs. (1)

After 1990-1992, the percentage of parasitism decreased as the number of TPBs had also decreased. (1)

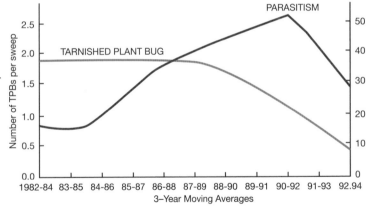

✓ CHECKIT!

1 Give two examples of a biotic factor.

2 Describe the effect of the loss of a tree species on the community.

3 A community has a population of peas, caterpillars, bees and sparrows. A new pathogen has decreased the number of bees in the community. Suggest the effect that this will have on the other species.

Adaptations

All organisms are adapted to live in their environment.

Adaptations enable species to survive in the conditions in which they normally live, for example, a cold climate. There are three types of adaptation:

1. Structural adaptations are adaptations to the body of the organism, such as the skeleton, body shape, body colouration or fur length. For example, dolphins have streamlined bodies to aid swimming in the water.

2. **Behavioural adaptations** are changes to a species' behaviour to help their survival. Some examples are: emperor penguins huddle together for warmth during the Antarctic winter; zebras stay together as a group when grazing so that their colouration confuses predators.

3. **A functional adaptation** is one that has occurred through natural selection over many generations in order to overcome a functional problem, such as birds having to eat seeds due to too much competition for other food sources, so the beak gradually becomes more adapted to seed eating.

Extremophiles

Some organisms are adapted to live in extreme environments, such as volcanoes or hot springs. These are usually microorganisms and are called extremophiles. Extremophiles can be found in habitats that have a high temperature, high pressure (under the sea) or a high salt concentration.

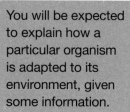

WORK IT!

Describe and explain how a camel is adapted to living in its environment. (4 marks)

Long eyelashes to keep the sand out of its eyes. (1) Large flat feet to support its weight on the sand. (1) Large fat store in hump to provide energy. (1) Does not sweat in high temperatures to keep water inside its body. (1)

CHECK IT!

1 Give an example of a structural adaptation in an Arctic animal.

2 What is the difference between a behavioural adaptation and a functional adaptation?

3 Explain how extremophiles can live in deep-sea hydrothermal vents.

Food chains

All organisms belong to a food chain or food web.

Photosynthetic organisms

The vast majority of food chains start with an organism that carries out photosynthesis. These organisms can be plants, algae, or bacteria that contain chlorophyll. These are called photosynthetic organisms. Light energy from the Sun is converted into biomass, which forms the basis of all food chains.

Food chains

Feeding relationships within a community can be represented with food chains. Photosynthetic organisms are known as producers. These are eaten by primary consumers, which in turn are eaten by secondary consumers. Secondary consumers are eaten by tertiary consumers. Food chains link together to form food webs.

One example of a food chain is shown below. The sweetcorn is eaten by crickets, which are eaten by lizards, which are eaten by snakes. The arrow represents the transfer of energy from one organism to another.

The lizards and snakes are predators because they kill and eat other animals. The crickets and the lizards are prey because they are eaten.

DO IT!

Draw an example of a food chain and expand it to form a food web.

WORKIT!

Describe the energy flow through a food chain that consists of cabbages and caterpillars. (3 marks)

Light energy from the Sun is absorbed by the cabbages by photosynthesis. (1)

The cabbages transfer the light energy into biomass. (1)

The biomass of the cabbages is eaten by the caterpillars and converted into their own biomass. (1)

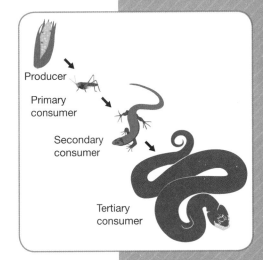

Producer

Primary consumer

Secondary consumer

Tertiary consumer

✓ CHECKIT!

NAILIT!

For Q3, think about the amount of energy being transferred.

1 What is a secondary consumer?

2 Draw and label a food chain to show the feeding relationship between grass, foxes and rabbits.

3 Suggest why there are often no more than four levels in a food chain.

Measuring species

Species are measured using a range of different sampling techniques.

Sampling techniques

It is difficult and time consuming to count every organism in a population. Sampling methods are quicker and use estimates of distribution and abundance of organisms in order to work out the size of the population.

Quadrats

Plants and slow-moving insects can be sampled using quadrats. This is a 1m by 1m square made of wire, which may be divided into smaller squares. The quadrat is placed on the ground and the number and type of each species is recorded. The quadrats are randomly placed in the area being investigated.

Transects

A transect is a long line made with string or a measuring tape, with quadrats placed at intervals, usually every metre, along it. The number and type of each species is recorded in each quadrat. Transects are usually used where the habitat changes over a short distance, for example, on a shoreline.

Animal traps

Animals can be trapped safely and then released again using a number of different traps, including nets, pooters and pitfall traps. Nets are used to sweep for larger insects. A pooter is a small jar with two straws and is useful for catching small insects. Suction is applied to one straw, so that the insect is sucked into the jar. A pitfall trap is a small beaker set into the ground with a raised lid to allow animals to climb in. These are useful for catching small ground invertebrates, such as woodlice.

DO IT!

Practice each of these techniques using equipment from your school.

Abundance of organisms

If you are using quadrats, the abundance of organisms is estimated by multiplying the average number of organisms found in the quadrats by the size of the area. The more quadrats used, the more reliable this estimate will be.

WORKIT!

24 daisies were found in 10 quadrats. The area measured 20 m by 20 m. Estimate the abundance of the daisies. (4 marks)

Average number of daisies per quadrat = 24 ÷ 10

$$= 2.4 \text{ (1)}$$

Area is 20 m × 20 m = 400 m² (1)

Estimated number of daisies = 2.4 × 400 (1)

$$= 960 \text{ daisies (1)}$$

If you are capturing and releasing animals, then you can use the capture, release, recapture method for measuring abundance. When you capture an animal, mark it in a safe way for the animal and release it. Then wait for a time and then capture animals in the same area again.

$$\text{Population size} = \frac{\text{number in first sample} \times \text{number in second sample}}{\text{number in second sample that are marked}}$$

WORKIT!

18 beetles were captured in an area of woodland and marked.
20 beetles were caught the second time, eight of these were marked.
Estimate the abundance of the beetles. (3 marks)

$$\text{Abundance of beetles} = \frac{18 \times 20}{8}$$

$$= \frac{360}{8}$$

$$= 45 \text{ beetles (3)}$$

CHECKIT!

1 Describe how to use a quadrat.

2 If five quadrats are used in a 100 m² area, and measure 25 buttercups, what is the abundance of buttercups in the area?

3 Explain why it is important to mark captured animals in a safe way.

Measuring the population size of a common species

In this practical, you will use sampling techniques to investigate the effect of a factor on the distribution of a common species.

WORKIT!

140 bluebells were found in 10 quadrats in an area of woodland. What is the mean number of bluebells per quadrat? (1 mark)

140 ÷ 10 = 14 bluebells per quadrat. (1)

Practical Skills

Investigation

Choose a local area in which to carry out your investigation. Use quadrats, a transect or animal traps to sample a common species.

Recording your findings

Record your results in a table. If using quadrats or a transect, record the number of individuals of a species in each quadrat, and then find the mean number of species per quadrat.

If using animal traps to count individuals of a species, mark your captured animals and release them. Then wait a period of time and capture animals in the same area.

Estimating population size

To estimate the population size of plants, you need to multiply the mean number of plants per quadrat by the size of the investigation area.

NAILIT!

To estimate the population size of animals, you need to know the number of individuals you captured each time, and how many were marked.

WORKIT!

In a woodland area measuring 100 m², there are 14 bluebells per quadrat. Estimate the population size of the bluebells. (2 marks)

Population size = 14 × 100 m² (1)

= 1400 bluebells (1)

Assumptions of capture–release–recapture

When estimating the population size of animals using this method, you must assume: no animals die or are born, no animals migrate and the marking of the animals has not affected their survival.

CHECKIT! ✓

1 Describe how to use an animal trap to estimate the population of an animal species.

2 If 12 snails are caught and marked, 11 are recaptured and 8 are marked, estimate the population size.

3 Explain why it must be assumed that there are no deaths, births or migration in a population when you estimate the population size.

The carbon cycle

Carbon is cycled through the ecosystem by biological and chemical processes.

How carbon is cycled through the ecosystem

Carbon dioxide in the air is absorbed by plants to use in photosynthesis. Plants convert carbon dioxide into sugars and the carbon becomes part of the biomass of the plant. When consumers eat plants, this carbon is passed to them.

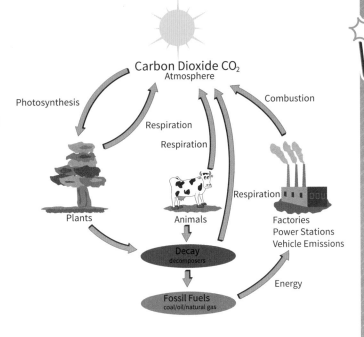

When plants and animals die their remains decay, and are decomposed by microorganisms, such as fungi and bacteria. These are known as the decomposers. Millions of years ago, some of their remains formed fossil fuels, such as coal or oil.

Carbon passes back into the atmosphere as carbon dioxide by:

• aerobic respiration of plants, animals and microorganisms

• combustion – the burning of fossil fuels.

WORKIT!

Explain the importance of the carbon cycle to living organisms. (4 marks)

Carbon dioxide is needed by plants in order to carry out photosynthesis. (1)

Photosynthesis produces sugars, which are used in respiration to provide energy for the plant. (1)

Sugars are eaten by animals and are used in respiration to provide energy for the animal. (1)

When dead matter is decomposed, the decomposers use the sugars in respiration to provide energy. (1)

✓ CHECKIT!

1 Name one way in which carbon dioxide returns to the atmosphere.

2 Explain why microorganisms are important in the carbon cycle.

3 Explain why there are increasing levels of carbon dioxide in the atmosphere.

The water cycle

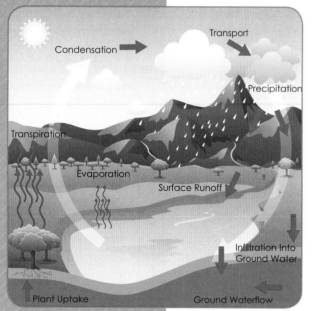

Water is cycled through the environment through the processes of evaporation and precipitation (rain).

Water in the oceans is warmed by the Sun, and evaporates into the atmosphere. The water droplets condense to form clouds and are transported inland by the wind. As the clouds rise over mountains, the water droplets are released as precipitation. The water moves under the land as ground water, and over the land into streams and rivers, and finally makes its way back to the oceans.

The importance of the water cycle

The water cycle is important to living organisms because it provides fresh water for plants and animals on land. Fresh water is needed for animals to drink, and for plants to use in photosynthesis. Water diffuses from the top surfaces of plants in a process called transpiration (see page 44).

SNAPIT!

Make your own copy of the water cycle diagram, including the key words and their definitions. Take a photo to revise from later.

WORKIT!

Describe and explain the role of plants in the water cycle. (3 marks)

Plants take up fresh water for use in photosynthesis. (1)

Water transpires from the surface of the plant. (1)

Water droplets return to the atmosphere and condense into clouds. (1)

CHECKIT!

1 Name two ways in which water can return to the atmosphere.

2 Explain why water evaporates form the surface of the oceans.

3 The land on one side of a mountain receives a large amount of rain, whereas the land on the other side of the mountain receives very little rain. Suggest how this is possible.

Biodiversity

Biodiversity is the variety of living organisms on the Earth, or within an ecosystem.

Waste management

The rapid growth of the human population and increasing standard of living, means that humans are using more and more resources. This means that more waste and chemical materials are produced, which in turn leads to increased pollution. Increasing levels of pollution lead to a decrease in biodiversity.

Land use

When humans clear land to build houses, or quarry for stone, or grow crops, or dump waste, it leaves fewer habitats for other organisms. This leads to decreased biodiversity.

Peat bogs have been dug up for many years to provide peat for compost. However, this destroys the beat bog habitat and decreases the biodiversity of the species that live there, including plants, animals and microorganisms. Burning peat also causes large amounts of carbon dioxide to be released into the atmosphere.

Large areas of tropical forest have been cut down in order to provide land for cattle and rice fields, and to grow crops for biofuels. This is called deforestation.

WORKIT!

Explain why peat bogs need to be protected as the human population and its food production increases. (3 marks)

As the global population increases, more food needs to be grown. (1)

Peat from peat bogs is a cheap and effective compost and increases the amount of food that can be produced. (1)

Peat bogs should be protected as destroying them decreases the biodiversity of the habitat. (1)

DO IT!

Write these examples of how human populations pollute the environment onto revision cards and learn them:

- in water – sewage, fertiliser leaching off from fields, toxic chemicals, e.g. oil spills

- in air – smoke and acidic gases from vehicle exhausts and power stations

- on land – decomposition of landfill and from toxic chemicals, e.g. pesticides.

 ## STRETCHIT!

Find out about some programs from organisations, such as the WWF, that are protecting species from extinction.

✓ CHECKIT!

1 What type of pollution do power stations produce?

2 Describe how using land for building houses decreases biodiversity.

3 Describe and explain the role of peat bogs in the carbon cycle.

Global warming

Global warming is the gradual increase in the overall temperature of the Earth's atmosphere.

Causes of global warming

Global warming is caused by the greenhouse effect. This is when carbon dioxide, methane and other greenhouse gases build up in the Earth's atmosphere. Radiation from the Sun warms the Earth, and this heat is reflected from the Earth's surface. The greenhouses gases in the atmosphere absorb the heat, increasing the temperature of the atmosphere.

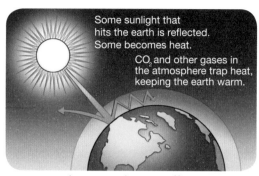

Some sunlight that hits the earth is reflected. Some becomes heat.

CO_2 and other gases in the atmosphere trap heat, keeping the earth warm.

The greenhouse effect

Greenhouse gases are increasing. They build up due to the respiration of organisms and the combustion of fossil fuels.

Biological consequences of global warming

Global warming has many consequences for the organisms on Earth:

- global weather patterns will change – causing flooding in some areas and drought in other areas. This will decrease available habitats, and food and water availability
- sea levels will rise – decreasing available habitats
- increased migration of organisms – species will move to more suitable habitats with enough available water and food
- increased extinction of species – some species will not be able to migrate or adapt quickly enough to the changing climate.

SNAPIT!

Draw your own version of this diagram and learn the process of global warming. Take a photo to help you revise later.

DOIT!

Make a revision card for each of the biological consequences.

WORKIT!

A species of mountain beetle has been found living higher up the mountain in recent years. Explain why this could have happened. (4 marks)

The mountain beetles feed on plants that grow at certain heights on the mountain. (1)

The plants have an optimum temperature that they like to grow in. (1)

As the temperature on the mountain increases, the plants cannot survive in the warmer areas. (1)

The plants begin to grow higher up the mountain, where it is cooler, and the beetles migrate to feed on them. (1)

CHECKIT! ✓

1 Name two greenhouse gases.

2 Describe the process of global warming.

3 Describe and explain human activities that could decrease the amount of greenhouse gases in the atmosphere.

Maintaining biodiversity

There are both positive and negative human interactions in an ecosystem that can impact biodiversity.

Positive interactions

- Conservation programmes – protecting species from extinction.
- Using renewable energy – reducing the amounts of greenhouses gases in the atmosphere and reducing the effects of global warming.
- Recycling waste – reusing resources to prevent pollution of the environment by dumping waste.
- Fish farming – breed fish for food rather than taking breeding adults out of the natural habitat.

Negative interactions

- Combustion of fossil fuels – increases the amounts of greenhouse gases in the atmosphere, and increases the effect of global warming.
- Pollution of the air, water and land – reduces the biodiversity of the polluted areas.
- Deforestation – removes habitats for organisms, and decreases photosynthesis, which contributes to global warming.
- Introduction of non-indigenous species – affects the number of native species.

Programmes to maintain biodiversity

Concerned people have put many programmes into place in order to reduce the negative effects of humans on ecosystems and biodiversity.

These include: Breeding programmes for endangered species, e.g. in zoos or wildlife reserves; protection and regeneration of rare habitats, e.g. conservation areas, to which people have limited access; reintroduction of hedgerows around fields where farmers only grow one type of crop; reduction of deforestation and carbon dioxide emissions.

DO IT!

Make a table of the positive and negative interactions of humans in an ecosystem.

WORKIT!

Explain and evaluate the conflicting pressures on maintaining biodiversity in a rare habitat where people live. (3 marks)

People need land to live on and grow food, which reduces the land available for other organisms. (1) People create waste which needs to be removed, otherwise it will pollute the land and make it unsuitable for other organisms. (1) If rare habitats are not protected, many species will become extinct. (1)

✓ CHECKIT!

1 Give one negative interaction that humans have with an ecosystem.

2 Explain how reintroducing hedgerows increases biodiversity.

3 The Galapagos Islands have many unique species and are visited by thousands of tourists each year. Suggest some ways that the impact of tourism to these islands can be reduced.

1 a Define the term 'population'.

  b What is a community?

  c Give two examples of an abiotic factor.

2 a How is carbon dioxide in the air used in plants?

  b Name two ways in which carbon dioxide is returned to the atmosphere.

  c Describe how water is recycled through the water cycle.

3 a Describe how global warming has affected:

   i   global temperatures

   ii water availability

   iii atmospheric gases.

  b What are the biological consequences of global warming?

4 a What is biodiversity?

  b Explain how human activity can affect biodiversity.

  c Give three ways that humans can help to maintain biodiversity.

5 Snowshoe hares are eaten by lynx, and this affects the number of hares and lynx in various years, in a pattern called a predator–prey cycle.

  a What type of competition is shown by the lynx?

  b Describe the predator–prey cycle shown by the hares and lynx.

6 a Some students were estimating the number of buttercups growing in the 10 m by 15 m school field. They used 10 quadrats and recorded their data in the table below:

	Quadrats										Total number of buttercups
	1	2	3	4	5	6	7	8	9	10	
Number of buttercups	1	2	0	10	4	6	2	1	6	2	

   i Complete the table.

   ii Use data from the table to estimate the abundance of the buttercups in the field.

  b The students then decided to estimate the abundance of snails, using the capture, release, recapture method. They caught and marked 25 snails the first time, and caught 18 snails the second time. Of these, 10 were marked. Estimate the abundance of the snails.

Atomic structure and the periodic table

Atoms, elements and compounds

All substances are made up of particles.

The atom is the smallest particle of an element that can take part in a chemical reaction.

There are over 100 elements and each element contains only one type of atom.

Each element is represented by its own chemical symbol and has its own atomic number.

A compound is formed when the atoms of two or more elements are joined by chemical bonds.

Chemical formulae are used to show which elements are in a compound and the number of atoms of each element that is present.

MATHS SKILLS

You will need to read the number of atoms in a compound from the chemical formula. Remember to leave out the number 1 if there is only one atom in the formula of a compound.

WORKIT!

The formula of iron(II) sulfate is $FeSO_4$. Identify the elements present in this compound and determine the number of atoms of each element. (1 mark)

The elements present are iron, sulfur and oxygen. There is 1 atom of iron, 1 atom of sulfur and 4 atoms of oxygen. (1)

The Roman numerals are not the number of atoms but the charge on the ion present in the compound.

NAILIT!

Some compounds have formulae containing brackets. When you work out the number of atoms in the compound you multiply the number of atoms inside the bracket by the number outside. For example, $Ca(NO_3)_2$ has 2 nitrogen atoms and 6 oxygen atoms.

✓ CHECKIT!

1 What type of atom would you find in the element sodium?

2 Explain why the chemical formula H_2 represents an element.

3 What type of substance is silver oxide?

4 What type of substance is represented by the following chemical formulae?

   a $CuBr_2$ **b** Mg

5 Name the elements present in the following compounds and give the number of atoms of each element.

   a $AgNO_3$ **b** $Fe(NO_3)_3$

Mixtures and compounds

DO IT!

Start a list of definitions and keep your list on a spreadsheet. You can begin this with definitions of an atom and a compound.

A physical property of a substance is one which you can measure or observe without changing the substance. Two examples of physical properties are melting point and appearance.

A chemical property of a substance is one which you can only observe by changing the substance.

A mixture is formed whenever elements or compounds are together but **not** joined chemically. An example of a mixture is iron mixed with sulfur. If the substances in a mixture are not joined chemically they can be easily separated. For example, the iron and sulfur can be easily separated using a magnet.

Mixtures have different physical properties to the substances that make them up. When they are separate, the substances in a mixture still keep their chemical properties.

There are different types of mixture and this means that there are different separation methods. See the Snap It! box on page 117.

The elements in a compound are joined chemically and therefore cannot be separated by physical means. For example, the compound iron sulfide cannot be separated into iron and sulfur using a magnet because they are combined chemically.

Practical Skills

The compulsory practicals that use these methods are:

1   The preparation of a soluble salt from an insoluble base.

2   The distillation of salt solution.

3   The separation of coloured substances using paper chromatography.

In some practicals, you will use more than one separation method. For example, in the separation of a soluble salt from an insoluble base you will first use filtration to remove any unreacted insoluble base and then crystallisation to get the salt from the solution.

NAIL IT!

Some of the marks (about 15%) will be allocated to questions about practical techniques. For some of the required practical experiments you use some of these methods. Make sure you know which separation method is used along with the apparatus that is required.

In the exam, you may be given data on the different boiling points or solubilities of different substances and asked to explain how you could separate them.

DO IT!

In the table below, there is an empty column labelled 'Example'.

For each separation method, give an example of where it is used. The answers are given below. Just put each example in the correct box.

Ethanol from ethanol and water; salt from salty water; chalk from chalk and water; water from salty water; colourings in sweets.

SNAP IT!

A table to show methods for separating different mixtures

Mixture	Method used and why	Apparatus used	Example
Separating an insoluble solid from a liquid.	Filtration because the insoluble solid cannot pass through the filter paper.	Filter funnel; filter paper and beakers.	
Separating the liquid from a solution of a solid in a liquid. The liquid is the distillate.	Simple distillation because the liquid has a much lower boiling point and so evaporates at a much lower temperature.	Flask; heating equipment and condenser.	
Separating two or more miscible liquids. (miscible means they can mix)	Fractional distillation because the liquid with the higher boiling point condenses on the column, the liquid with the lower boiling point carries on up as a vapour.	Flask; heating equipment; fractionating column and condenser.	
Separating coloured substances.	Paper chromatography which relies on the substances having different attractions for the paper and the solvent.	Container and chromatography paper.	
Separating the dissolved solid from a solution.	Crystallisation which depends on the big differences in boiling points between the solvent and the dissolved solid.	Evaporating basin and heating equipment.	

WORKIT!

The boiling points of two substances X and Y along with water are shown in the table below.

Substance	Boiling point/°C
X (X is a solid at room temperature)	1800
Y (Y is a liquid at room temperature)	67
Water	100

a   Explain how you could get X from a solution of X in water. (2 marks)

Crystallisation. (1) The water has a much lower boiling point and can be evaporated off to give solid X. (1)

b   Explain how you could get Y from a solution of Y in water. (3 marks)

Fractional distillation. (1) Y and water are miscible (they mix), otherwise Y would not dissolve in water. (1) Their boiling points are close so fractional distillation is needed. (1)

CHECKIT! ✓

1   Why is a mixture of iron and sulfur easy to separate but it is very difficult to separate iron from sulfur in iron sulfide?

2 a   Explain how you could separate a mixture of chalk and salt. Check the Snap It! on page 117 for ideas.

b   The table below shows the solubilities of two solid substances, Q and R, in petrol and water.

Substance	Q	R
Solubility in petrol	Soluble	Insoluble
Solubility in water	Insoluble	Insoluble

Explain how you could use filtration to separate a mixture of Q and R.

Scientific models of the atom

From the ancient Greeks up to the end of the 19th century, atoms were thought to be indivisible.

Joseph John Thomson discovered the electron and he suggested that the atom is a positive ball (the plum pudding) with negatively charged electrons (the currants) dotted around inside it.

A few years later, Ernest Rutherford showed that the atom has a central nucleus which contains positively charged protons with electrons orbiting around it.

It is now thought that the electrons are in energy levels or shells around the nucleus.

James Chadwick discovered the neutron (which is neutral) in the nucleus. This means that there are three sub-atomic particles – the electron, the proton and the neutron.

NAIL IT!

The most important experiment is Rutherford's experiment. When Rutherford and his team fired high-energy positively charged alpha particles at gold foil they expected these particles to pass straight through. What they saw was that most of the particles did pass straight through but some were deflected or rebounded straight back.

Think about Rutherford's reasoning on his experimental results. The nucleus must be positive to repel the positive alpha particles and very dense because it had to withstand their high energy. As most of the alpha particles passed through the foil, most of the atom must be empty space.

SNAP IT!

The diagram below shows the model of the atom we now use.

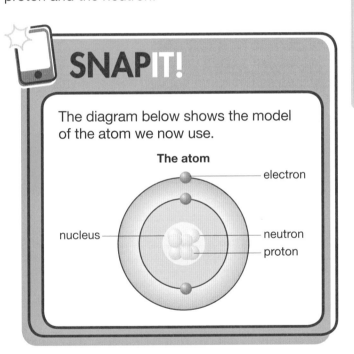

The atom

electron, nucleus, neutron, proton

DO IT!

Copy and complete the table below to show how scientists' vision of the atomic model atom has changed over time, as new evidence became available.

Scientist	What they discovered	Comments

CHECK IT!

1 What does indivisible mean?

2 Name the three main sub-atomic particles.

3 Describe the structure of the atom.

Atomic structure

NAIL IT!

You may need to know the size of an atom:

Radius of
an atom = 0.1nm

= 1×10^{-10} m

Radius of
a nucleus $=$ 1/10 000 of an atom
or 1×10^{-14} m

Make sure you can write numbers in standard form.

In an atom, the protons and neutrons are in the central nucleus and the electrons move around the nucleus in electron shells.

The charges and masses of the three sub-atomic particles are very small and because of this, we use their relative charges and relative masses. See the table below for a summary of their properties.

Each of the atoms of an element contains the same number of protons. This is called the atomic number.

For each element the atomic number is fixed and cannot change.

In a neutral atom, the number of protons equals the number of electrons.

Ions are charged atoms and are formed when atoms of an element react with atoms from another element.

An ion is positive if electrons are lost and the number of positive charges on the ion is equal to the number of electrons lost.

An ion is negative if electrons are gained and the number of negative charges is equal to the number of electrons gained.

The mass number of an atom is the sum of the number of protons and neutrons in the nucleus.

This table shows the properties of sub-atomic particles

Name of sub-atomic particle	Where is it in the atom?	Relative charge	Relative mass
Proton	In the nucleus	+1	1
Neutron		0	1
Electron	In shells around the nucleus	−1	Very small

DO IT!

Draw a diagram of an atom, label each sub-atomic particle in your diagram and write a short note to describe its relative mass and its relative charge.

Use the words electron, proton, neutron, atomic number, mass number, nucleus and neutral in your notes.

SNAP IT!

Atoms of any element can be represented as follows:

mass number
atomic number **X**

For example, $^{27}_{13}$Al

MATHS SKILLS

You can work out the number of each sub-atomic particle in an atom by using its atomic number and mass number.

In a **neutral atom** the atomic number $=$ number of protons $=$ number of electrons

The mass number $=$ number of protons $+$ number of neutrons $=$ atomic number $+$ number of neutrons

This means that: the number of neutrons $=$ mass number $-$ atomic number

In a **positive ion** the number of electrons $=$ atomic number $-$ the number of charges on the ion

In a **positive ion** the number of electrons $=$ atomic number $+$ the number of charges on the ion

NAIL IT!

You must remember that the number of protons (the atomic number) in atoms of the same element never changes. If its atomic number were to change, it would be a different element.

WORKIT!

1 An atom of potassium has the atomic number 19 and mass number 39. What are the number of electrons, protons and neutrons in an atom of potassium? (1 mark)

The number of protons and electrons are equal to the atomic number $= 19$

The number of neutrons $=$ mass number $-$ atomic number
$= 39 - 19 = 20$ (1)

2 How many electrons are there in a calcium ion Ca^{2+}? [Atomic number of calcium $= 20$] (1 mark)

For Ca^{2+} number of electrons $= 20 - 2 = 18$ (1)

3 How many electrons are there in N^{3-} ion? [Atomic number of nitrogen $= 7$] (1 mark)

For N^{3-} number of electrons $= 7 + 3 = 10$ (1)

CHECKIT!

1 An oxygen atom contains 8 protons. Why can't it have 9 protons?

2 Explain why scientists have concluded that most of the mass of an atom is in the nucleus.

3 A phosphorus atom (symbol P) has an atomic number 15 and a mass number 31. Show how you could represent the phosphorus atom.

4 How many electrons are there in an Al^{3+} ion? [Atomic number of aluminium $= 13$]

5 An atom of sodium can be represented as shown below. Give the number of protons, electrons and neutrons in a sodium atom. $^{23}_{11}Na$

Isotopes and relative atomic mass

The atomic number of an element cannot change. It is fixed. Therefore the number of protons is also fixed.

However, mass number of an element can have different values and so the number of neutrons must also vary in number.

Atoms of the same element with different mass numbers are called isotopes.

Isotopes of an element have the same number of protons but different numbers of neutrons.

Examples of isotopes are the three naturally occurring isotopes of magnesium:

a $^{24}_{12}Mg$ **b** $^{25}_{12}MG$ **c** $^{26}_{12}Mg$

These all have 12 protons in the nucleus but atom **a** has 12 neutrons, **b** has 13 neutrons and **c** has 14 neutrons in the nucleus.

For any element there are different amounts of each isotope and this has to be taken into account when calculating the relative atomic mass of an element.

The units of atomic mass are atomic mass units (amu).

NAILIT!

The relative atomic mass is a weighted average which means that we don't just add up the mass numbers of the isotopes and find the average. We have to take into account the abundance of each isotope. If you consider the two isotopes of chlorine $^{35}_{17}Cl$ and $^{37}_{17}Cl$, the average is (35+37)/2 =36 but this does not take into account that there is more of the $^{35}_{17}Cl$ isotope and that is why the relative atomic mass is 35.5 atomic mass units – the average is nearer to 35.5. This means that when working out the relative atomic mass its value should be near the most abundant isotope.

MATHS SKILLS

The amount of an isotope in terms of its percentage is its percentage abundance.

When you work out the relative atomic mass of an element, you start off by saying 'Let there be 100 atoms'. The percentage abundance of an isotope is the number of atoms out of the 100 which are that isotope.

You then multiply the percentage abundance of the isotope by its mass number to give the mass due to its atoms.

Repeat this for the other isotopes and add up all the masses due to all the isotopes.

Finally, divide the total mass by 100 to get the average mass which is the relative atomic mass.

DOIT!

The mass numbers for the isotopes of chromium are shown below. Estimate the relative atomic mass of chromium to the nearest whole number and then look up the relative atomic mass in the periodic table.

Element	Mass number for each isotope with percentage abundance in brackets
Chromium	50 (4.31%); 52 (83.76%); 53 (9.55%) and 54 (2.36%)

WORKIT!

1 One example of isotopes is chlorine with its two isotopes $^{35}_{17}Cl$ and $^{37}_{17}Cl$. The isotope $^{35}_{17}Cl$ makes up 75% of the atoms and the $^{37}_{17}Cl$ isotope 25%. Calculate the relative atomic mass of chlorine. (3 marks)

Let there be 100 atoms. 75 of these are the $^{35}_{17}Cl$ isotope and they have a total mass of 75 × 35 = 2625 amu

The total mass of the $^{37}_{17}Cl$ isotope = 25 × 37 = 925 amu   (1)

The total mass of 100 atoms of all the isotopes = 2625 + 925 amu = 3550 amu (1)

The average mass = 3550/100 = 35.5 amu = the relative atomic mass of chlorine.   (1)

> Remember that amu stands for atomic mass units.

2 There are three naturally occurring isotopes of magnesium $^{24}_{12}Mg$ (78.6% of total); $^{25}_{12}Mg$ (10.11% of total) and $^{26}_{12}Mg$ (11.29% of total). What is the relative atomic mass of magnesium? (3 marks)

> Even though the percentages are not whole numbers just use the same method as for chlorine. One way of checking your answer is to estimate which number the relative atomic mass would be nearest to. In this case the most abundant isotope is magnesium-24 and therefore you expect the relative atomic mass would be nearer to 24 than the others.

Let there be 100 atoms. Mass of magnesium-24 isotope = 78.6 × 24 amu = 1886.4 amu

Mass of magnesium-25 isotope = 10.11 × 25 amu = 252.8 amu   (1)

Mass of magnesium-26 isotope = 11.29 × 26 amu = 293.5 amu   (1)

The total mass of 100 atoms = 2432.7 amu

Therefore the relative atomic mass is the average mass of each atom = 2432.7/100 = 24.3 amu (to 3 significant figures).   (1)

As expected, this is near to 24 which is the mass number of the most abundant isotope.

✓ CHECKIT!

1 a Explain why the two atoms represented by $^{38}_{18}Ar$ and $^{40}_{18}Ar$ are isotopes.

b Give the numbers of electrons, protons and neutrons in these two atoms.

2 a There are two naturally occurring isotopes of copper. Their mass numbers and percentage abundance are given in this table.

Mass number	Percentage abundance/%
63	69%
65	31%

Use this data to calculate the relative atomic mass of copper to 3 significant figures.

b The atomic number of copper is 29. How many electrons, protons and neutrons are there in each isotope of copper?

The development of the periodic table and the noble gases

The periodic table we have today was originally developed by Dmitri Mendeleev.

At the time, Mendeleev organised the elements in order of their atomic weights and he placed elements with similar properties in groups.

He did not place elements where they did not fit and left spaces for elements that he thought had not yet been discovered.

He also predicted the properties of these missing elements. When they were discovered his predictions were found to be very accurate.

Today, the elements are placed in order of their atomic number. If the elements were placed in order of their relative atomic mass the elements iodine and tellurium would be placed in the wrong groups. The same applies to argon and potassium.

Vertical columns of elements are called groups and horizontal rows of elements are called periods.

The number of electrons in the outer shell of an element is its group number.

The number of occupied electron shells is the period number.

The first of the noble gases to be discovered was argon. This new element had completely different properties to those elements already discovered.

Due to Mendeleev's concept of grouping elements with similar properties, discoverers of argon then started looking for other elements with similar properties. This led to the discovery of the other noble gases: helium (in the Sun) and neon, krypton and xenon from the fractional distillation of liquid air.

The noble gases are very unreactive because they have full outer electron shells which are stable electron arrangements.

As you go down the group, the noble gases become denser and their boiling points increase.

DO IT!

Write a short description of the present-day periodic table using the words below:

- groups
- similar
- period
- columns
- electron shells
- properties
- atomic number
- horizontal rows.

Neon gas is often used in lights

SNAPIT!

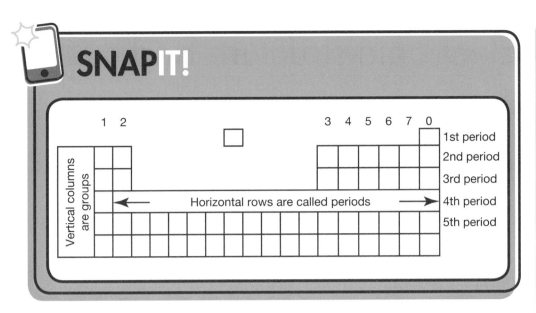

The Periodic table of the elements

NAILIT!

- Once you have located an element in the periodic table you will probably need to select information about the element.
- Remember, the top number above the symbol is the relative atomic mass and the number below the symbol is the atomic number.
- The first row of elements consists of hydrogen and helium. These are easily missed because the hydrogen is placed on its own because it does not fit into a particular group.
- The groups you should concentrate on are the ones featured in this book. These are group 1– the alkali metals, group 7 – the halogens and group 0 – the noble gases.

CHECKIT!

1 What are the names given to vertical columns and horizontal rows in the periodic table?

2 Name the element that is in group 3 and period 3 of the periodic table.

3 Explain why Mendeleev left spaces in his original periodic table.

4 Look at the pairs of elements potassium and argon and tellurium and iodine. Use these examples to explain why we do not arrange the elements in order of their atomic mass.

Electronic structure and the periodic table

The electrons in atoms are arranged in electron shells.

The first shell can take a maximum of 2 electrons, the second shell holds 8 electrons and the third shell can also hold 8 electrons.

The number of electrons in the outer shell of an element is its group number.

The number of occupied electron shells is the period number.

The electronic structures with full outer shells are particularly important because they are very stable. These structures are the electronic structures of the noble gases − 2 and 2,8 and 2,8,8.

DO IT!

Suppose you were given the atomic number of an element. Describe to a revision partner two ways you could locate an element in the periodic table. (Hint: number of protons and electrons)

Alternatively, record a short MP3 of your explanation and check its accuracy by playing it back and comparing it with your notes or your textbook.

SNAP IT!

The diagram below gives two ways to represent the electronic structure of an element. In this case the elements carbon and chlorine.

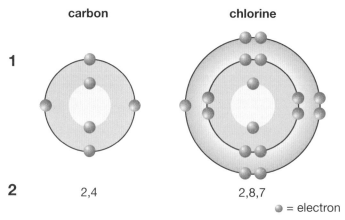

carbon   chlorine

1

2   2,4   2,8,7

= electron

As you can see, one way shows the electrons in their orbits. The second way is a shorthand way and just gives the number of electrons in each shell.

NAIL IT!

Electronic structure is very important. If you can work out the electronic structures of different atoms you can then go on to work out the type of bond formed between them.

The electronic structures with full outer shells are particularly important because they are stable arrangements. These are sometimes called the noble gas arrangements.

You only need to work out the electronic structures of the first 20 elements, i.e. those for the elements hydrogen to calcium.

WORKIT!

Phosphorus has 15 electrons. Only 2 electrons go into the first shell, leaving 13 electrons. Only 8 can go into the second shell leaving 5 electrons for the outermost shell which can take up to 8.

Write out the electronic structure of phosphorus. (1 mark)

The electronic structure or electron arrangement of phosphorus is written as 2,8,5.   (1)

If we use the periodic table we can check this is right by looking at the third period (third row) and group 5. If we look here we find phosphorus.

You will need to work out the electronic structure of an element. You do this by filling the shells up starting from the first or innermost shell which takes up to 2 electrons. You then fill up the second shell until it is full and so on.

Match heads contain phosphorous

✓ CHECKIT!

1 Describe where electrons are found in an atom.

2 Give the maximum number of electrons that can be held in the second electron shell.

3 An element has 19 electrons. Give its electron arrangement and describe its position on the periodic table.

Metals and non-metals

DOIT!

Choose a metal and a non-metal from the periodic table. Look up the properties of both elements on the internet and check to see if they have all the physical properties listed in the table and record if they form a positive or negative ion.

Don't forget to add the definitions we have used here to your spreadsheet of definitions recommended at the beginning of the guide.

Metals are found on the left-hand side of the periodic table and non-metals on the right-hand side.

Most elements (about 92 out of 118) are metals.

The two types of element are different in their appearance, electrical conductivity, malleability and ductility. These differences are shown in the table below.

When metals react with non-metals their atoms lose electrons to form positive ions.

When non-metals react with metals their atoms gain electrons to form negative ions.

Physical property	Metals	Non-metals
Electrical conductivity	Electrical conductors	Electrical insulators
Malleability and ductility	Malleable and ductile	The solids are brittle. They snap when you try to bend or stretch them.
Appearance	All shiny	The solids are dull in appearance.

A table that shows the differences between the physical properties of metals and non-metals

SNAPIT!

Dividing line

Non Metals

Metals

Transition metals

NAILIT!

Malleability means it can be hammered into shape and ductility means it can be drawn into wires.

Learn the differences in physical properties. If you learn the properties of metals then the properties of non-metals are usually the opposite.

CHECKIT!

1 Calcium is a metal. Describe where you would find calcium in the periodic table.

2 Classify the elements sulfur and sodium as metals or non-metals. You should explain your answer by referring to some of their physical properties.

3 Give the charge on a sodium ion and explain how it forms.

4 Black phosphorus is a form of phosphorus that conducts electricity. By referring to the periodic table explain why this is unusual.

Group 1 – the alkali metals

The alkali metals are all in group 1 because they have one electron in their outer electron shell.

They are typical metals because they are shiny and are good electrical conductors.

But, unlike most other metals, they are soft and have low densities. The first three, lithium, sodium and potassium, are less dense than water.

They all react with water to give hydrogen gas and an alkaline solution of the metal hydroxide. For example, sodium gives hydrogen and sodium hydroxide when it reacts with water.

sodium(s) + water(l) → hydrogen(g) + sodium hydroxide(aq)

$2Na(s) + 2H_2O(l) \rightarrow H_2(g) + 2NaOH(aq)$

When they react they all lose their outer electron to form a +1 ion (e.g. Na^+). This gives them a stable full outer shell of electrons.

They get more reactive as you go down the group because it gets easier to lose their outer electron. See the table below.

SNAPIT!

Li	Hardest in group but can be cut with a knife, tarnishes (goes dull) in air. When added to water, it floats and fizzes because hydrogen is released and leaves an alkaline solution of lithium hydroxide (LiOH).	They all have 1 electron in their outer electron shell and this is why they are in group 1.
Na	Easier to cut with a knife than lithium, tarnishes quickly. When added to water, it moves around on the surface of the water and gets hot enough to melt, giving hydrogen gas and alkaline sodium hydroxide (NaOH) solution.	They are all soft and easily cut with a knife. They all react with air and water and this is why they are stored under oil.
K	Even softer and tarnishes immediately. With water, it whizzes around the surface and melts, and the hydrogen given off burns producing a lilac flame. The solution remaining contains potassium hydroxide (KOH).	As you go down the group, their reactions with water get more and more violent. Each reaction produces hydrogen and an alkaline solution of the metal hydroxide.
Rb	Look at videos from the internet to see their reaction with water.	
Cs		

DOIT!

You can find videos of the reactions of the alkali metals on the internet. They will show what happens when they are added to water. Just do a search for 'alkali metals reactions with water'.

NAILIT!

There are often questions on the group 1 elements in the periodic table. You need to know the word and symbol equations for their reactions with water.

You should also be able to predict the reactions of rubidium and caesium using what you know about the first three elements in the group.

WORKIT!

What is the electron arrangement of sodium? (1 mark)

Sodium has the atomic number 11. This means that it has 11 electrons. 2 electrons go in the first shell, 8 in the second and this leaves 1 for the outer third shell. Its electronic structure is 2,8,1.   (1)

MATHS SKILLS

You should be able to work out the electron arrangements of the first three alkali metals. Remember 2 electrons can go in the first shell, 8 in the second and 8 in the third.

A piece of sodium

NAILIT!

Make sure you can explain why the elements get more reactive as you go down the group:

- They all lose their outer single electron when they react to give a stable electron arrangement.
- This gets easier as you go down the group for two reasons:
  - The outer electron gets further from the positive nucleus and so it feels less of an attractive force and can leave the atom more easily.
  - As you go down the group there are more electron shells between the nucleus and the outer electron. This also lowers the attractive force from the nucleus.

CHECKIT! ✓

1 a Give the electron arrangements of lithium (atomic number 3) and potassium (atomic number 11).

b Use your answers to explain why they are in the same group of the periodic table.

2 List four physical properties of sodium.

3 Write the word equation and balanced symbol equation for the reaction of potassium with water.

4 Why is sodium more reactive than lithium?

5 a Predict the observations you would make if rubidium was added to water.

b Write the balanced symbol equation for the reaction of rubidium with water.

6 Explain why the alkali metals are always found in compounds, never uncombined.

Group 7 – the halogens

The halogens are all in group 7 because they have 7 electrons in their outer electron shell.

As elements they exist as molecules of 2 atoms (diatomic molecules) e.g. Cl_2 and Br_2.

When they react with metals, they all gain 1 electron to form a -1 ion. These ions are called halide ions.

As you go down the group the:

- elements get heavier as the relative molecular masses increase
- elements get less reactive because it gets harder to gain an extra electron (see Snap It! box below)
- more reactive halogens displace less reactive ones from solutions of their salts (see next sub-topic).

SNAPIT!

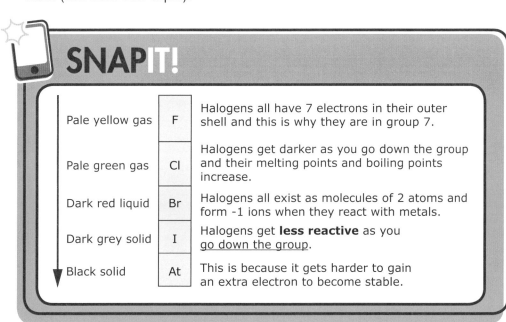

Pale yellow gas	F	Halogens all have 7 electrons in their outer shell and this is why they are in group 7.
Pale green gas	Cl	Halogens get darker as you go down the group and their melting points and boiling points increase.
Dark red liquid	Br	Halogens all exist as molecules of 2 atoms and form -1 ions when they react with metals.
Dark grey solid	I	Halogens get **less reactive** as you go down the group.
Black solid	At	This is because it gets harder to gain an extra electron to become stable.

DOIT!

Using the Snap It! box, describe the properties of the elements as you go down the group. Predict the properties of astatine and then look it up on the internet. How many did you get right?

NAILIT!

There are often questions on the group 7 elements in the periodic table.

You should also be able to predict the reactions of fluorine and astatine using what you know about the middle three elements in the group: chlorine, bromine and iodine.

Of the group 7 elements, you will only be asked to work out the electron arrangements (or draw the electron structure) for fluorine and chlorine. However, you should be able to work out that the electron arrangement of iodine and bromine also end in 7.

One common error is to use chloride instead of chlorine. Chloride is used when chlorine is part of a compound. You can have sodium chloride but not sodium chlorine. The element 'chloride' does not exist.

NAIL IT!

Exam questions often need you to use different techniques and areas of knowledge and these link together different areas of the specification. For example, in the same question you may need to work out relative atomic mass and numbers of sub-atomic particles.

WORKIT!

Bromine is in group 7. It has two isotopes $^{79}_{35}Br$ (50.5%) and $^{81}_{35}Br$ (49.5%).

a Calculate the relative atomic mass of bromine. (3 marks)

The first step to remember is to let there be 100 atoms and then work out the mass of each isotope in the 100 atoms.

Mass of bromine – 79 isotopes = 50.5 × 79 = 3990 amu

amu = atomic mass units

Mass of bromine – 81 isotopes = 49.5 × 81 = 4010 amu   (1)

The total mass of 100 atoms = 3990 + 4010 = 8000 amu   (1)

This makes the relative atomic mass = 80   (1)

b Calculate the number of electrons, protons and neutrons in the bromine-79 isotope. (1 mark)

The atomic number is 35 and this means that there must be 35 electrons and 35 protons.   (1)

The mass number = 79 and the number of neutrons = 79 – 35 = 44

c How many electrons are there in a Br⁻ ion? (1 mark)

Because the ion is 1– it must have 1 extra electron compared with the neutral atom. The number of electrons = 35 + 1 = 36 electrons.   (1)

NAIL IT!

Make sure you can explain why the elements get less reactive as you go down the group:

- They all gain an extra electron when they react to give a stable electron arrangement.
- This gets harder as you go down the group for two reasons:
  - The outer electron shell gets further from the positive nucleus and so any electron feels less of an attractive force and this makes it harder to gain an electron.
  - As you go down the group, there are more electrons between the nucleus and the outer electron. This also lowers the attractive force from the nucleus and makes it harder to gain an electron.

CHECKIT!

1 Give the number of outer shell electrons a bromine atom has.

2 When chlorine reacts with sodium what ion is formed, Cl⁺ or Cl⁻?

3 Draw the electron arrangement of a chloride ion.

Displacement reactions in group 7

A displacement **reaction** is a chemical **reaction** in which a more reactive element displaces a less reactive element from its compound.

In group 7 a displacement reaction takes place when a more reactive halogen takes the place of another less reactive halogen in a compound.

A halide is a compound formed between a halogen and another element.

The more reactive halogen displaces a less reactive halogen ion in a solution of a metal halide (a salt).

For example, when bromine is added to a solution of sodium iodide, the more reactive bromine displaces the iodide ion to form sodium bromide and iodine.

This is because bromine accepts an electron more easily than iodine so the iodide ion donates its extra electron to a bromine atom.

DO IT!

Write word equations using the results table in the Snap It! box and use these to explain the observations for each reaction. Then give the order of reactivity from your observations.

SNAPIT!

The results table below shows what happens in some of the halogen displacement reactions. Try and work out the order of reactivity. Check your answer online to see if you were correct.

Halogen	Sodium chloride solution	Sodium bromide solution	Sodium iodide solution
Chlorine	X	turns yellow/ pale orange	turns brown
Bromine	No change	X	turns brown
Iodine	No change	No change	X

In solution, chlorine is very pale green; bromine is pale yellow to orange and iodine is brown or purple.

Note: X is placed where there is no experiment. For example, you cannot displace a chloride using chlorine!

WORKIT!

The word equation for reacting sodium bromide with chlorine is:
sodium bromide + chlorine → sodium chloride + bromine

Write a balanced equation for this reaction. (2 marks)

Write symbols $NaBr + Cl_2 \rightarrow NaCl + Br_2$ (1)

Balance equation $2NaBr + Cl_2 \rightarrow 2NaCl + Br_2$ (1)

MATHS SKILLS

You should be able to balance the equations for displacement.

First, write down the word equation, then the symbol equation, and then balance it.

Practical Skills

When you do displacement reaction experiments, you must always add the halogen to water as a control. This means you can compare it with the reaction mixture to see if there is a change.

Sometimes cyclohexane is added to the mixture and shaken. Any halogen formed dissolves more in the cyclohexane than the water. This helps because in water, bromine and iodine are not very different in colour (depending on their concentrations). In cyclohexane, bromine is still orange/red in colour whilst iodine is purple. So this can be used to confirm the identity of the halogen produced.

STRETCH IT!

For the Higher Tier, you could be asked to write ionic equations for the displacement reactions that take place. The simple rule for these is that the salts exist as ions and the halogen molecules do not. For example, the reaction between chlorine and potassium iodide solution gives iodine and potassium chloride solution.

Symbol equation: $2KI + Cl_2 \rightarrow 2KCl + I_2$

The potassium iodide and potassium chloride can be written as ions:

$\cancel{2K^+} + 2I^- + Cl_2 \rightarrow \cancel{2K^+} + 2Cl^- + I_2$

The K^+ ions do not change and can be cancelled out because it does not take part in the reaction. The K^+ ion is called a **spectator ion** (it just looks on).

So the final equation is $2I^-(aq) + Cl_2(g) \rightarrow 2Cl^-(aq) + I_2(s)$ ◄── The letters in brackets are state symbols, these are explained on page 138.

For more information on ions, see pages 138 and 139.

NAIL IT!

When you write the formulae for the halogens remember they are diatomic molecules, X_2. So, for example, chlorine is Cl_2 and bromine is Br_2. The general formula of all the sodium halides is NaX.

CHECK IT!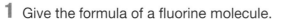

1 Give the formula of a fluorine molecule.

2 Describe the observations when chlorine is added to an aqueous solution of sodium iodide.

3 a Write the word and balanced symbol equations for the reaction between bromine and sodium iodide (NaI).

  b Describe what you would see happen when the bromine is added to the sodium iodide. Explain this observation.

  H c Write the ionic equation for this reaction.

4 When bromine is added to a solution containing sodium chloride no change is observed.

  a What would be the control for this reaction?

  b Explain the observation.

1 a Balance the symbol equation below for the reaction between sodium and chlorine.

$$Na(s) + Cl_2(g) \rightarrow NaCl(s)$$

  b i How can you tell from the symbols in the equation that this is a reaction between two elements?

   ii Describe the appearance of each reactant.

   iii In the product the two types of particle formed are Na^+ and Cl^-. What is the name given to these charged particles?

  c Explain why sodium chloride solution in water is a mixture.

2 In the periodic table, what is a group and what is a period?

3 Sodium has the atomic number 11. Explain why sodium cannot have the atomic number 12.

4 Sulfur has the atomic number 16. Describe its position in the periodic table.

5 a A metal X forms a green compound and a blue compound. Explain where you would find X in the periodic table.

  b Predict four physical properties of the element X.

6 a When group 7 elements react with metals they form ions. What is the charge on these ions?

  b Explain why the group 7 elements get less reactive as you go down the group.

7 Magnesium has the atomic number 12 but has atoms with three different mass numbers, 24, 25 and 26.

  a What does the atomic number 12 tell you about an atom of magnesium?

  b How is it possible for magnesium atoms to have three mass numbers?

  c Gallium has two isotopes, gallium-69 (60%) and gallium-71(40%). Use these figures to calculate the relative atomic mass of gallium.

8 Describe two things that Mendeleev did to make his version of the periodic table work.

9 Explain why the elements in group 1 get more reactive as you go down the group.

10 Explain why the noble gases are so unreactive.

11 a When the element argon was discovered it had very different properties to any other element. Why do you think scientists went on to look for other elements like it?

   b Why was this element so hard to discover?

12 The element astatine is in group 7. It is so rare that most of its chemistry has been guessed at from its position below iodine in group 7.

   a Why do we think it is black in colour?

   b The symbol for astatine is At. What is the formula of an astatine molecule?

   c Predict the formula of sodium astatide.

   d What is the charge on an astatide ion?

Bonding, structure and the properties of matter

Bonding and structure

Strong attractive forces between the particles in a substance make its melting and boiling point high. This is because more energy is needed to overcome these forces.

You can work out the state of a substance at room temperature or any other temperature using its melting and boiling points (see Work It! box below).

When a substance melts, attractive forces between the solid particles are broken as the particles break away from the solid lattice and become liquid particles. The temperature remains constant until all these bonds are broken.

In the same way, when a liquid boils the particles overcome the attractive forces between the liquid particles to become a gas. The temperature stays constant at the boiling point until all these attractive forces are overcome.

The state of a substance at room temperature and pressure (RTP) can be shown in a symbol equation by using the state symbol for the substance.

These state symbols are (s) for solids; (l) for liquids and (g) for gases. The extra symbol (aq) is for substances dissolved in water to give an aqueous solution.

DO IT!

Imagine yourself as a particle. Write a short account of what it would feel like if you were in a solid, a liquid and a gas. What would happen to you if you were a particle in the solid and you were heated until the solid melted?

WORKIT!

A substance X melts at 250°C and boils at 890°C. What state is X in at a) 140°C and b) 723°C? Explain your answers. (2 marks)

a 140°C is below the melting point of X so it hasn't melted and is a solid. (1)

b 723°C is above the melting point so X would have melted but it is below the boiling point and this means that it has not boiled and at this temperature X is a liquid. (1)

MATHS SKILLS

You could be asked to work out what state a substance is in at a given temperature. You will be given the melting point and boiling point of the substance.

To work out the state you have to answer two questions:

• At this temperature has it melted?
• At this temperature has it boiled?

SNAPIT!

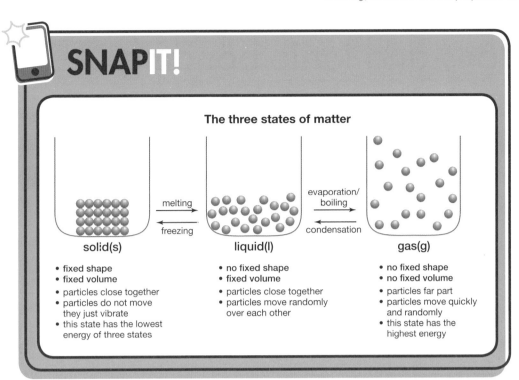

The three states of matter

solid(s) — melting → — freezing ← liquid(l) — evaporation/boiling → — condensation ← gas(g)

solid(s)
- fixed shape
- fixed volume
- particles close together
- particles do not move they just vibrate
- this state has the lowest energy of three states

liquid(l)
- no fixed shape
- fixed volume
- particles close together
- particles move randomly over each other

gas(g)
- no fixed shape
- no fixed volume
- particles far part
- particles move quickly and randomly
- this state has the highest energy

NAILIT!

The three states of matter can be represented by a simple model, as shown in the Snap It! box above. This particle model can help to explain melting, boiling, freezing and condensing. However, for Higher Tier, you will be expected to understand the limitations of this model:

- The model assumes that the particles are spheres.
- Many particles are not spheres, e.g. polymers.
- In the model there are no forces between the particles, which is not correct!

CHECKIT!

1 Which state of matter could be described as having a fixed volume but no fixed shape?

2 If a substance has a low melting point, what can you say about the forces of attraction between its particles?

3 Describe what happens to the particles in a substance as it melts, in terms of movement and arrangement.

4 A substance Y has a melting point of −25°C and a boiling point of 135°C.

   a In which state of matter is Y at i) 20°C and ii) 250°C?

   b Which state symbol could you use for Y in a symbol equation?

H 5 The particle theory states that particles are solid spheres. Give two reasons why this statement is incorrect.

Ions and ionic bonding

Ions are formed when metal atoms transfer their outer electrons to the outer shells of non-metal atoms. This electron transfer forms charged particles called ions.

Metal atoms form positive ions because they have lost negatively charged electrons. The positive charge on the ion is the same as the number of electrons that are lost.

Non-metal atoms form negative ions because they have gained negatively charged electrons. The negative charge on the ion is the same as the number of electrons that are gained by the atom.

The elements in groups 1 and 2 of the periodic table are metals and they form positive ions.

Each group 1 atom forms a 1+ ion by losing 1 electron and each group 2 atom forms a 2+ atom by losing 2 electrons.

The elements in groups 6 and 7 are non-metals and they form negative ions.

Each group 6 atom gains 2 electrons to form a 2− ion and each group 7 atom gains 1 electron to form a 1− ion.

All the ions formed have the same electronic structure as the nearest noble gas because these are stable electron arrangements.

The ionic bond is the electrostatic attraction between the positive and negative ions.

You can work out the formula of an ionic compound by balancing the charges on the ions (see Work It! box on page 139).

DOIT!

The atomic number of lithium is 3 and fluorine's is 9. Draw diagrams of the atoms including their nuclei before reaction and diagrams of the ions after reaction. Use your diagrams to explain why the lithium ion is Li$^+$ and the fluoride ion is F$^-$.

SNAPIT!

The **dot-and-cross** diagrams below show the outer electronic structures of some atoms and ions before and after electron transfer.

In the third example, each chlorine can accept just 1 electron and the magnesium has to lose 2 electrons. This means that each magnesium has to react with 2 chlorines.

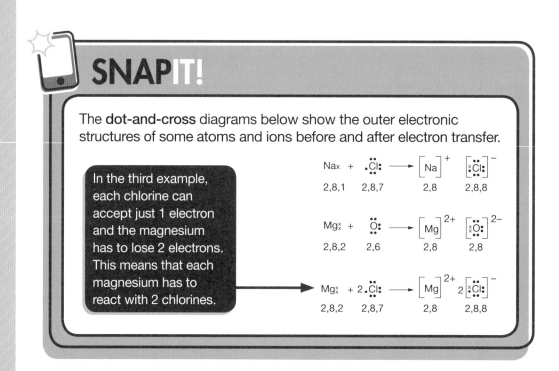

NAILIT!

When you draw the ions and their electronic structures, you only have to give the outer electrons. For the positive metal ions, you do not have to draw any dots or crosses because they have lost their outer electrons.

You must also remember to draw the electrons gained by the non-metal atom as different to the ones already there. For example, if you look at the chloride ion in the Snap It! box on page 138 the electron gained from the sodium is written as an × when the chlorine electrons are given as •.

You can also identify whether or not a compound is ionic by looking at the elements. A compound is ionic if one element comes from either group 1 or 2 and the other one comes from either group 5, 6 or 7.

As far as hydrogen is concerned, it resembles group 7 elements because it has to gain 1 electron to be stable.

WORKIT!

a   Calculate the formula of the compound sodium oxide. (2 marks)

Sodium is in group 1 which means that its ion is Na^+. Oxygen is in group 6 which means the oxide ion is O^{2-}.

To make the charges add up to zero we need 2 of the Na^+ ions and 1 of the O^{2-} ion. (1)

The formula is therefore Na_2O. (1)

A dot-and-cross diagram could also be used to find the formula. The sodium atom needs to lose 1 electron and the oxygen needs to gain 2 electrons. Therefore to make this happen there needs to be 2 sodium atoms combining with 1 oxygen and the formula is therefore Na_2O.

b   Calculate the formula of strontium fluoride. (2 marks)

Strontium is in group 2 and this means that its ion is Sr^{2+}. Fluorine is in group 7 so its ion is F^-. To make the charges add up to zero we need 1 strontium ion and 2 fluoride ions. (1) The formula is SrF_2. (1)

MATHS SKILLS

When you work out the formula of an ionic compound, you have to make the charges on the ions add up to zero. Let's suppose you have a compound containing M^{2+} ions and X^- ions. To make the charges add up to zero we need 2 of the 1− ions and 1 of the 2+ ions. This means that the formula is MX_2.

✓ CHECKIT!

1   The list below shows the formulae of six compounds. From the list choose the three ionic compounds.

   $LiCl$ CS_2 NH_3 $BaBr_2$ CO_2 NaH

2   Draw dot-and-cross diagrams for the three ionic compounds you have chosen.

3   Explain how group 1 elements form 1+ ions.

4   Give the formula of a sulfide ion.

5   Explain how the ions in NaCl stay together.

6   Give the formula of the ionic compounds potassium sulfide and magnesium iodide.

The structure and properties of ionic compounds

The ions in ionic compounds are held together because of the strong electrostatic attraction between the oppositely charged ions.

The electrostatic forces around each ion extend in all directions so each ion attracts several oppositely charged ions around it. This results in a giant lattice where the ions are regularly arranged in a repeating pattern.

The diagrams below showing the structure of sodium chloride illustrate this.

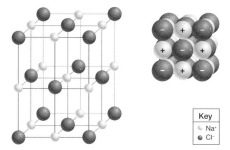

Key
◦ Na⁺
● Cl⁻

Even though there are many ions in a giant ionic lattice, the numbers of each ion are in the same ratio as they are in the formula of the compound. For example, in NaCl the number of sodium and chloride ions are equal in number. In $CaBr_2$, there are twice as many bromide ions as there are calcium ions.

Ionic compounds have high melting and boiling points because the ionic bonds are **strong** and in the giant lattice there are lots of them to break. You need a lot of energy to break these bonds.

Ions are charged particles and when they move they can carry an electric current.

In solid form ionic compounds do not conduct electricity because the ions are fixed in a lattice and do not move and therefore they cannot carry the current.

If an ionic compound is dissolved in water or is in molten form, then the ions are no longer fixed in a lattice and can move. This means that the ionic compound will conduct electricity.

DOIT!

You may be asked to complete part of an ionic lattice by placing the ions in their places on the grid. The one most usually asked is NaCl because it is 1:1 in terms of positive and negative ions

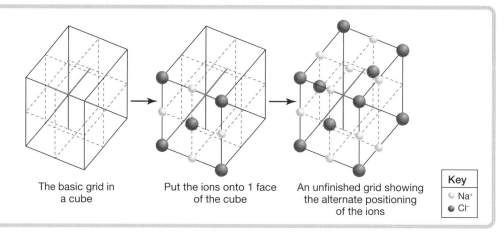

The basic grid in a cube

Put the ions onto 1 face of the cube

An unfinished grid showing the alternate positioning of the ions

Key
◦ Na⁺
● Cl⁻

NAILIT!

With a few exceptions, all giant structures have high melting and boiling points because they have lots of bonds that have to be broken. The attraction between ions increases as the charges on the ions increase. For example, the attraction between magnesium ions Mg^{2+} and oxide ions O^{2-} is greater than the attraction between sodium ions Na^+ and chloride ions Cl^-. This means that the melting point of magnesium oxide is higher than that of sodium chloride.

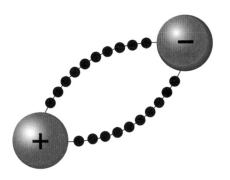

SNAPIT!

Property	Ionic compounds	Explanation
Melting and boiling point	High	Strong electrostatic attraction between the ions and there are lots of bonds to be broken.
Electrical conductivity of solid	Poor	The ions cannot move and cannot carry the current.
Electrical conductivity of liquid	Good	The ions can move and can carry the current.

✓ CHECKIT!

1 Describe what keeps the ions together in an ionic lattice.

2 Explain why the melting point of sodium chloride is very high.

3 Explain why sodium chloride does not conduct electricity as a solid but it does conduct electricity when it is dissolved in water.

4 What can you say about the number of fluoride ions compared to magnesium ions in the ionic lattice of magnesium fluoride (MgF_2)?

5 Explain which of potassium bromide (KBr) and calcium oxide (CaO) has the higher melting point.

Covalent bonds and simple molecules

A covalent bond is formed when a pair of electrons is **shared** between the atoms of two non-metal atoms.

The number of covalent bonds formed by an atom is equal to the number of electrons it needs to gain.

After forming the bonds each atom has the same electronic structure as the nearest noble gas.

Covalent bonds are strong and require a lot of energy to break.

Covalent bonding can be represented by dot-and-cross diagrams or straight lines drawn between atoms. For example:

HCl can be represented by H——Cl or H $\overset{\bullet\bullet}{\underset{\bullet\bullet}{\overset{\times}{C}l}}$

CH$_4$ can be represented by

$$H—\overset{\displaystyle H}{\underset{\displaystyle H}{\overset{|}{\underset{|}{C}}}}—H$$

or H $\overset{\bullet\,x}{\underset{\bullet\,x}{\overset{\times}{C}}}$ H with H above and below

Covalently bonded substances can exist as simple molecules, giant covalent structures or polymers.

Simple molecules are neutral particles and have no charges or free electrons. This means they cannot carry an electric current either in liquid or solid form.

Simple molecular substances have weak intermolecular forces between the molecules. These weak attractive forces need little energy to break them. This means that simple molecular substances have low melting and boiling points and quite often, they are liquids or gases at room temperature.

The intermolecular forces increase as the molecular size **increases** and this means that the melting and boiling points also increase.

NAIL IT!

Important! When a simple molecular substance melts or boils, the weak intermolecular forces are broken **not** the strong covalent bonds.

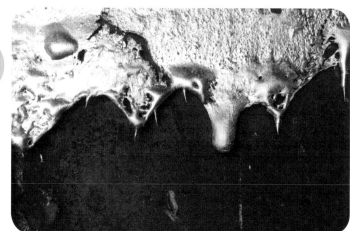

SNAPIT!

See the examples below:

Nitrogen 2,5
needs to gain 3
electrons so forms
3 covalent bonds
per atom

+   3 x H   ⟶

Hydrogen 1
needs to gain 1
electron so forms
1 covalent bond per
atom

Ammonia NH_3

    +   ⟶

Oxygen 2,6
needs to gain
2 electrons so
forms 2 covalent
bonds per atom

Carbon 2,4
needs 4 electrons
so forms 4 covalent
bonds per atom and
because it needs to gain
more electrons it is the
central atom of the
molecule

Carbon dioxide CO_2

$O = C = C$

Each oxygen shares two
pairs of electrons with
the carbon and this means
there are double covalent
bonds between the carbon
and each oxygen

NAILIT!   H

If you are doing the Foundation exam, concentrate on drawing dot-and-cross diagrams for compounds containing hydrogen. For example, H_2O and NH_3.

If you are doing the Higher Tier exam, you may be asked to draw dot-and-cross diagrams for compounds containing hydrogen, as well as for more complicated covalent bonding such as that shown in carbon dioxide where multiple bonds are involved (see Snap It! box above).

Remember, only the outer electrons are shown in dot-and-cross diagrams. The element which has to gain the most electrons is always the central atom in a molecule.

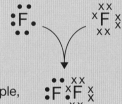

Some elements also exist as simple diatomic molecules. These elements are H_2, O_2, and N_2 and all the halogens (see group 7 on page 131) F_2, Cl_2, Br_2, I_2 and At_2. This is because they can form stable electron arrangements by bonding with each other. For example, fluorine, F_2, which is shown to the right.

CHECKIT!

1 Explain what is meant by a covalent bond.

2 What type of element has atoms that form covalent bonds?

3 Draw a molecule of water (H_2O) using a dot-and-cross diagram and using straight lines for the covalent bonds.

4 Draw dot-and-cross diagrams and line diagrams for: **a** HF   **b** CF_4

5 CF_4 is a compound with a simple molecular structure. Predict some of its physical properties such as melting point, electrical conductivity, and so on.

Diamond, graphite and graphene

All of the carbon atoms in diamond, graphite and graphene are linked together by strong covalent bonds.

Diamond and graphite exist as giant covalent structures.

In diamond, each carbon atom is covalently bonded to four other carbon atoms in a giant repeating pattern.

Each covalent bond is strong and difficult to break and this makes diamond **very hard**.

As diamond has strong covalent bonds in a giant structure it has very high melting and boiling points.

There are no charged particles in diamond or any free electrons to carry an electric current so it does not conduct in the liquid or solid state.

In graphite, the carbon atoms are in layers of hexagonal rings where each carbon atom is covalently bonded to three other carbons. This means that each carbon has a spare electron which is free to move through the layers so graphite conducts electricity both as a solid and as a liquid.

The covalent bonds in the layers are strong but between the layers there are only weak intermolecular forces. These are easy to break and so the layers can slide over each other easily, making graphite very soft and slippery. This means it can be used in industry as a lubricant.

Graphite also has high melting and boiling points because to melt graphite all of its strong covalent bonds have to be broken and this requires lots of energy.

Graphene has an identical structure to a single layer of graphite. This means that it is one atom thick.

It also means that graphene has identical properties to one layer of graphite, making it strong, flexible, transparent and a good electrical conductor. These properties give it lots of potential uses in, for example, electrical circuits, medical applications, displays and solar cells.

DO IT!

Redraw this table. Compare the physical properties of simple molecular and giant covalent structures.

	Physical property	Simple molecular	Giant covalent
Diamond			
Graphite			
Graphene			

SNAPIT!

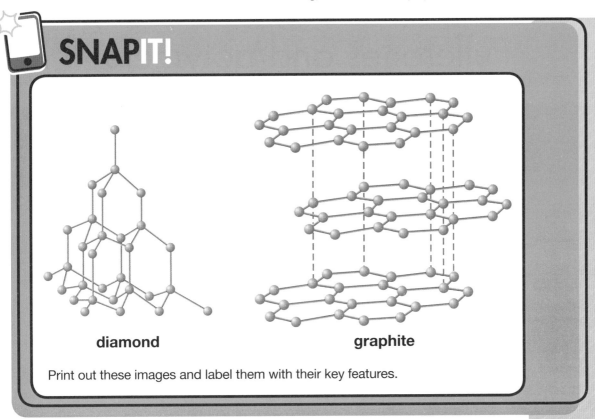

diamond   **graphite**

Print out these images and label them with their key features.

Diamonds

Graphite

NAILIT!

You should be able to recognise the diagrams of diamond and graphite and understand that they are examples of giant covalent structures.

Describing or recognising the structures of diamond and graphite is not a high-level skill and would only get you a few marks in a question. The same applies to simply describing the properties. It is using the structure to explain the bulk properties that is a high-level skill.

CHECKIT!

1 Describe the structure found in graphite and diamond.

2 Explain why the melting points of graphite, diamond and graphene are very high.

3 Explain why graphite is very soft.

4 Explain why diamond does not conduct electricity but graphite does.

5 Graphene is one layer of graphite. Describe this layer.

6 Silicon dioxide has a structure which is similar to diamond. Predict the properties of silicon dioxide.

Fullerenes and polymers

The smallest example of a spherical fullerene is buckminsterfullerene which is a large but simple molecular substance with the formula C_{60}. These spheres contain mostly 6-carbon rings but there are also small numbers of 5- and 7-carbon rings.

These spherical fullerenes can be used to trap drugs and deliver them to parts of the body. They are also used in industry as lubricants and catalysts.

Carbon nanotubes are cylindrical fullerenes which have a high length to diameter ratio. Due to the strong covalent bonds between the carbons in the layers they have a high tensile strength. They are good electrical and heat conductors.

Polymers are very large molecules which are made up of long chains of carbon atoms joined by strong covalent bonds. As the chains are very long, the intermolecular forces between each chain are quite large and this means polymers are solids.

DOIT!

You may have used a key in biology to identify plants or animals. You can use a key to identify lots of different things. A key will usually ask questions based on easily identifiable features of something.

Draw out a key for carbon structures which can be used to identify them using the structures and properties. The key should include graphite, diamond, polymers, graphene and fullerenes (spherical and nanotubes).

SNAPIT!

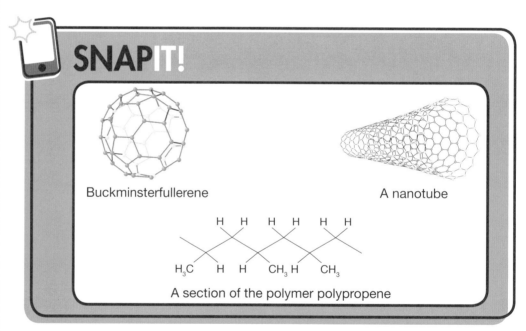

Buckminsterfullerene

A nanotube

A section of the polymer polypropene

CHECKIT!

1 a How many carbon atoms there are in the smallest molecule of the buckminsterfullerene series of structures?

 b This substance has a simple molecular structure. Predict some of its physical properties.

2 Describe the properties of nanotubes.

3 Explain why nanotubes are used to reinforce tennis rackets.

4 Give a use for buckminsterfullerene.

Giant metallic structures and alloys

Metals form a giant lattice of positive metal ions in a sea of delocalised electrons.

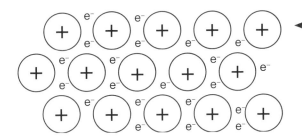

> This diagram shows how the delocalised electrons can move around the positive metal ions.

The electrostatic attraction between the positive ions and the delocalised or free electrons is called a metallic bond. These metallic bonds are usually strong and need a lot of energy to overcome them, which is why most metals have high melting and boiling points.

The metallic bond operates in both pure metals and in alloys.

The layers of metal ions can slide over each other without disrupting the structure. This means that metals can be bent, shaped (they are malleable) and drawn into wires (they are ductile).

The delocalised electrons can carry an electric current so metals are good electrical conductors both as solids and liquids.

Metals are also good heat or thermal conductors because the electrons can transfer the heat along the metal.

Alloys are mixtures of metals.

Introducing different-sized metal atoms into a metal lattice makes it harder for the layers to slide over each other so alloys are harder than pure metals.

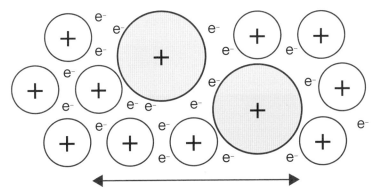

More difficult for metal ions to slide over each other

Car wheels are often made from alloys of magnesium or aluminium

DO IT!

Make a summary table for the different types of structures. Use the following headings:

Type of structure	Example	Particles present	Melting point and boiling point	Electrical conductivity	
				As solid	As liquid
Giant ionic					
Simple molecular					
Giant covalent					
Giant metallic					

NAIL IT!

A common question is recognising the type of structure from the properties of a substance. These properties are usually melting points, boiling points and electrical conductivity in the solid and liquid states.

SNAP IT!

Property	For metals	Explanation
Melting and boiling point	Usually high	Strong metallic bonds in a giant lattice
Electrical conductivity as solid	Good	Delocalised electrons carry the current
Electrical conductivity as liquid	Good	Delocalised electrons carry current
Thermal (heat) conductivity	Good	Delocalised electrons transfer the heat energy
Malleability and ductility	Malleable (can be shaped) and ductile (can be drawn into wires)	Layers of metal ions can slide over each other without changing the structure

CHECK IT! ✓

1 Describe a giant metallic lattice.

2 Explain why metals are good electrical conductors in both the solid and liquid form.

3 Explain the term alloy.

4 Explain why alloys are often harder than pure metals.

1 Explain how a lithium ion is formed from a lithium atom. [Atomic number of lithium = 3]

2 When magnesium (atomic number 12) reacts with chlorine (atomic number 17) the ionic compound magnesium chloride is formed.

a Showing the outer electrons only, draw dot-and-cross diagrams of magnesium and chloride ions.

b Give the formula of magnesium chloride.

c Explain why magnesium chloride has a high melting point.

d Why is solid magnesium chloride a poor electrical conductor?

3 Carbon (atomic number 6) and hydrogen (atomic number 1) combine to form methane (CH_4).

a Draw a dot-and-cross diagram to show the bonding in methane.

b Explain why methane is a gas at room temperature even though the bonds between the carbon and hydrogen are strong.

4 These diagrams show the structures of diamond and graphite.

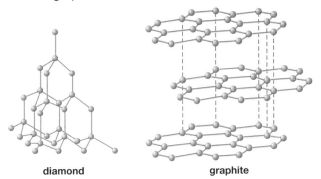

diamond   graphite

a What type of structure do they both have?

b Why do they both have high melting points?

c Why is graphite slippery?

d Explain why solid graphite is a good electrical conductor but solid diamond is a poor electrical conductor.

5 The table below shows the properties of four substances. The letters are not their symbols.

Using the data in the table name the type of structure present in A to D.

Substance	Melting point /°C	Electrical conductivity	
		As solid	As liquid
A	−55	Poor	Poor
B	2015	Poor	Good
C	1897	Good	Good
D	2567	Poor	Poor

6 Using your knowledge of types of structure, explain the following observations:

a Methane has a lower melting and boiling point than potassium chloride.

H b Magnesium oxide has a higher melting point than potassium chloride.

c Brass (an alloy of copper and zinc) is harder than pure copper.

d Sodium conducts electricity in both the solid and liquid state but sodium chloride only conducts electricity in the liquid state or in a solution.

149

Quantitative chemistry

Conservation of mass and balancing equations

In chemical equations the reactants are written on the left-hand side and the products on the right-hand side.

Chemical formulae are used to represent the substances in the reactants and the products.

The law of conservation of mass states that no atoms are lost or gained during a chemical reaction.

Therefore, when you represent chemical reactions by symbol equations there must be the same number of each type of atom on both sides of the equation. This means that the equation must be **balanced**.

One consequence of this is that the mass of the reactants is equal to the mass of the products.

If the chemical container is open and gas is consumed or produced in a reaction, the mass appears to go up or down respectively but in terms of the atoms taking part it really has not changed.

Example:

$CaCO_3(s) + 2HNO_3(aq) \rightarrow Ca(NO_3)_2(aq) + H_2O(l) + CO_2(g)$

The mass appears to diminish because the CO_2 gas comes off from the reaction.

DO IT!

Take the following equation and explain to a revision partner why the equation has to be balanced and show them how it should be balanced. If working alone, write down your notes and check against your textbook.

$Fe_2O_3 + Al \rightarrow Fe + Al_2O_3$

SNAP IT!

The diagram below shows apparatus that could be used to prove that the law of conservation of mass is true. Your teacher may have demonstrated this in class. There is a bung in the top of the flask so that no substances can enter or leave. Once the reactants have mixed and the reaction has started the top-pan balance shows that the mass does not change.

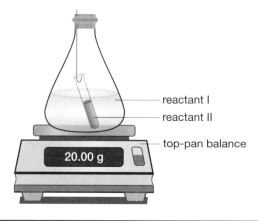

reactant I
reactant II
top-pan balance
20.00 g

WORKIT!

The reaction between ammonia and copper(II) oxide gives nitrogen, copper and water as products. Write a balanced symbol equation for this reaction. (2 marks)

The unbalanced symbol equation may be written as
$NH_3 + CuO \rightarrow N_2 + Cu + H_2O$. (1)

NH_3 and N_2 are both gases so are followed by (g); CuO and Cu are solids, followed by (s) and H_2O is a liquid (l).

The number of nitrogen and hydrogen atoms are unequal and so the equation needs balancing. To balance it you can balance the nitrogen atoms first to give 2 on each side. Then the hydrogen atoms – 6 on each side – followed by the oxygen atoms – 3 on each side and the copper atoms – 3 on each side.

The final equation is
$2NH_3(g) + 3CuO(s) \rightarrow N_2(g) + 3Cu(s) + 3H_2O(l)$ (1)

NAILIT!

Balancing equations and being able to interpret them is one of the vital skills you need to get a good grade in chemistry. One thing that you may have to do is to balance equations that have in them negative acid groups such as nitrate (NO_3^-) and sulfate (SO_4^{2-}) or positive ammonium (NH_4^+). You must remember that if you have more than one of these groups in a compound then brackets are placed round the group.

For example, calcium nitrate contains the Ca^{2+} ion and the NO_3^- ion. Calcium nitrate's formula is therefore $Ca(NO_3)_2$.

When balancing equations like this, treat the nitrate group like a single atom except when there is more than one, then they need brackets round them.

MATHS SKILLS

A balanced symbol equation has the same number of each type of atom in the reactants as there are in the products.

When you balance a chemical equation the first thing you must remember is **not to change the formulae!**

You can only write numbers in normal-sized writing **before the formulae.**

Stage 1 – work out which atoms are not equal in number on both sides of the equation. Balance these by putting numbers in the correct places.

Stage 2 – see which atoms are now wrong in number and balance them and continue until the atoms are equal in number on both sides.

CHECKIT!

1 When magnesium is heated with aluminium oxide, magnesium oxide and aluminium are formed.

   a In this reaction, identify which are the reactants and which are the products.

   b If 72 g of magnesium completely reacts with 103 g of aluminium oxide, what is the total mass of aluminium and magnesium oxide formed?

2 a Write the word equation for the following **unbalanced** chemical reaction:

   $HCl + CaCO_3 \rightarrow CaCl_2 + H_2O + CO_2$

   b Write out the balanced symbol equation.

   c Add state symbols to the equation.

   d If this reaction was carried out in an open container the measured mass would decrease. Explain why.

151

Relative formula masses

The relative atomic mass (symbol = A_r) of an element is the weighted average mass of its naturally occurring isotopes.

You calculate the relative formula mass (symbol = M_r) of a compound by adding up all the relative atomic masses of all the atoms present in the formula of the compound.

The elements hydrogen, oxygen, nitrogen, chlorine, bromine, iodine and fluorine exist as diatomic molecules. This means that in equations their relative formula masses are twice their relative atomic masses.

All other elements are represented by just their symbols in an equation. So their relative formula mass is equal to their relative atomic mass.

Using the law of conservation of mass we can say that in a chemical reaction the sum of the relative formula masses of the reactants is equal to the sum of the relative formula masses of the products.

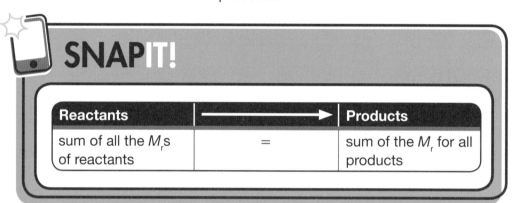

SNAP IT!

Reactants		Products
sum of all the M_rs of reactants	=	sum of the M_r for all products

For example, when calcium carbonate (formula mass = 100) is heated it decomposes to calcium oxide (formula mass = 56) and carbon dioxide. This means that the formula mass of carbon dioxide equals 44.

MATHS SKILLS

In a chemical formula, the number to the right-hand side of an atom is the number of that type of atom in the compound. For example, in carbon dioxide CO_2 there is 1 carbon atom and 2 oxygen atoms.

Also, when there is a group of atoms such as hydroxide and there are brackets around the group in the formula, then you multiply whatever is inside by the number outside it.

For example, calcium hydroxide $Ca(OH)_2$ has 1 calcium, 2 oxygens and 2 hydrogens.

WORKIT!

a Find the relative formula mass of aluminium oxide Al_2O_3. (1 mark)

There are 2 aluminium atoms and 3 oxygen atoms. The relative
formula mass = $(2 \times 27) + (3 \times 16) = 102$. (1)

b Find the relative formula mass of calcium nitrate $Ca(NO_3)_2$. (1 mark)

In this formula there is 1 calcium atom and 2 times whatever is inside the
brackets. This means there are 2 nitrogen atoms and 6 oxygen atoms so the
relative formula mass = $(1 \times 40) + (2 \times 14) + (6 \times 16) = 164$. (1)

NAILIT!

You do not have to remember relative atomic masses of elements. You will be given a periodic table showing them. Make sure you use the right number. The relative atomic mass is the top one.

CHECKIT!

1 Give the relative formula masses of the following compounds:

  a CaO

  b $MgCl_2$

  c KNO_3

  d $Al_2(SO_4)_3$

2 **a** The balanced equation below represents the reaction taking place when silver carbonate is heated:

    $2Ag_2CO_3(s) \rightarrow 4Ag(s) + 2CO_2(g) + O_2(g)$

  b If 552 g of silver carbonate is heated, 88 g of carbon dioxide and 32 g of oxygen are formed. Calculate the mass of silver produced by the reaction.

  c What law are you applying in your calculation?

The mole

The **amount** of a chemical substance is measured in moles (symbol = mol).

The number of atoms, ions or molecules in a mole is equal to Avogadro's constant (N_A). The value of this is 6.02×10^{23}.

This number applies to all particles. For example, there are the same number of CH_4 molecules in 1 mole of methane as there are sodium atoms in 1 mole of the element sodium.

The mass of 1 mole of any substance is its relative formula mass expressed in grams.

The number of moles of a substance is equal to its mass in grams divided by its relative formula mass.

SNAPIT!

The equations for moles and number of particles

$m = n \times M_r$ number of particles = $n \times N_A$

$n = m \div M_r$

$M_r = m \div n$

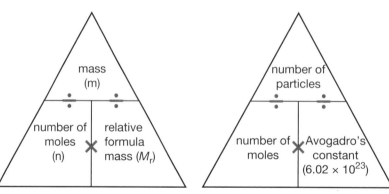

There are two ways of learning the equations – either using the triangles or learning one equation and then rearranging it.

Consider the left-hand triangle. You have to use the operations shown in the triangle. This means that the top quantity m = the product of the two bottom quantities n × M_r. The bottom left-hand quantity n = the top quantity m divided by the bottom left, and so on.

MATHS SKILLS

Use the calculation triangles to construct formulae for your calculations.

$n = m/M_r$ $m = n \times M_r$ no. of particles = $n \times N_A$

When using Avogadro's constant you need to know how to use your calculator.

Type in 6.02 then press the [x10^x] button and then 23. You will get 6.02×10^{23}

WORKIT!

How many moles of carbon dioxide (CO_2) are there in 2.2g of the substance? (2 marks)

The relative formula mass of carbon dioxide = 44   (1)

see left-hand triangle on page 154

This means that $n = m/M_r = 2.2/44 = 0.05\,mol$ (1)

How many carbon dioxide particles are there in the same mass of substance? (2 marks)

Number of CO_2 particles = number of moles × Avogadro's constant
$$= 0.05 \times 6.02 \times 10^{23}$$ (1)
$$= 3.1 \times 10^{22}$$ (1)

NAILIT!

Know the relationships:
$n = m/M_r$
$m = n \times M_r$
number of particles = $n \times N_A$.

You need to be able to write down large numbers like Avogadro's constant in **standard form**.

The same applies to very small numbers. For example 0.00041 is expressed as 4.1×10^{-4}.

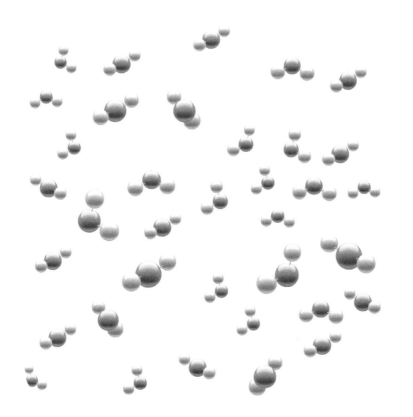

✓ CHECKIT!

H 1 How many particles of a substance are found in 1 mole of the substance.

H 2 Describe how you would find the mass of one mole of a compound.

H 3 a Calculate the relative formula mass of sulfur dioxide (SO_2).

   b Calculate how many moles there are in 1.6g of sulfur dioxide.

   c Calculate how many sulfur dioxide molecules there are in 1.6g of the substance.

Reacting masses and using moles to balance equations

H

You can tell how many moles of a substance react or are produced in a reaction from the numbers in front of the formulae in the balanced symbol equation.

When you are given the masses that react you can convert these into moles and then use the equation to find the masses of the substances produced by using the ratios in the equation.

The same applies to finding the mass of reactants needed to produce a certain mass of products.

If you know the masses of reactants and products in a reaction you can convert these to moles and use the results to balance the equation.

DOIT!

Have a go at calculating one of the Work It! questions from page 157 yourself. Record yourself explaining how you would work it out and compare to the solution given.

SNAP**IT!**

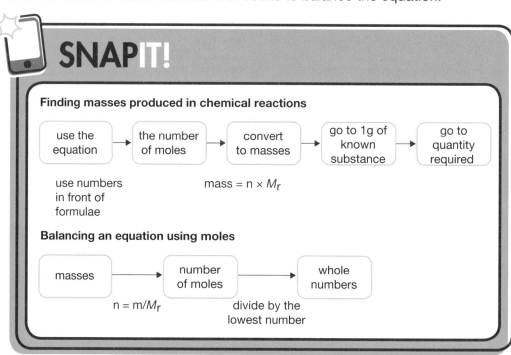

Finding masses produced in chemical reactions

use the equation → the number of moles → convert to masses → go to 1g of known substance → go to quantity required

use numbers in front of formulae

$mass = n \times M_r$

Balancing an equation using moles

masses → number of moles → whole numbers

$n = m/M_r$ divide by the lowest number

NAIL**IT!**

When you answer this type of numerical question and you are using a calculator, it is sometimes tempting to just write down the answer. This is not a good idea because if you make a mistake using the calculator and give the wrong answer without any working you will not get any marks. Always show your working and you will get credit for the correct approach.

You should also make sure your calculator will give an answer that enables you to quote your answer to 3 significant figures and will give the answer in standard form. For example:

$45 \div 4789$

If you use scientific mode there is no problem. You can ask for 3 figures and you get 9.40×10^{-3}. So make sure you have it in this setup.

MATHS SKILLS

1 To work out the **mass of a substance produced** see the first worked example.

2 To see how to **balance an equation** see the second worked example.

(H)

WORKIT!

1 Mass of a substance produced.

Calculate the mass of silver formed when 23.2 g of silver oxide is heated. (3 marks)

Note: Formula masses $Ag_2O = 232$; $Ag = 108$

1.	Write the balanced equation.	$2Ag_2O$	⟶ 4Ag	$+ \quad O_2$
2.	Use the equation to find the number of moles of reactants and products.	2 moles of silver oxide	give 4 moles of silver	You are not asked about the oxygen.
3.	Convert these moles to masses.	$2 \times 232g =$ 464g	give $4 \times 108g$ = 432g	Remember a mole is the M_r in g.
4.	IMPORTANT: Go down to 1g of reactant to make calculation easier.	1g	give 432/464g (1)	Divide both sides by 464.
5.	Go to the desired quantity.	23.2g	give $\dfrac{23.2 \times 432}{464}g$ (1) $= 21.6g$ (1)	This will give 23.2 times whatever 1g gives.

2 Balancing equations using moles.

In the reaction between aluminium oxide and magnesium, 10.2 g of aluminium oxide reacts with 7.2 g of magnesium to give 5.4 g of aluminium and 12.0 g of magnesium oxide. Use these results to find the balanced equation for the reaction. [M_r $Al_2O_3 = 102$; $Mg = 24$; $Al = 27$ and $MgO = 40$] (2 marks)

Write word equation	Aluminium oxide +	magnesium ⟶	aluminium	+ magnesium oxide
Write down the masses	10.2g	+ 7.2g	5.4g	12.0g
Convert these masses to moles	$\dfrac{10.2}{102} = 0.1\,mol$	$\dfrac{7.2}{24} = 0.3\,mol$	$\dfrac{5.4}{27} = 0.2\,mol$	$\dfrac{12.0}{40} = 0.3\,mol$ (1)
Divide by the lowest number which is 0.1 to give simplest whole-number ratio	$\dfrac{0.1}{0.1} = 1$	$\dfrac{0.3}{0.1} = 3$	$\dfrac{0.2}{0.1} = 2$	$\dfrac{0.3}{0.1} = 3$
Write out the balanced equation	Al_2O_3	+ 3Mg ⟶	2Al	+ 3MgO (1)

✓ CHECKIT!

H 1 Calculate the mass of silver that could be formed from 6.35 g of copper when copper is added to silver nitrate solution:

$Cu(s) + 2AgNO_3(aq) \rightarrow Cu(NO_3)_2(aq) + 2Ag(s)$

H 2 When 68 g of silver nitrate ($AgNO_3$) is heated it decomposes to form 43.2 g of silver (Ag), 18.4 g of nitrogen dioxide (NO_2) and 6.4 g of oxygen O_2.

Use this data to write a balanced equation for the reaction. You must show all your working.

[Relative formula masses: $AgNO_3 = 170$; $Ag = 108$; $NO_2 = 46$; $O_2 = 32$]

H

Limiting reactant

If there are two reactants in a chemical reaction and there is an excess (more than needed) of one of them, then the other reactant is called the limiting reactant.

The amounts of the products formed depend on the amount of the limiting reactant.

The limiting reactant is identified by comparing the number of moles of each reactant with those required by the balanced symbol equation.

At the end of the reaction, the limiting reactant is completely used up and the reactant in excess remains along with the products.

SNAPIT!

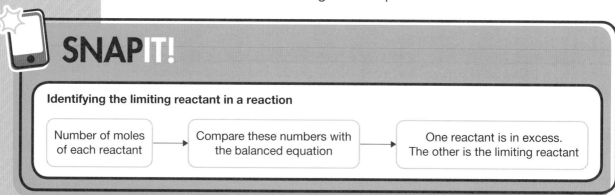

Identifying the limiting reactant in a reaction

Number of moles of each reactant → Compare these numbers with the balanced equation → One reactant is in excess. The other is the limiting reactant

MATHS SKILLS

To identify the limiting reactant, you work out the number of moles of each reactant and then compare with what is needed using the chemical equation for the reaction.

WORKIT!

The equation for the reaction between zinc and hydrochloric acid is
$Zn(s) + 2HCl(aq) \rightarrow ZnCl_2(aq) + H_2(g)$

If 0.10 mol of zinc is added to 0.3 mol of hydrochloric acid what is the limiting reactant? Explain your answer. (3 marks)

From the equation 2 mol of hydrochloric acid react with 1 mol of zinc, and this means that 0.1 mol of zinc needs exactly 0.2 mol of hydrochloric acid to react. (1)

This means that 0.3 mol of hydrochloric acid is in excess and the limiting reactant is zinc. (1) Because there is 0.1 mol of zinc this means that we will get 0.1 mol of hydrogen and 0.1 mol of zinc chloride. (1)

CHECKIT! ✓

H 1 Magnesium and sulfuric acid react as follows:

$Mg(s) + H_2SO_4(aq) \rightarrow MgSO_4(aq) + H_2(g)$

1.00 mol of magnesium is added to 0.90 mol of sulfuric acid.

a Identify the limiting reactant.

b Explain your answer.

c How many moles of hydrogen are produced in the reaction?

Concentrations in solutions

In a solution the substance being dissolved is called the solute and the liquid in which it dissolves is called the solvent.

A solution in water (a solvent) is described as an aqueous solution.

The greater the number of moles of solute dissolved in a solution, the more concentrated it is. If the amount of water is increased, then the solution becomes more dilute.

To compare the concentrations of solutions we use the amount of solute dissolved in $1\,dm^3$ ($1000\,cm^3$).

The concentration of a solution can be expressed in two different ways:

1. In grams of solute per dm^3 of solution. Units can be written as g per dm^3 or g/dm^3.

2. In moles of solute per dm^3 of solution. Units can be written as mol per dm^3 or mol/dm^3.

DOIT!

Get all the 'calculation triangles' together and put them onto a sheet which you can put on your wall. Every so often when you take a break, look at the sheet. Gradually the information will sink in!

SNAPIT!

The equation triangles are shown below

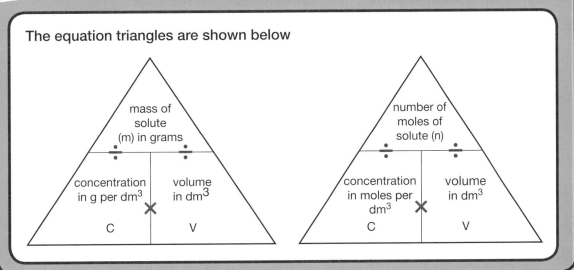

MATHS SKILLS

Make sure you can use the formulae from these triangles:

$C = m/V$

$m = C \times V$

$V = m/C$

$C = n/V$

$n = C \times V$

$V = n/C$

NAILIT!

Always convert the volume of a solution to dm^3 if it is given as cm^3.

$1\,cm^3 = 1/1000\,dm^3$

$= 1 \times 10^{-3}\,dm^3$

$= 0.001\,dm^3$

NAILIT!

If memorising the equation triangles is not your way of learning, then think of how we express concentration. The units of concentration are mol per dm³. Which is saying $C = n/V$. From there you can rearrange the equation to find V and n. The same applies to the formula $C = m/V$.

NAILIT!

As with all the other topics where calculations are made, make sure you show your working. Also, when the question is longer, highlight the part of the question that tells you what you have to find. See Check It! box below as an example.

WORKIT!

A solution contains 0.400 g of sodium hydroxide (M_r of NaOH = 40) in 100 cm³ of solution. What is the concentration of the solution in both g per dm³ and moles per dm³? (3 marks)

Use the equations $C = m/V$; $n = m/M_r$ and then $C = n/V$.

$C = m/V$; $V = 100/1000 \, dm^3 = 0.100 \, dm^3$.

The concentration $= 0.400/0.1 \, g = 4.00 \, g$ per dm³.   (1)

Use $n = m/M_r = 0.4/40 \, mol = 0.01 \, mol$.   (1)

The concentration $= 0.0100/0.100 \, mol/dm^3$
$= 0.100 \, mol/dm^3$.   (1)

CHECKIT! ✓

1   A solution of hydrochloric acid contains 0.1 mol in 500 cm³.

What is its concentration?

a   in mol/dm³

b   in g/dm³

2   A solution has a concentration of 0.2 mol/dm³. Calculate many moles of solute there are in 250 cm³ of the solution.

For this set of questions you will need your periodic table. You will also need to use the following information:

Avogadro's constant (N_A) = 6.02×10^{23}; 1 mole of any gas occupies $24\,dm^3$ at RTP.

You should also refer to the formulae sheets that you have been making during this topic. Have your list of formulae ready.

1 a Express the following numbers in standard form:

 i 0.0833   ii 223 000
 iii 856.1   iv 0.0000453

  b Write the following numbers in standard form and to three significant figures.

 i 4   ii 0.06572
 iii 0.04550   iv 0.0004389567900

2 a Balance the following equations:

 i $H_2(g) + Cl_2(g) \rightarrow HCl(g)$
 ii $Na(s) + Br_2(l) \rightarrow NaBr(s)$
 iii $K(s) + N_2(g) \rightarrow K_3N(s)$
 iv $Mg(s) + AgNO_3(aq) \rightarrow Mg(NO_3)_2(aq) + Ag(s)$
 v $Na(s) + O_2(g) \rightarrow Na_2O(s)$

  b Explain why these equations have to be balanced.

3 Calculate the relative formula masses of the following compounds:

  a K_2O b $Ca(OH)_2$ c $Mg(NO_3)_2$ d $Al(OH)_2$
  e potassium sulfate   f copper(II) chloride
  g silicon dioxide

H 4 a Calculate the relative formula mass of CO_2.

  b How many moles of carbon dioxide are present in 4.4 g of CO_2?

  c How many CO_2 molecules are there in 4.4 g of CO_2?

  d What is the volume of this mass of CO_2 at RTP?

5 The equations below show two ways by which carbon dioxide can be prepared:

 i $2HCl(aq) + CaCO_3(s) \rightarrow CaCl_2(aq) + H_2O(l) + CO_2(g)$
 ii $C(s) + O_2(g) \rightarrow CO_2(g)$

  a Calculate the atom economies of both methods.

  b Why is atom economy important to chemists?

H 6 The equation below represents the reaction between sodium hydroxide and hydrochloric acid:

$HCl(aq) + NaOH(aq) \rightarrow NaCl(aq) + H_2O(l)$

In one experiment on this reaction, $20.0\,cm^3$ of hydrochloric acid reacted with $30.0\,cm^3$ of NaOH which had a concentration of 1 mole per dm^3. What is the concentration of the acid?

H 7 Magnesium and hydrochloric acid react as follows:

$Mg(s) + 2HCl(aq) \rightarrow MgCl_2(aq) + H_2(g)$

6 g of magnesium are added to $200\,cm^3$ of hydrochloric acid with a concentration of 1 mole per dm^3.

  a How many moles of magnesium are there in 6 g of the metal?

  b How many moles of hydrochloric acid were used?

  c Explain which of the two reactants was the limiting reactant.

Chemical changes

Metal oxides and the reactivity series

Most metals react with oxygen to form metal oxides.

Gain of oxygen is oxidation and loss of oxygen is reduction.

The reactivity series (see Snap It! box on page 163) is an arrangement of the metals in order of their reactivity. The non-metals carbon and hydrogen are also included in the series.

When metals react they lose electrons to form positive ions. The more reactive the metal the more easily it loses electrons.

Metals that react with water at room temperature give metal hydroxides and hydrogen gas is produced. For example, when calcium is added to water, calcium hydroxide is formed along with hydrogen.

$$Ca(s) + 2H_2O(l) \rightarrow Ca(OH)_2(aq) + H_2(g)$$

When metals react with dilute acids, metal salts are produced and hydrogen gas is given off. For example, when magnesium is added to hydrochloric acid the salt formed is magnesium chloride and hydrogen is also given off.

$$Mg(s) + 2HCl(aq) \rightarrow MgCl_2(aq) + H_2(g)$$

More reactive metals will displace less reactive metals from solutions of metal salts.

For example, magnesium is more reactive than iron. This means that if magnesium is added to a solution of iron(II) sulfate the iron is displaced to give a solution of magnesium sulfate and iron.

$$Mg(s) + FeSO_4(aq) \rightarrow Fe(s) + MgSO_4(aq)$$

NAILIT!

Metal salts are formed when the hydrogen in an acid molecule is replaced by a metal atom. For example, sulfuric acid, H_2SO_4, has two replaceable hydrogens and either one or both of them can be replaced. This means that salts of sulfuric acid include $NaHSO_4$, Na_2SO_4 and $MgSO_4$.

Crystals of iron(II) sulfate

SNAPIT!

The Reactivity Series

This diagram shows the relative reactivity of some common metals. The way to remember the order is simply to say **PoSLiCaMZIC**. The letter(s) representing the metals are not necessarily their usual symbols.

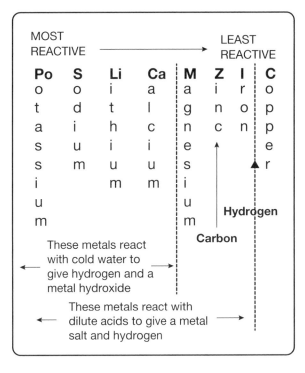

MOST REACTIVE → LEAST REACTIVE

Potassium **S**odium **Li**thium **Ca**lcium **M**agnesium **Z**inc **I**ron **C**opper

Hydrogen

Carbon

These metals react with cold water to give hydrogen and a metal hydroxide

These metals react with dilute acids to give a metal salt and hydrogen

DOIT!

If you do not like the **phonetic** method of remembering the reactivity series (PoSLiCaMZIC), make up your own mnemonic.

✓ CHECKIT!

1 What is formed when a metal atom reacts?

2 Which is more reactive, carbon or copper?

3 List what is formed when lithium is added to water.

4 Write the word equation and balanced symbol equation for the reaction between hydrogen (H_2) and copper(II) oxide (CuO).

5 Write the word equation and balanced symbol equation for the reaction between magnesium (Mg) and copper sulfate ($CuSO_4$).

6 What is formed when dilute hydrochloric acid is added to:

a zinc metal

b copper metal?

Extraction of metals and reduction

Metals in the Earth's crust are found in rocks called ores. These ores contain enough of the metal to make it worthwhile to use them.

Unreactive metals such as gold and silver can be found uncombined with other elements and don't need chemical reactions to extract them.

If a metal is less reactive than carbon then it can be extracted by heating the metal oxide with carbon. The metal oxide is reduced by losing its oxygen.

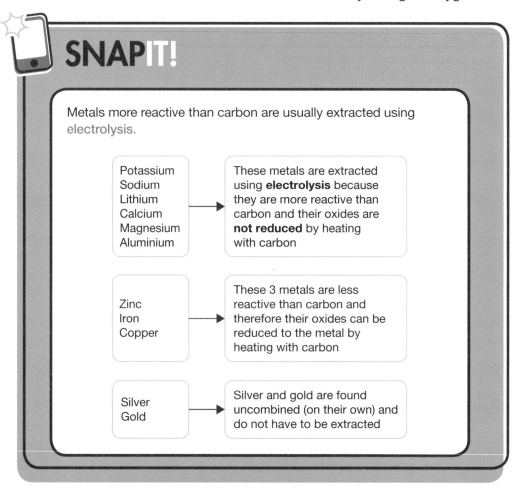

SNAPIT!

Metals more reactive than carbon are usually extracted using electrolysis.

Potassium Sodium Lithium Calcium Magnesium Aluminium	→	These metals are extracted using **electrolysis** because they are more reactive than carbon and their oxides are **not reduced** by heating with carbon
Zinc Iron Copper	→	These 3 metals are less reactive than carbon and therefore their oxides can be reduced to the metal by heating with carbon
Silver Gold	→	Silver and gold are found uncombined (on their own) and do not have to be extracted

Gold and silver can be found uncombined with other elements, while other elements are found in ores

STRETCH IT!

In compounds, metals exist as positively charged ions. This means that when they react they lose electrons. Another definition of oxidation is that **Oxidation Is Loss** of electrons.

The more reactive the metal the more easily they lose electrons and the less easily they take them back.

In electrolysis, electrons are gained by metal ions to form metal atoms. This is another version of reduction. **Reduction Is Gain** of electrons.

We remember this definition of oxidation and reduction by using the mnemonic **OILRIG**.

Oxidation **I**s **L**oss of electrons and **R**eduction **I**s **G**ain of electrons.

Examples of oxidation involving metals losing electrons are:

$Fe \rightarrow Fe^{2+} + 2e^-$ and $Al \rightarrow Al^{3+} + 3e^-$

Examples of reduction are the reverse of those above and $Na^+ + e^- \rightarrow Na$.

NAIL IT!

When metals oxides are reduced by carbon to the metal, carbon dioxide is usually formed.

For example, copper(II) oxide and carbon form copper and carbon dioxide.

$2CuO(s) + C(s) \rightarrow 2Cu(s) + CO_2(g)$

When carbon is heated with the oxide of a more reactive metal, there is no reaction.

$Al_2O_3(s) + C(s) \rightarrow$ No reaction

CHECK IT!

1 What term is used for the following:

a a metal gaining oxygen

b a metal losing oxygen?

2 In the reaction shown below which metal is reduced and which metal is oxidised?

$2Al(s) + Fe_2O_3(s) \rightarrow Al_2O_3(s) + 2Fe(s)$

3 a Lead is a more reactive metal than copper but less reactive than iron. Suggest a method for extracting lead from its ore.

b Barium is more reactive than calcium. Suggest a method for extracting barium from its ore.

The reactions of acids

Acids are a group of substances with similar properties. This means that if you know the properties of one acid you can predict the properties of the others.

The reason all acids have similar properties is that they all contain hydrogen, and when in aqueous solution they all give hydrogen ions (H^+).

Hydrochloric acid (HCl) and sulfuric acid (H_2SO_4) react with metals to form salts and hydrogen.

Alkalis are soluble metal hydroxides; bases are insoluble metal oxides and hydroxides.

Examples of alkalis are sodium hydroxide (NaOH) and potassium hydroxide (KOH). Examples of insoluble bases are copper(II) oxide (CuO) and magnesium hydroxide ($Mg(OH)_2$).

Hydrochloric acid (HCl), nitric acid (HNO_3) and sulfuric acid (H_2SO_4) are all neutralised by alkalis and bases. Each neutralisation reaction forms a salt and water.

$$acid(aq) + alkali(aq) \rightarrow salt(aq) + water(l)$$

$$acid(aq) + base(s) \rightarrow salt(aq) + water(l)$$

All three of the acids above react with metal carbonates to give a salt, water and carbon dioxide gas. So the general equation is:

$$acid(aq) + carbonate(s) \rightarrow salt(aq) + water(l) + carbon\ dioxide(g)$$

For example, calcium carbonate ($CaCO_3$) reacts with nitric acid (HNO_3) to produce the salt calcium nitrate ($Ca(NO_3)_2$), carbon dioxide (CO_2) and water (H_2O).

$$CaCO_3(s) + 2HNO_3(aq) \rightarrow Ca(NO_3)_2(aq) + CO_2(g) + H_2O(l)$$

H

The reaction between acids and metals is a redox reaction. The metal atoms lose electrons and are oxidised. The hydrogen (H^+) ions are reduced because they gain electrons. For example:

$$Mg(s) + 2H^+(aq) \rightarrow Mg^{2+}(aq) + H_2(g)$$

Note that the negative ions associated with the acid do not appear in these equations because they do not take part in the reaction. They are spectator ions ('they just look on').

H

DOIT!

Make up an exam question about the reactions of acids for a friend or revision partner.

- Each part of the question should be worth a certain number of marks.
- Write a mark scheme for your question.

NAILIT!

One common problem associated with the writing of symbol equations involving acids is the writing of the formulae of the salt because of the negative ions in acids like sulfuric acid and nitric acid.

- The key to this is realising that the acids contain replaceable hydrogen ions.
- For example sulfuric acid is H_2SO_4. It contains two replaceable hydrogen ions. Each of these has one positive charge. Therefore the sulfate ion must have two negative charges to balance out the positive charges from the hydrogen ions and its formula is SO_4^{2-}.
- If sulfuric acid forms a salt, then we can use this information to write the correct formulae for the salt.
- For example, magnesium sulfate contains the magnesium Mg^{2+} ion (magnesium is in group 2). Therefore the formula of magnesium sulfate is $MgSO_4$ as the +2 on the magnesium ion cancels out the -2 on the sulfate ion.
- Similarly, the nitrate ion is NO_3^- [only one replaceable hydrogen ion in nitric acid (HNO_3)]. Therefore in magnesium nitrate we need two nitrate ions to cancel out the charges on the Mg^{2+}. Therefore magnesium nitrate is written $Mg(NO_3)_2$.

SNAPIT!

This table shows salts formed by metals with different acids. Remember to find the formula of the salt the charges on the ions have to be balanced (see Ions and ionic bonding on page 138.)

Metal in base; alkali etc.	Formula of metal ion	Acid	Ion from acid	Name of salt formed	Formula of salt
Magnesium	Mg^2	Nitric (NHO_3)	NO_3^-	magnesium nitrate	$Mg(NO_3)_2$
Copper	Cu^{2+}	Sulfuric (H_2SO_4)	SO_3^{2-}	copper(II) sulfate	$CuSO_4$
Zinc	Zn^{2+}	Hydrochloric (HCl)	Cl^-	zinc chloride	$ZnCl_2$
Sodium	Na^+	Sulfuric (H_2SO_4)	SO_4^{2-}	sodium sulfate	Na_2SO_4

CHECKIT!

1 What is meant by the terms:

  a alkali

  b base?

2 Name the salts formed when sodium hydroxide reacts with:

  a hydrochloric acid

  b nitric acid.

3 When copper carbonate is added to sulfuric acid, the mixture fizzes/effervesces. Explain this observation.

4 Nitric acid forms sodium nitrate ($NaNO_3$).

  a Give the formula of the nitrate ion.

  b Give the formula of magnesium nitrate.

5 **Complete** and **balance** the following symbol equations:

  a $MgO(s) + H_2SO_4(aq) \rightarrow$

  b $MgCO_3(s) + H_2SO_4(aq) \rightarrow$

  c $Mg(OH)_2(s) + HCl \rightarrow$

 6 a Give the ionic equation for the reaction between an acid and magnesium.

  b Explain why this is a redox reaction.

The preparation of a soluble salt

A soluble salt is prepared by the neutralisation reaction between an insoluble base and an acid.

If a metal carbonate is used, then the only difference is that carbon dioxide is given off.

The metal ion needed to make the salt comes from the base and the negative ion in the salt comes from the acid.

The general equation for the reaction shows that the reaction gives a clear solution because there is only water and an aqueous solution of a salt formed.

base(s) + acid(aq) → salt(aq) + water(l)

Therefore when the insoluble base stops dissolving we know all the acid has been used up.

Practical Skills

The skills/ techniques used are heating, filtration, evaporation and crystallisation.

Stage	What is done/action	Explanation of what is done
1	Heat up the acid.	Higher temperature means faster reaction.
2	Add the solid base.	Neutralisation reaction takes place between acid and base.
3	Stop adding the base when it stops dissolving or if a carbonate is added it stops fizzing.	When the base stops dissolving or the carbonate stops fizzing, it means that the acid has been used up.
4	Filter off the unreacted base or carbonate.	Remove unreacted insoluble base or carbonate, leaving just the aqueous solution of the salt.
5	Heat the salt solution on a steam bath.	Steam bath evaporates the water in the salt solution slowly so we get crystals and not powder.
6	Stop heating when solid salt starts to appear. You now have a saturated solution.	Saturated solution means that no more will dissolve in the solution at the higher temperature.
7	Leave the saturated solution to cool so that crystallisation can take place.	At a lower temperature the solid is less soluble so more crystals appear.

SNAPIT!

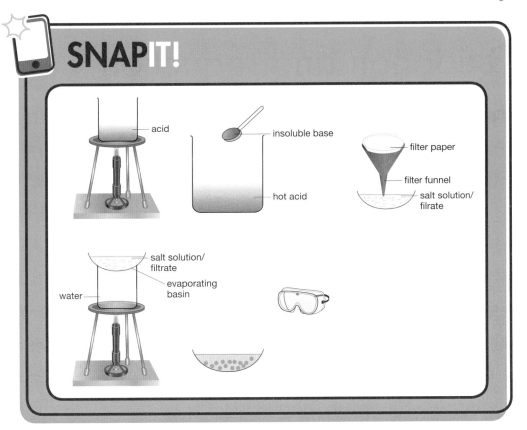

acid

insoluble base

hot acid

filter paper

filter funnel

salt solution/ filrate

salt solution/ filtrate

evaporating basin

water

DOIT!

Write an account to go with the diagrams in the Snap It! box.

NAILIT!

The reasons for not getting as much salt as theoretically possible are as follows:

- Some acid is lost when the insoluble base is added as spitting takes place.
- Some salt solution is left in the filter paper.
- Not all of the salt solution crystallises.

MATHS SKILLS

You might be asked to calculate the percentage yield for the preparation and account for any losses.

$$\text{Percentage yield} = \frac{\text{actual yield (what you get)}}{\text{theoretical yield (what you should get)}} \times 100\%$$

CHECKIT!

1 What reactants could you use to prepare the following soluble salts?

   a Copper sulfate

   b Zinc nitrate

   c Magnesium chloride

2 a Describe how an unreacted base is separated from the salt solution.

   b Why is this method of separation used?

3 In a salt preparation 4.5 g of the salt were prepared when 5 g was the theoretical yield. Calculate the percentage yield.

4 The preparation of magnesium nitrate can be made using two different reactions and the equations for these are shown below:

   A $MgO(s) + 2HNO_3(aq) \rightarrow Mg(NO_3)_2(aq) + H_2O(l)$

   B $MgCO_3(s) + 2HNO_3(aq) \rightarrow Mg(NO_3)_2(aq) + H_2O(l) + CO_2(g)$

   a Suggest two observations that suggest reaction B has been completed.

   b Assuming that magnesium nitrate is the desired product, calculate the atom economy for each method.

   [Relative formula masses MgO = 40; $MgCO_3$ = 84; HNO_3 = 63; $Mg(NO_3)_2$ = 148; H_2O = 18]

Oxidation and reduction in terms of electrons

NAILIT!

When you write the ionic equations <u>leave out</u> the non-metal ions and just put in the metal atoms and ions present. For example, the reaction between zinc metal and an aqueous solution of copper(II) sulfate:

$Zn(s) + Cu^{2+}(aq) \rightarrow Zn^{2+}(aq) + Cu(s)$

Note that the zinc has lost electrons and become more positive by forming a Zn^{2+} ion (oxidised) and at the same time the copper(II) ion has lost its +2 charge by accepting electrons and forming a neutral copper atom (reduced).

In displacement reactions, a reactive metal will displace a less reactive one from an aqueous solution of its salt.

As far as electrons are concerned, Oxidation Is Loss of electrons and Reduction Is Gain of electrons (remembered as OILRIG).

Ionic equations for displacement reactions will only involve metal atoms and positive metal ions.

The ions of the less reactive metal gain electrons from the atoms of the more reactive metals.

This means that the less reactive metal ions are reduced and the more reactive metal atoms are oxidised.

The negative ions from the salt are spectator ions – they are not involved in the reaction and can be removed from the equation.

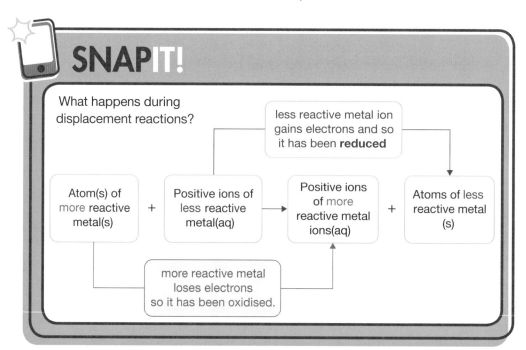

SNAPIT!

What happens during displacement reactions?

less reactive metal ion gains electrons and so it has been **reduced**

Atom(s) of more reactive metal(s) + Positive ions of less reactive metal(aq) → Positive ions of more reactive metal ions(aq) + Atoms of less reactive metal (s)

more reactive metal loses electrons so it has been oxidised.

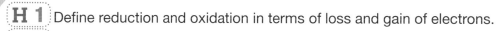

CHECKIT!

H 1 Define reduction and oxidation in terms of loss and gain of electrons.

H 2 a Write the following reaction as an ionic equation:

$Mg(s) + ZnCl_2(aq) \rightarrow MgCl_2(aq) + Zn(s)$

b Explain which species has been oxidised and which has been reduced in part **a**.

The pH scale and neutralisation

An aqueous solution is one formed when a substance dissolves in water. Its state symbol is (aq).

Acids are a group of substances with similar properties. This is because all aqueous solutions of acids produce H^+ ions.

In the same way aqueous solutions of alkalis all give OH^- ions.

In neutralisation reactions H^+ and OH^- ions react to give water. The ionic equation for this reaction is as follows:

$$H^+(aq) + OH^-(aq) \rightarrow H_2O(l)$$

The pH scale runs from 0 to 14 and is a measure of the acidity or alkalinity of a solution.

Acid solutions have a pH value less than 7. The lower the pH the more concentrated the H^+ ions are.

Neutral solutions such as pure water have a pH of 7.

SNAPIT!

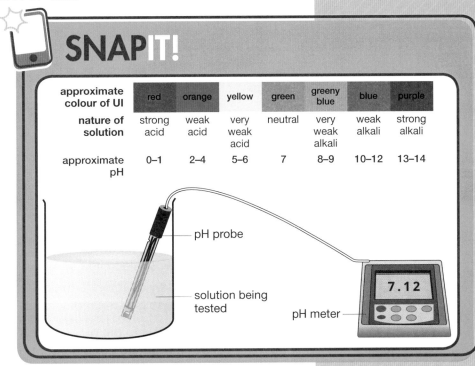

approximate colour of UI	red	orange	yellow	green	greeny blue	blue	purple
nature of solution	strong acid	weak acid	very weak acid	neutral	very weak alkali	weak alkali	strong alkali
approximate pH	0–1	2–4	5–6	7	8–9	10–12	13–14

pH probe

solution being tested

7.12

pH meter

Alkaline solutions have a pH value greater than 7 and the higher the pH the greater the concentration of OH^- ions.

The pH of a solution can be measured accurately using a pH probe or more approximately using the colour of universal indicator (UI) in the solution.

A pH probe measures the concentration of hydrogen ions accurately and the pH is read from a pH meter either digitally or on a scale.

CHECKIT!

1 Name the ions that would be found in aqueous solutions of both hydrochloric acid and sulfuric acid.

2 Name the ions that would be found in aqueous solutions of both sodium hydroxide and calcium hydroxide.

3 Classify the reaction which takes place when hydrochloric acid is neutralised by an alkali.

4 The pH of a solution of hydrochloric acid is 1. Ethanoic acid of the same concentration has a pH of 3. Explain what conclusions you can draw from this data.

Strong and weak acids

A concentrated solution of an acid has a greater amount (in moles) of the acid dissolved in the same volume of water than a dilute solution.

A strong acid like hydrochloric acid is completely ionised in aqueous solution.

A weak acid like ethanoic acid is only partially ionised in aqueous solution.

This means that 1000 molecules of the strong acid, hydrochloric acid, will all ionise to give 1000 H^+ ions in aqueous solution. On the other hand, only 4 out of 1000 molecules of the weak acid, ethanoic acid, will ionise in aqueous solution.

Important: strong is not the same as concentrated and weak is not the same as dilute.

If the pH value of a solution decreases by 1 unit, then the concentration of the H^+ ions increases by 10 times or 1 order of magnitude.

DOIT!

Use your calculator to express concentrations (mol per dm^3) in standard form. The numbers you need to look at are 1, 0.1, 0.001, etc.

Describe what happens to the order of magnitude n, in 1×10^{-n} as the concentration decreases 10-fold each time.

MATHS SKILLS

Understand what is meant by order of magnitude.

An increase of 1 order of magnitude means that the value has increased 10 times. A decrease by 1 order of magnitude means that the value has decreased 10 times to one-tenth of the original value.

An increase of 100 times is 2 orders of magnitude and so on.

WORKIT!

As the pH decreases by 1 unit, the hydrogen ion concentration **increases by a factor of 10** or **1 order of magnitude**. On the other hand, if the pH increases by 1, then the hydrogen ion concentration decreases by a factor of 10 or has decreased by 1 order of magnitude to one-tenth of what it was before.

What happens to the pH as the concentration of H^+ ions goes from 0.1 mol per dm^3 to 0.001 mol/dm^3? (2 marks)

In going from 0.1 mol/dm^3 to 0.001 mol/dm^3 the concentration decreases 100 times. (1)

A decrease in the hydrogen ion concentration of 100 times is a decrease of 2 orders of magnitude and this means the pH value goes up by 2. (1)

CHECKIT!

H 1 Explain the difference between a strong acid and a weak acid.

H 2 Explain the difference between a dilute and a concentrated solution.

H 3 If the pH of a solution goes down by 3, explain what has happened to the concentration of the H^+ ions in the solution.

The basics of electrolysis and the electrolysis of molten ionic compounds

Electrolysis is the splitting up of an ionic compound using electricity.

Electrolysis takes place when the ionic compounds are in the liquid state or in aqueous solution because the ions are free to move and carry the current.

The liquid that is decomposed by electrolysis is called the electrolyte.

When ions lose or gain electrons at the electrodes they are discharged.

During electrolysis, negative ions move towards the positive electrode (the anode (+)) and lose electrons to form non-metallic elements.

Positive ions move towards the negative electrode (cathode (−)) and gain electrons to form metallic elements (or hydrogen).

For example, molten potassium iodide forms iodine at the anode (+) and potassium at the cathode (−).

H The reactions at electrodes can be represented using ionic half equations:

Generally at cathode (−) $M^{n+} + ne^- \rightarrow M$

At anode (+) $2X^{m-} \rightarrow X_2 + 2me^-$

These equations show that at the cathode the metal ions gain electrons and this is reduction. Also, at the anode the non-metal ions lose electrons and this is oxidation.

SNAPIT!

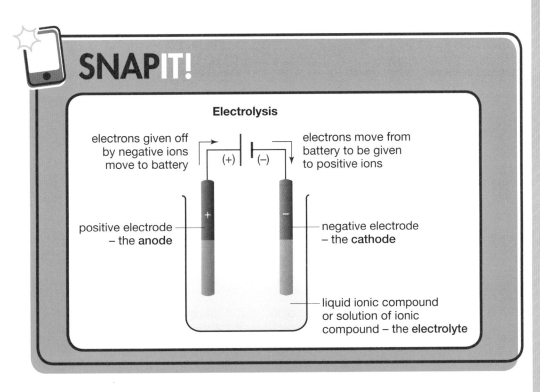

Electrolysis

electrons given off by negative ions move to battery

(+)   (−)

electrons move from battery to be given to positive ions

positive electrode – the **anode**

negative electrode – the **cathode**

liquid ionic compound or solution of ionic compound – the **electrolyte**

DOIT!

Write a short account of what happens to the ions and the electrons at each electrode during electrolysis. You could print out the diagram in the Snap It! box and complete it by putting in the ions and what happens at each electrode.

NAIL IT!

If you are doing the Higher Tier exam, then you should learn how to balance ionic half equations.

Look back to your notes on ionic compounds – remember that group 1 elements form +1 ions, e.g. Na^+; group 2 elements form +2 ions: group 6 elements form 2– ions; and group 7 elements form 1– ions.

MATHS SKILLS H

You need to balance the charges on both sides of the equation.

WORKIT!

Foundation Tier What is formed at the cathode (−) and anode (+) when molten calcium chloride is electrolysed? (1 mark)

Calcium is formed at the cathode and chlorine at the anode. (1)

Higher Tier   Give the equations for the reactions at the electrodes. (2 mark)

At the cathode (−) $Ca^{2+} + 2e^- \rightarrow Ca$ (1)

At the anode (+) $2Cl^- \rightarrow Cl_2 + 2e^-$ (1)

Flakes of calcium chloride

CHECKIT!

1 What is the term used for the following:

   a the liquid that is electrolysed

   b the negative electrode

   c the positive electrode?

2 When molten magnesium chloride is electrolysed, name the substances formed at:

   a the anode

   b the cathode.

H 3 a Write ionic equations for the reactions taking place at both electrodes when molten magnesium chloride is electrolysed.

   b Explain why the reaction at the anode is oxidation.

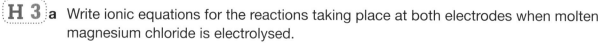

The electrolysis of aqueous solutions

Water ionises to give hydrogen ions ($H^+(aq)$) and hydroxide ions ($OH^-(aq)$).

At the cathode (−), if the $H^+(aq)$ ion is discharged then hydrogen (H_2) gas is produced.

At the anode (+), if the hydroxide ion is discharged then oxygen is produced.

The presence of these two ions from the water means that there is a choice from two products at each electrode.

At the cathode, we have a choice between the metal ion and the hydrogen ion.

If the metal present is more reactive than hydrogen, then we get **hydrogen from the discharge of the hydrogen ions**.

At the anode (+), we have a choice between the discharge of the hydroxide ion and the other negative ion.

We get oxygen from the discharge of the hydroxide ion unless the other negative ion is a group 7 halide ion (Cl^-, Br^- or I^-). If so we get the halogen which is Cl_2, Br_2 or I_2.

NAILIT!

Remember that the product at the anode is always a diatomic molecule. The elements that form diatomic molecules are H_2, O_2, N_2, Cl_2, Br_2, I_2 and F_2.

 STRETCHIT!

At the Higher Tier, you are asked to write half equations for the reactions at the electrodes. When aqueous solutions are electrolysed, you may be asked to write the half equations for the discharge of the hydrogen (H^+) and the hydroxide (OH^-) ions.

The half equations for the reactions at each electrode are shown below.

At the negative cathode (−)

$$2H^+(aq) + 2e^- \rightarrow H_2(g)$$

And at the positive anode (+)

$$4OH^-(aq) \rightarrow 2H_2O(l) + O_2(g) + 4e^-$$

This means that at the cathode, hydrogen ions gain electrons and are reduced. At the same time at the anode, hydroxide ions lose electrons and are oxidised.

DOIT!

Choose a solution. Draw a diagram showing both electrodes and the ions that would be attracted to those electrodes. Write notes on your diagram to explain what happens to the ions.

WORKIT!

1 How can we predict which positive ion is discharged at the cathode (–) during the electrolysis of an aqueous solution? (2 marks)

The answer is hydrogen unless the metal is less reactive than hydrogen. This means that in your reactivity series list (PoSLiCaMZIC) only copper ions (C in the list) would be discharged. (1)

Cu^{2+} ions accept electrons to give Cu. (1)

2 How can we predict which non-metal ion is discharged at the anode during electrolysis of an aqueous solution? (2 marks)

If the ion other than hydroxide (OH^-) is a halide ion (Cl^-, Br^- or I^-) then the halide ion is discharged to give the halogen Cl_2, Br_2 or I_2. (1) If there is any other negative ion (such as SO_4^{2-} and NO_3^-) then we get oxygen given at the anode from the discharge of the hydroxide (OH^-) ion. (1)

 Practical Skills

A required practical is to investigate what happens when aqueous solutions are electrolysed using inert electrodes.

NAILIT!

When asked which products you would get from the electrolysis of an aqueous solution you should also be aware of what remains. For example, if you have a solution of copper(II) sulfate you will get copper at the cathode because copper is less reactive than hydrogen. At the anode, you get oxygen from the hydroxide ion. This means that hydrogen ions and sulfate ions are left behind, which together make sulfuric acid.

At the Higher Tier, you may be asked to write the half equations for the reactions at the electrodes.

WORKIT!

When aqueous potassium chloride solution is electrolysed, what is formed at each electrode and what solution remains after the electrolysis? (3 marks)

Potassium is more reactive than hydrogen and this means that at the cathode (–) we will get hydrogen (H_2) as the product. (1)

Chloride (Cl^-) is a halide ion so we will get chlorine (Cl_2) formed at the anode (+). (1)

This means that the ions left behind are potassium (K^+) and hydroxide (OH^-) ions and the solution that remains is potassium hydroxide (KOH) solution. (1)

CHECKIT! ✓

1 a Give the formulae of the four ions present in an aqueous solution of sodium chloride.

  b Suggest the products at each electrode when aqueous sodium chloride solution is electrolysed.

  c What solution remains after the electrolysis?

  H d Write the half equations for the reactions at each electrode and explain whether oxidation or reduction has taken place.

The extraction of metals using electrolysis

If a metal is more reactive than carbon then the metal oxide cannot be reduced to the metal by heating with carbon.

Electrolysis is used to extract metals **more reactive** than carbon. An example is aluminium.

To extract aluminium, the electrolyte used is aluminium oxide. Aluminium oxide has a very high melting point so to save energy and lower the operating temperature the aluminium oxide is dissolved in a compound called cryolite.

Carbon is used for both electrodes. Aluminium ions are discharged at the cathode (–) to give molten aluminium metal which is run off.

At the anode (+), oxide ions are discharged to form oxygen gas. This oxygen gas then reacts with the carbon anode to form carbon dioxide. The carbon anode (+) burns away and loses mass. This means that the anode (+) has to be replaced at regular intervals.

DOIT!

Write a brief description of how aluminium is extracted from aluminium oxide. Your description should include materials used for electrodes, how energy is saved and what happens at each electrode.

SNAPIT!

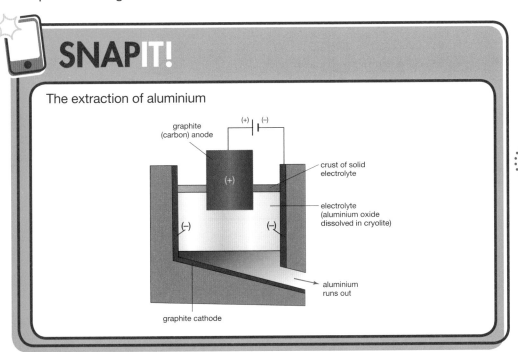

The extraction of aluminium

graphite (carbon) anode

crust of solid electrolyte

electrolyte (aluminium oxide dissolved in cryolite)

aluminium runs out

graphite cathode

NAILIT!

At the cathode aluminium ions are **reduced** by gaining electrons.

$Al^{3+} + 3e^- \rightarrow Al$

At the anode (+) oxide ions are **oxidised** by losing electrons.

Remember OILRIG.

CHECKIT!

1 List the metals that are extracted using electrolysis.

2 Justify the use of electrolysis to extract aluminium from its ore.

3 Write a brief description of the electrolysis of aluminium oxide.

H 4 Sodium is extracted from molten sodium chloride. Write the half equations for the reactions at each electrode.

Investigation of the electrolysis of aqueous solutions

Part of the investigation is to predict the products formed by the electrolysis of various solutions. See page 175.

You can make predictions about the identity of the solution that remains because you know that it is formed from the ions that do not react.

Universal indicator can be used to identify the pH of the solution.

The apparatus you will use is similar to that shown in the Snap It! box below. The gas is collected by downward displacement of the solution used.

The volume of gas produced depends on the identity of the gas. If hydrogen and oxygen are formed, then the volume of hydrogen is twice that of the volume of oxygen.

The volume of chlorine gas is less than predicted because chlorine dissolves in water.

The products at the electrodes are identified by chemical tests:

- Hydrogen gas 'pops' when a lighted splint is placed in the gas.
- Oxygen gas relights a glowing splint.
- Chlorine bleaches blue litmus paper or UI in the solution.
- Bromine turns the solution yellow/orange and iodine will turn it brown.

Practical Skills

This practical tests the skills of planning and predicting, carrying out, making observations and analysing the results.

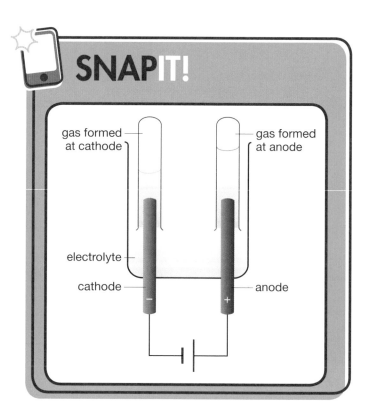

SNAPIT!

gas formed at cathode

gas formed at anode

electrolyte

cathode

anode

DOIT!

Describe how the apparatus shown in the Snap It! box is used in the investigation.

WORKIT!

If an aqueous solution of sodium chloride is electrolysed, what would be the predictions and what would be seen? (6 marks)

Predictions			Observations		
At cathode	At anode	Solution	At cathode	At anode	Solution
Hydrogen formed because sodium is more reactive than hydrogen.	Chloride is a halide ion so chlorine would be formed from the discharge of the chloride ion.	The ions left behind are sodium and hydroxide ions.	The gas pops with a lighted splint which shows hydrogen.	The gas formed bleaches the UI in the solution. This shows chlorine is formed.	The remaining solution turns UI purple around the cathode. This is because the ions remaining are sodium and hydroxide ions which form the alkali sodium hydroxide. (6)

In practicals, make sure you record your observations!

NAILIT!

There are several areas of chemical knowledge that are used in this practical so make sure you are able to answer questions on them.

1 The electrolysis of aqueous solutions.

2 Testing for gases.

3 Formulae of ions.

4 The colour of UI in different types of solution.

✓ CHECKIT!

1 In an investigation, the gas at the cathode popped with a lighted splint and the gas at the anode relit a glowing splint. Name these two gases.

2 Explain why the solution around the anode turns yellowy orange when aqueous potassium bromide is electrolysed.

3 When sodium iodide solution is electrolysed the following reaction takes place:

$H_2O(l) + 2NaI(aq) \rightarrow H_2(g) + I_2(aq) + 2NaOH(aq)$

The iodine turns the solution brown. Describe how you would identify the hydrogen and the sodium hydroxide solution.

For additional questions, visit
www.scholastic.co.uk/gcs

1 The five metals aluminium, copper, iron, magnesium and zinc are in the reactivity series.

   a Put them in order of reactivity from the **least reactive** to **the most reactive**.

   b Complete the following word equations for each reaction :

      i   zinc(s) + copper(II) sulfate(aq) →

      ii aluminium(s) + magnesium oxide(s) →

      iii aluminium(s) + iron(III) oxide(s) →

   c When burning magnesium is lowered into a test tube containing carbon dioxide, the magnesium continues burning.

      At the end of the reaction there are traces of a black solid and a white solid on the sides of the test-tube.

      i   Write the word equation for the reaction including state symbols.

      ii Write the balanced symbol equation for the reaction including state symbols.

      iii Explain why this is a redox reaction.

      iv I Identify the white solid formed.

          II Identify the black solid formed.

2 When hydrochloric acid solution is added to zinc metal, the zinc disappears and the mixture effervesces but when the same acid is added to copper metal there is no reaction.

   a Explain the effervescence and describe how you can test for the product that causes the effervescence.

   b Explain why there is no reaction between the acid and copper.

   H c i   Write the ionic equation for the reaction between zinc metal and the acid. Note that the formula for the zinc ion is Zn^{2+}.

      ii Explain why this is a redox reaction.

3 In the electrolysis of aqueous potassium bromide solution, there is a gas produced at the cathode and at the anode the solution turns yellow-orange in colour.

   a Name the products produced at both electrodes.

   b What solution remains after the electrolysis?

4 a A solution of ammonia turns Universal Indicator blue. What does this tell you about the ammonia solution?

   b Universal Indicator turns yellow in phenol solution. What does this tell you about phenol?

Energy changes

Exothermic and endothermic reactions

An exothermic reaction is where the reaction gives out heat energy to its surroundings. This results in an increase in the temperature of the surroundings.

Examples of exothermic reactions are the burning of fuels in combustion reactions and most neutralisation and oxidation reactions.

Everyday uses of exothermic reactions are hand warmers and self-heating cans of food.

An endothermic reaction is where the reaction takes in heat energy from its surroundings. This results in a decrease in the temperature of the surroundings.

Examples of endothermic reactions are thermal decomposition (breaking up a compound using heat) and the reaction of citric acid with sodium hydrogen carbonate.

Sports injury packs use endothermic reactions.

SNAP IT!

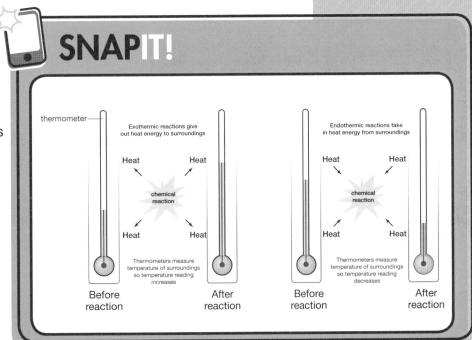

thermometer

Exothermic reactions give out heat energy to surroundings

Heat   Heat

chemical reaction

Heat   Heat

Thermometers measure temperature of surroundings so temperature reading increases

Before reaction   After reaction

Endothermic reactions take in heat energy from surroundings

Heat   Heat

chemical reaction

Heat   Heat

Thermometers measure temperature of surroundings so temperature reading decreases

Before reaction   After reaction

DO IT!

Write a brief summary of the diagram in the Snap It! box above.

What do the changes in the thermometers mean?

CHECK IT!

1 a In a chemical reaction the temperature increased. What type of reaction is this.

   b The table below shows the results from two experiments:

Experiment	Initial temperature/°C	Final temperature/°C	Temperature change
I	20		+25
II	20	15	

Complete the table and then classify each reaction as either exothermic or endothermic.

2 List some types of reaction that are endothermic reactions.

3 List some uses of exothermic reactions.

Investigation into the variables that affect temperature changes in chemical reactions

When you are carrying out an investigation, it is often useful to draw a flow chart of what you are going to do. Discuss it with a practical partner or review it yourself to identify where you might be doing things in the wrong order or points where a piece of apparatus is needed. Practical work is improved by good organisation.

This is one of the required practicals and you could be questioned on it in the exam. This particular practical places emphasis on you putting forward a hypothesis and making predictions based on it. It also requires the accurate use of appropriate measuring apparatus and the need to make and record a range of measurements. When investigating temperature changes, you should know what factors need to be controlled to make a fair test.

Exothermic reactions are accompanied by a temperature rise and **endothermic** reactions give a temperature decrease.

Reactions you could investigate include combustion, neutralisation reactions between acids and alkalis, and acids reacting with either metals or carbonates.

Factors affecting the temperature changes are amount of reactant, surface area (lumps or powder) of solids, concentrations of solutions and the reactivity of different metals with acids.

Relevant pieces of apparatus are thermometers or temperature probes and data loggers; measuring cylinders; spirit burners for combustion experiments – see diagram below; top-pan balance and reaction containers such as beakers and test tubes. The container that is used to measure the heat change for a reaction is called a calorimeter. A polystyrene cup can be used or a lagged container like a beaker. The lagging reduces heat loss through the sides of the container.

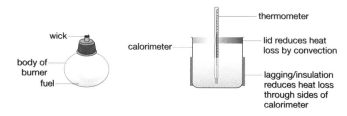

Practical Skills

One required practical is to investigate the variables that affect temperature changes in reacting solutions. For this practical you should:

- Put forward a hypothesis about what you think would happen based on your knowledge of the chemistry involved.
- Identify variables that could affect the temperature change and explain how you would make it a fair test.
- Identify safety factors involved in the experiment.
- Know what measurements you would make.
- Record your results in a suitable form.

If you are taking the Higher Tier exam, then you should be aware that if you are investigating different substances then equal masses does not mean equal amounts, and masses should be converted to moles using $n = m/M_r$.

WORKIT!

A student was given some magnesium ribbon, magnesium powder and hydrochloric acid. They were asked to investigate the effect of the surface area of the magnesium on the temperature change caused by the reaction. They carried out a fair test and measured the starting temperature for both as 20°C. In the reaction with the magnesium ribbon, the final temperature was 28°C and with the powder it was 42°C.

a List the apparatus she would need and give the use of each piece. (3 marks)

- A small beaker for the reaction container which is very well insulated and a thermometer (or temperature data logger) to measure the temperature changes.   (1)

- A measuring cylinder to measure the volume of acid.   (1)

- A spatula for adding the powder and a top-pan balance for weighing out the magnesium ribbon and powder.   (1)

b Describe how they would make it a fair test. (2 marks)

The volume and concentration of the acid should be the same for both experiments and so should the mass of the ribbon and powder. (1) The containers should be identical or use the same one for both experiments.   (1)

c Show how they might record their results. (2 marks)

Experiment	Starting temperature/°C	Final temperature/°C	Temperature change/°C
Magnesium powder	20	42	22
Magnesium ribbon	20	28	8

(2)

d Make any conclusions possible from their results. (2 marks)

The reaction with the powder gave a greater temperature rise. (1) This is because it had the greater surface area and reacted more quickly.   (1)

CHECKIT!

NAILIT!

A common error when drawing up a table of results is to forget to put in the units.

1 List some factors that could affect the temperature changes in a chemical reaction.

2 A student added measured amounts of magnesium powder to separate calorimeters containing 100 cm³ of 2 mol/dm³ hydrochloric acid and then measured the temperature changes. The results are shown below:

Experiment	1	2	3	4	5	6
Mass of magnesium/g	1.00	2.00	3.00	4.00	5.00	6.00
Temperature change/°C	5.00	10.0	15.0	20.0	24.0	24.0

  a Explain the results obtained for experiments 1 to 4.

  b Suggest an explanation for the results from experiments 5 and 6.

3 List some reactions that could be investigated in terms of temperature changes.

4 Using their reactions with hydrochloric acid, describe how you could place the metals copper, iron, magnesium and zinc in order of reactivity.

Reaction profiles

A reaction profile shows how the energy changes from reactants to products.

In a reaction profile for an exothermic reaction the products are lower in energy than the reactants because energy is released to the surroundings during the reaction.

In a reaction profile for an endothermic reaction, the products are higher in energy than the reactants because energy is taken in from the surroundings during the reaction.

Chemical reactions occur when reacting particles collide with enough energy to react. This energy is called the activation energy (E_a).

The activation energy is the minimum energy required for a reaction to occur. The activation energy is a barrier to reaction.

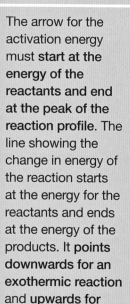

DOIT!

Draw reaction profiles for:

• a combustion reaction
• a thermal decomposition.

NAILIT!

The arrow for the activation energy must **start at the energy of the reactants and end at the peak of the reaction profile**. The line showing the change in energy of the reaction starts at the energy for the reactants and ends at the energy of the products. It **points downwards for an exothermic reaction and upwards for an endothermic reaction.**

SNAPIT!

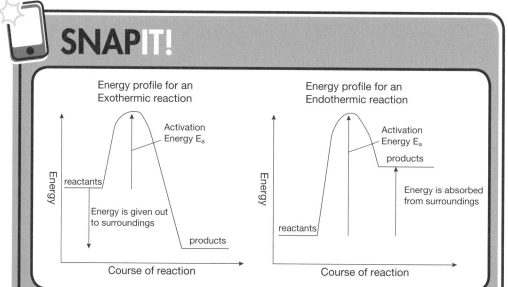

Energy profile for an Exothermic reaction

Activation Energy E_a

Energy

reactants

Energy is given out to surroundings

products

Course of reaction

Energy profile for an Endothermic reaction

Activation Energy E_a

products

Energy

Energy is absorbed from surroundings

reactants

Course of reaction

CHECKIT!

1 Draw and label the reaction profile for the combustion of carbon to give carbon dioxide.

2 a i Define what is meant by a catalyst.

 ii Explain how a catalyst speeds up a reaction.

 b i Draw a reaction profile for an endothermic reaction.

 ii On the same graph, draw the profile you would get for a catalyst for the same endothermic reaction.

The energy changes of reactions

Energy is absorbed when chemical bonds are broken in a chemical reaction.

Energy is given out when chemical bonds are formed in a chemical reaction.

Bond energies are used to calculate the energy needed to break bonds and the energy given out by making bonds.

The energy given out or taken in during a chemical reaction is measured in kJ/mol.

The energy difference between the energy needed to break the bonds and the energy given out by their formation is the energy change in the reaction.

If the energy needed to break the bonds in the reactants is less than the energy given out by the formation of new bonds in the products, then the reaction is exothermic.

If the energy needed to break the bonds in the reactants is greater than the energy given out by the formation of new bonds in the products, then the reaction is endothermic.

DO IT!

Methane and oxygen react as follows:

$$CH_4(g) + 2O_2(g) \rightarrow CO_2(g) + 2H_2O(l)$$

NAIL IT!

It is a common error to miscalculate the number of bonds in a molecule. For example, in the Do It! box there are three carbons in the propane (C_3H_8) molecule and it is tempting to put this as 3 C-C bonds when in fact there are only 2 C-C bonds (C-C-C). Also remember that in a water molecule there are 2 O-H bonds and in carbon dioxide there are 2 C=O bonds.

DO IT!

For the reaction between methane and oxygen, draw a reaction profile and show the bonds broken in the reactants and the bonds formed in the products.

To help you, the following example of the burning of propane in oxygen to give water and carbon dioxide:

$$C_3H_8 \quad + \quad 5O_2 \quad \longrightarrow \quad 3CO_2 \quad + \quad 4H_2O$$

H—C—C—C—H + 5O=O ⟶ 3O=C=O + 4O (with H,H)

with H H H on top and H H H on bottom of the three carbons

bonds broken		bonds made	
2C — C	5O=O	6C=O	8O — H
8C — H			

MATHS SKILLS

You will need to read data from a table and carry out simple multiplications, additions and subtractions.

WORKIT!

Calculate the energy change for the combustion of propane shown in the equation below. The required bond-energies are shown in the table. (3 marks)

You will be supplied with this information in the exam.

C_3H_8 + $5O_2$ $\longrightarrow$ $3CO_2$ + $4H_2O$

H   H   H
| | |
H—C—C—C—H + 5O=O → 3O=C=O + 4O⟨H H
| | |
H   H   H

Bond	Bond energy (kJ per mol)
C–C	350
C–H	415
O=O	500
C=O	800
O–H	465

bonds broken
2C—C   5O=O
8C—H

bonds made
6C=O   8O—H

Bonds broken: heat energy taken in	Bonds made: heat energy given out
2 C–C bonds need 2 × 350 kJ = 700 kJ of energy to break them	6 C=O bonds give out 6 × 800 kJ of energy when they form = 4800 kJ
8 C–H bonds need 8 × 415 = 3320 kJ of energy to break them	8 O–H bonds give out 8 × 465 kJ of energy when they are formed = 3720 kJ
5O = O bonds need 5 × 500 = 2500 kJ of energy to break them	
Total energy taken in to break bonds = 700 + 3320 + 2500 kJ = 6520 kJ (1)	Total energy given out when bonds are made = 4800 + 3720 kJ = 8520 kJ (1)

There is more energy given out than taken in so the reaction is exothermic.
The energy of reaction = 8520 − 6520 = 2000 kJ/mol   (1)

CHECKIT!

H 1 When methane reacts with oxygen the following complete combustion reaction takes place:

$CH_4(g) + 2O_2(g) \rightarrow 2H_2O(l) + CO_2(g)$

a Draw the bonds present in the molecules.

b Use the values in the table above to calculate:

  i the energy taken in to break bonds

  ii the energy given out when bonds are formed

  iii the energy change for the reaction.

c Is the reaction exothermic or endothermic? Explain your answer.

1 What happens to the temperature in a reaction when the reaction is endothermic?

2 When limestone is heated it undergoes thermal decomposition to give calcium oxide (known as quicklime) and carbon dioxide. The reaction between calcium oxide and water can be used to heat up meals. The relevant equations are shown below:

A $CaCO_3(s) \rightarrow CaO(s) + CO_2(g)$

B $CaO(s) + H_2O(l) \rightarrow Ca(OH)_2(s)$

Classify reactions A and B as either exothermic or endothermic. In each case explain your answer.

3 The **incomplete** table below shows the results from an experiment on the reactivity of three metals, X, Y and Z. The same amounts of all three metals were reacted with hydrochloric acid and the temperature change in each reaction was measured.

Metal	Starting temperature	Final temperature	Temperature change
X	22		1
Y	22		19
Z	22	29	

a i Complete the table.

   ii Apart from the missing temperature figures, what else is missing from the table?

b Describe how you would ensure the results are valid.

c List the apparatus for this experiment.

d Suggest the order of reactivity of the three metals and explain your answer.

4 The diagram opposite shows the reaction profile for a chemical reaction.

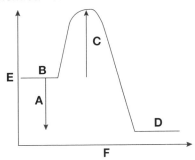

a Write down the correct labels for A to F.

b What type of reaction is represented by the reaction profile above.

c Explain what is meant by the term 'activation energy'.

H 5 The equation below shows the structures of the reactants and products of the chemical reaction between ethene and chlorine.

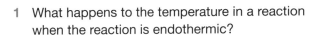

ethene   chlorine   dichloroethane

a List the bonds that are broken and calculate the energy required to break all these bonds.

b List the bonds that are made and calculate the energy given out by their formation.

c i Calculate the energy change for the reaction and the units.

   ii Explain whether it is exothermic or endothermic.

Table of bond energies

Bond	Bond energy (kJ/mol)
C=C	610
C–H	415
Cl–Cl	245
C–C	350
C–Cl	345

Rates of reaction and equilibrium

Ways to follow a chemical reaction

The main ways of following a chemical reaction to measure its rate are:

- measuring the volume of gas produced over a period of time
- measuring the loss in mass of the reactants over a period of time
- measuring how long it takes for a cross to be obscured when a solid is formed in a reaction between two solutions.

When you investigate the factors that affect the rate of a chemical reaction there are several variables that can be changed, measured or controlled. For example, if you were investigating the effect of temperature on the rate of reaction between marble chips and acid, you could measure the volume of gas produced over time. To make it a fair test you would keep the concentration of acid and the surface area of the marble chips constant for all the experiments.

The table below summarises facts about variables using this example.

Type of variable	Description	In the example it is...	Where it is plotted on a graph
Independent	The variable whose effect you have chosen to measure	Temperature	On the horizontal or x-axis
Dependent	The variable that you measure to find the effect of changing the independent variable	The volume of gas	Up the vertical or y-axis
Control variables	The variables you keep constant to make it a fair test	Concentration of acid and surface area of the marble chips	These are not plotted but could be noted in the title of the graph

SNAPIT!

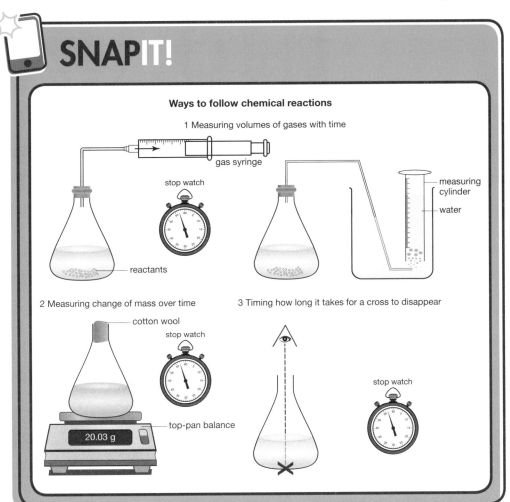

Ways to follow chemical reactions

1 Measuring volumes of gases with time

gas syringe

stop watch

reactants

measuring cylinder

water

2 Measuring change of mass over time

cotton wool

stop watch

top-pan balance

20.03 g

3 Timing how long it takes for a cross to disappear

stop watch

DOIT!

Using the diagrams in the Snap It! box, describe briefly how you would make the measurements in each method and create a suitable results table for each experiment.

NAILIT!

If you are asked to pick an appropriate method for following a chemical reaction to see how quickly it is going, then the equation for that reaction will help you.

You can use the state symbols in the equation to help you decide on the method to use. For example, if you see the (g) state symbol in the products, then you know a gas is produced.

If a gas is given off, then measurement of gas volumes using the gas syringe or displacement of water are obvious alternatives. If the gas given off is carbon dioxide, then measuring the loss in mass is also a possibility.

When two solutions react to form a solid then the mixture goes cloudy and you can time how long it takes to obscure a cross.

If a solid disappears during a reaction then you can time how long it takes for this to happen.

STRETCHIT!

If you use the disappearing cross technique then you measure how long it takes to obscure the cross so that you cannot see it.

The longer the time taken for this to happen, the slower the reaction is. If you want to have an idea of how quick the reaction is, you measure the time but use 1/time when you plot your results because 1/time tells you how quick the reaction is.

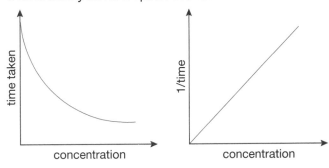

time taken

concentration

1/time

concentration

CHECK IT!

1 Suggest which of the three methods shown on page 189 could be used to measure the rate of the following reactions. In some cases more than one method can be used.

   a $2HCl(aq) + CaCO_3(s) \rightarrow H_2O(l) + CaCl_2(aq) + CO_2(g)$

   b $2HCl(aq) + Mg(s) \rightarrow H_2(g) + MgCl_2(aq)$

2 A student was investigating the effect of concentration on the rate of the reaction between marble chips and hydrochloric acid. They decided to measure the volume of gas over a period of time.

   In this investigation name the:

   a the independent variable

   b the dependent variable.

3 The equation below shows the reaction between hydrochloric acid and sodium thiosulfate solution.

   $2HCl(aq) + Na_2S_2O_3(aq) \rightarrow SO_2(g) + 2NaCl(aq) + H_2O(l) + S(s)$

   The rate of this reaction is usually followed by timing how long it takes to obscure a cross.

   a Give some problems that might make this method inaccurate.

   b i You are given a small electric lamp, a power pack, some black card, a light sensor and a data logger. Using a simple diagram explain how you could measure the rate of the same reaction using this apparatus.

   ii Explain how this method would be better than 'obscuring the cross' for measuring the rate.

You can measure the rate of reaction by measuring change in mass over time

Calculating the rate of reaction

The rate of a chemical reaction can be looked at in two different ways. Either using the reactant being used up, in which case the

$$\text{Mean rate of reaction} = \frac{\text{Amount of reactant used up}}{\text{Time taken}}$$

Or in terms of the product formed in which case the

$$\text{Mean rate of reaction} = \frac{\text{Amount of product formed}}{\text{Time taken}}$$

When a graph is plotted to show the change in a product or reactant you will probably get a curve. The rate at any time can be found by drawing a tangent to the curve and measuring its slope.

The steeper the gradient of the tangent the faster the rate of reaction.

When the gradient is zero, no reaction taking place and the reaction is complete.

DO IT!

Sketch a graph showing the course of a reaction and on it draw a few tangents. Write a brief description about why the tangents show that as time goes on the reaction gets slower.

WORK IT!

In an experiment, two substances were reacted together and one of the products, a gas, was collected. The reaction was finished after 63 seconds and the total volume of gas collected was 50 cm³.

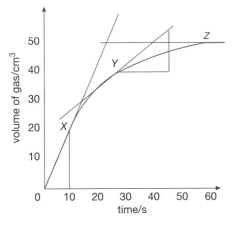

a) Calculate the average rate of the reaction. (1 mark)

a) The average rate of reaction = 50/63 = 0.794 cm³/s. (1)

b) Calculate the rate of reaction at points X and Y on the graph. (2 marks)

b) At X the tangent's gradient = 20/10 = 2 cm³/s. (1)

At Y the tangent's gradient = 15/20 = 0.750 cm³/s. (1)

c) Where is the rate of reaction zero? Explain why. (2 marks)

c) At Z (1) the gradient = 0 because the reaction has finished. (1)

MATHS SKILLS

You will need to:

- Plot a graph or graphs and compare the results.

- Determine the slope of a linear graph.

- Understand that a tangent is a line that touches a curve at a point but does not cross it anywhere else.

- Draw tangents to a graph, calculate their gradients and use them as a measure of the rate of a reaction.

- Express results to three significant figures.

- Express very small or very large numbers in base form.

- Know that a straight line between concentration and rate shows that the rate is proportional to the concentration.

STRETCH IT!

You may be asked to express the rate in mol per s. Remember for a gas at RTP 1 mol of any gas occupies $24\,000\,cm^3$.

At **X** the rate $= 2\,cm^3/s$. $2\,cm^3 = 2/24\,000\,mol/s$

$= 8.33 \times 10^{-5}\,mol/s$

At **Y** the rate $= 0.750\,cm^3/s = 0.750/24\,000\,mol/s = 3.13 \times 10^{-5}\,mol/s$

NAIL IT!

For Graph I, the reaction giving line A is faster than the reaction responsible for line B, this is shown by the fact the line obtained is steeper at the beginning than the one representing reaction B.

Similarly, for Graph II, reaction C is faster than reaction D and this is why its line is steeper at the beginning than the line for reaction D. The line levels off as the reaction slows down and it is horizontal because the reaction has finished.

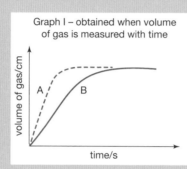

Graph I – obtained when volume of gas is measured with time

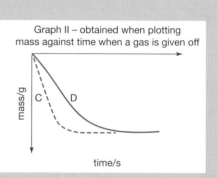

Graph II – obtained when plotting mass against time when a gas is given off

CHECK IT!

1 Describe how you can find the mean rate of a reaction using the amount of product formed in a reaction.

2 Two students carried out an investigation into the reaction between excess calcium carbonate and $100\,cm^3$ of hydrochloric acid.

$$CaCO_3(s) + 2HCl(aq) \rightarrow CaCl_2(aq) + H_2O(l) + CO_2(g)$$

They measured the volume of carbon dioxide produced at various time intervals. The results are shown in the table below.

Time/s	0	5	10	15	20	25	30	40	50	60	70	80
Volume of CO_2/cm^3	0	27	41	50	57	62	66	72	77	79	80	80

a Plot these results on a suitable graph.

b i Draw tangents to the graph at 0 s, 10 s and 30 s.

ii Find the rate of reaction in cm^3/s at each of these times.

iii Explain why the rate decreases with time.

c Explain why the gradient of the tangent drawn at 80 s is zero.

H d i Calculate how many moles of carbon dioxide were produced in the reaction.

ii Calculate the number of moles of hydrochloric acid present.

iii Calculate the concentration of the hydrochloric acid.

The effect of concentration on reaction rate and the effect of pressure on the rate of gaseous reactions

Particles have to collide in order to react.

When you increase the concentration of a solution in a chemical reaction the rate of the reaction also increases.

This is because when the concentration increases, the reacting particles get more crowded as there are more of them in a given volume, so they collide more frequently and there are more successful collisions.

During a chemical reaction, the concentration of the reactant particles in the solution decreases as they react. This means that they become less crowded and collide less **frequently** and the reaction slows down.

When a graph is plotted to show how a reaction is progressing, the slope of the tangent to the graph at any time shows us how quickly the reaction is at that time. The quicker the reaction the steeper the slope of the tangent.

As the concentration of the reactant decreases during a reaction, the reaction gets slower and this means that the slope of the graph gets less steep and the graph becomes a curve.

In reactions that involve gases increasing the pressure, it also makes the reactant particles more crowded; they collide more frequently and react more quickly so the rate increases.

NAILIT!

When you talk about collisions saying just 'more collisions' will not get you the marks. There could be more collisions but over a much longer time. Using the word **frequently** is very important.

SNAPIT!

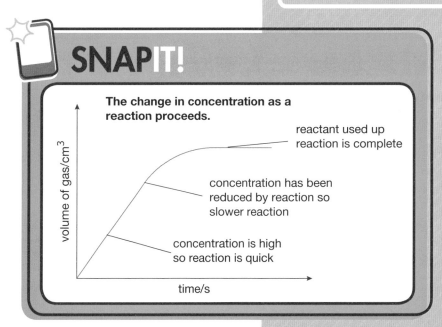

The change in concentration as a reaction proceeds.

- reactant used up reaction is complete
- concentration has been reduced by reaction so slower reaction
- concentration is high so reaction is quick

volume of gas/cm^3 (y-axis)

time/s (x-axis)

CHECKIT!

1 Describe what happens to the rate if the concentration of a reacting solution decreases.

2 Using collision theory, explain why increasing the concentration has an effect on the rate.

3 When calcium carbonate reacts with excess hydrochloric acid, carbon dioxide gas is produced.

   a Sketch a graph of volume of carbon dioxide gas produced over time.

   b On the same scales sketch a graph showing the reaction using the same concentration of acid but only **half the amount** of calcium carbonate.

Rates of reaction – the effect of surface area

DOIT!

Imagine a cube with sides of 1 cm. It has 6 faces. What is its total surface area? If the cube was broken up into smaller cubes with sides 0.1 cm what is the new surface area? Note that the number of smaller cubes is 1000.

If a solid is broken up into smaller pieces its surface area increases.

Smaller pieces have a larger surface area to volume ratio than larger pieces. This means that a powdered solid has a greater surface area than a lump.

If the experiment involves a comparison between the reaction rate given by a powdered reactant and a lump then a line graph cannot be drawn and the results are expressed using bar charts.

A larger surface area means that more solid particles are exposed to collisions with other reactant particles because there are more points of contact.

This means that there are more frequent collisions and the rate of reaction increases.

So increasing the surface area of a solid reactant increases the rate of reaction.

SNAPIT!

Graphs to show effects of surface area

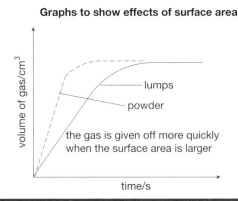

the gas is given off more quickly when the surface area is larger

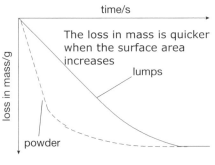

The loss in mass is quicker when the surface area increases

The steepness of the slopes of the graphs show how quickly the reaction is proceeding at any time. For both methods, the initial (starting) slope of the graph is steeper for the powder showing that it is producing a faster reaction. Therefore the greater the surface area the faster the rate.

CHECKIT!

1 Which has the larger surface area – lumps or powder?

2 Explain in terms of collision theory what effect increasing the surface area has on the rate of reaction.

3 A student wanted to find the effect of surface area on the rate of the reaction between calcium carbonate (used as marble chips) and hydrochloric acid.

$$CaCO_3(s) + 2HCl(aq) \rightarrow CaCl_2(aq) + H_2O(l) + CO_2(g)$$

a Explain how they could vary the surface area.

b List two ways they could follow the reaction rate.

c Suggest how they could display their results.

The effects of changing the temperature and adding a catalyst

The activation energy is the **minimum** energy needed for a reaction to take place.

If colliding particles do not have an energy greater than the activation energy then they will not collide with enough energy to break bonds and react.

When the temperature is raised particles move around more quickly and have more kinetic energy.

This means that there are more **frequent collisions** because they collide more frequently (more collisions per second).

There are more frequent effective collisions because the particles have a greater chance of colliding with an energy greater than the activation energy.

Adding a catalyst to a reaction speeds up the reaction but it is unchanged chemically at the end of the reaction.

A catalyst lowers the activation energy for a reaction by providing an alternative pathway for the reaction which has a lower activation energy.

This means that there will be even more frequent effective collisions and therefore a faster rate of reaction.

Enzymes are biological catalysts and carry out reactions in living organisms.

The formulae for catalysts are not included in chemical equations because they are unchanged chemically by the reaction. They can be written on the arrow between the reactants and products. For example:

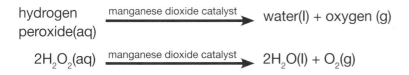

hydrogen peroxide(aq) $\xrightarrow{\text{manganese dioxide catalyst}}$ water(l) + oxygen (g)

$2H_2O_2(aq) \xrightarrow{\text{manganese dioxide catalyst}} 2H_2O(l) + O_2(g)$

NAIL IT!

For the variables concentration and surface area, doubling either of them will double the rate. This does not work with temperature. An approximate effect is that the rate for some reactions doubles if the temperature goes up by 10°C.

DO IT!

Manganese dioxide is a catalyst for the decomposition of hydrogen peroxide to give water and oxygen gas.

You are given some hydrogen peroxide, manganese dioxide, wooden splints, test tubes and filtration apparatus. Explain how you could show that the manganese dioxide can be reused as a catalyst. Hint: the wooden splints are for testing for the presence of a gas.

SNAPIT!

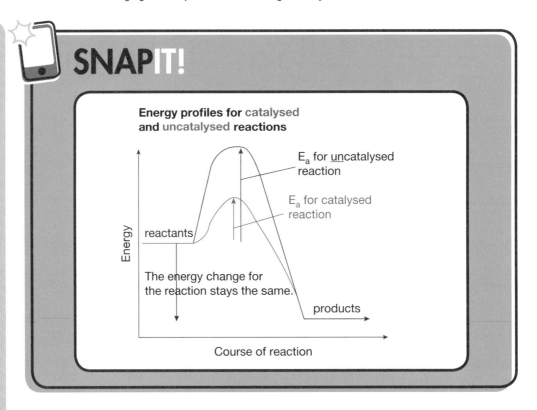

Energy profiles for catalysed and uncatalysed reactions

E_a for uncatalysed reaction

E_a for catalysed reaction

reactants

The energy change for the reaction stays the same.

products

Course of reaction

Energy

CHECKIT!

1 Describe the effect of increasing the temperature on the rate of reaction.

2 Using collision theory, explain the effect of raising the temperature on reaction rate.

3 a Define a catalyst.

   b Explain how a catalyst speeds up a chemical reaction.

4 The diagram below shows the energy levels for the reactants and products in an **endothermic** reaction.

   Complete the diagram by:

   a Drawing the reactions profiles for an **uncatalysed** reaction and a **catalysed** reaction.

   b Labelling the two activation energies and the energy of reaction.

Energy

Course of reaction

Investigation into how changing the concentration affects the rate of reaction

This is one of the required practicals and you could be questioned on it in the exam. This particular practical places emphasis on you putting forward a hypothesis and making predictions based on it. It also requires the accurate use of appropriate measuring apparatus and the need to make and record a range of measurements using apparatus that measures gas volumes or changes in turbidity. When investigating concentration you should know what factors need to be controlled to make a fair test.

The hypothesis is that increasing the concentration of a reactant in solution increases the rate of reaction. The concentration is your independent variable and is plotted along the horizontal axis (*x*-axis).

This can be tested by using several different reactions and the practical method requires both the measurement of **gas volume with time** and either a change in **turbidity** or **colour with time**. This is the dependent variable and is plotted up the vertical (*y*-axis).

SNAPIT!

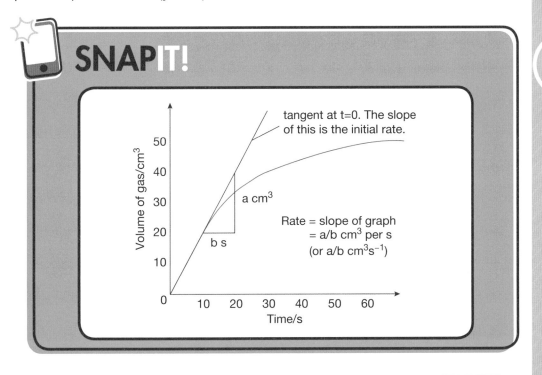

tangent at t=0. The slope of this is the initial rate.

a cm^3

b s

Rate = slope of graph
= a/b cm^3 per s
(or a/b cm^{3}s^{-1})

NAILIT!

If the plot of concentration against rate is a straight line then you can say that the **rate of reaction is proportional to the concentration.**

MATHS SKILLS

You should be able to draw tangents to curves, calculate the slopes of these tangents and express the results in terms cm^3/s or g/s.

You must also remember that the independent variable is plotted on the horizontal axis.

Practical Skills

The practical skills that you might be tested on are:

1 The variables you would keep constant to make it a fair test. These are the **control variables**.

When measuring the effect of concentration other variables must be kept constant. These variables are:

- the **surface area** and **mass** of any solid reactant
- the **temperature** of the reactant solution
- the **acid** used if one of the reactants is an acid
- the **volume** of acid used.

2 The apparatus you would use to carry out the experiment

Apparatus used for measuring gas volumes, loss in mass and turbidity is shown in the rates of reaction topic on page 189.

A **colorimeter** can be used to follow changes in colour in a reaction or if the turbidity changes with time then this change can be followed using a **turbidity meter** or a **light meter**.

3 Measuring the rate.

The simplest way is to measure the volume of gas given off in a measured time like 1 minute or the loss in mass in 1 minute.

Another way is to measure the volume of gas with time or mass with time. Plot the results on a graph and then measure the initial rate (see Snap It! for gas volume on page 197 as an example).

Measure how long it takes to give a certain volume of gas or change in mass. If this used then the time tells you how slow the reaction is. To give a measure of rate you plot 1/time.

> **DO IT!**
>
> Write short notes or record an MP3 file in which you describe how you would carry out this investigation.

CHECKIT! ✓

1 The reaction between sodium thiosulfate solution and hydrochloric acid produces sulfur as a solid and this turns the solution cloudy. The rate of reaction is measured by the time it takes to obscure a cross on a piece of paper. The results table below shows the results obtained in one experiment.

Volume of $Na_2S_2O_3$ solution/cm³	Volume of hydrochloric acid/cm³	Volume of water/cm³	Time taken to obscure cross/s	Reaction Rate/s⁻¹
5	10	35	128	
10	10	30	64	
20	10		32	
30	10		22	
40	10		16	

Complete the gaps in the table.

What variables are kept constant in this experiment?

Plot a graph of the volume of sodium thiosulfate solution against the rate of reaction and explain what this tells you about the effect of its concentration on the reaction rate. [HINT: The volume of sodium thiosulfate solution is the independent variable.]

Reversible reactions

A reversible reaction is one which can go both ways. This means that as well as reactants forming products, the products can also react to give the reactants.

The reaction where the reactants form the products is called the forward reaction.

The reaction where the products form the reactants is called the reverse reaction.

The symbol used for a reversible reaction is $\rightleftharpoons$.

An example of reversible reactions is the thermal decomposition of ammonium chloride to give ammonia gas and hydrogen chloride gas.

$$NH_4Cl(s) \underset{cool}{\overset{heat}{\rightleftharpoons}} NH_3(g) + HCl(g)$$

If the forward reaction is endothermic then the reverse reaction is exothermic. The opposite is also true.

Example:

$$CuSO_4 5H_2O(s) \underset{exothermic}{\overset{endothermic}{\rightleftharpoons}} CuSO_4(s) + 5H_2O(l)$$

A dynamic equilibrium is when the rate of the forward reaction is equal to the rate of the reverse reaction.

The forward and reverse reactions do not stop they are going on all the time and that is why it is a dynamic equilibrium.

At the same time the concentrations of the reactants and products remain constant.

DO IT!

Write a short account of the experiments shown in the Snap It! box below. How can you show that they are reversible reactions?

SNAP IT!

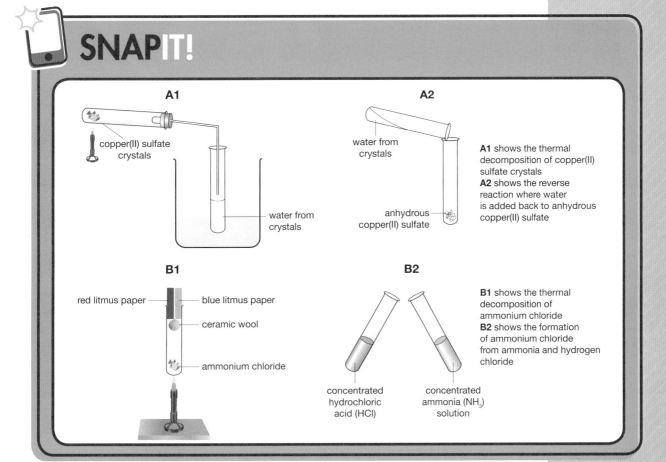

A1

copper(II) sulfate crystals

water from crystals

A2

water from crystals

anhydrous copper(II) sulfate

A1 shows the thermal decomposition of copper(II) sulfate crystals
A2 shows the reverse reaction where water is added back to anhydrous copper(II) sulfate

B1

red litmus paper — blue litmus paper

ceramic wool

ammonium chloride

B2

concentrated hydrochloric acid (HCl)

concentrated ammonia (NH_3) solution

B1 shows the thermal decomposition of ammonium chloride
B2 shows the formation of ammonium chloride from ammonia and hydrogen chloride

CHECKIT!

1 What is the symbol that tells you a reaction is reversible?

2 What is meant by the term reversible reaction?

3 Two compounds A and B react to form Y and Z in a reversible reaction.

   a Complete the equation for the reaction.

   A + B

   b When A is added to B the temperature goes up. What can you say about the forward and reverse reactions?

   c At the beginning when A is added to B why can't there be any reverse reaction?

   d At **equilibrium** which of the following alternatives is/are correct?

   i Both the forward and reverse reactions stop.

   ii There are equal amounts of reactants and products.

   iii The rate of the forward equals the rate of the reverse reaction.

   iv The concentrations of the reactants and products remain constant.

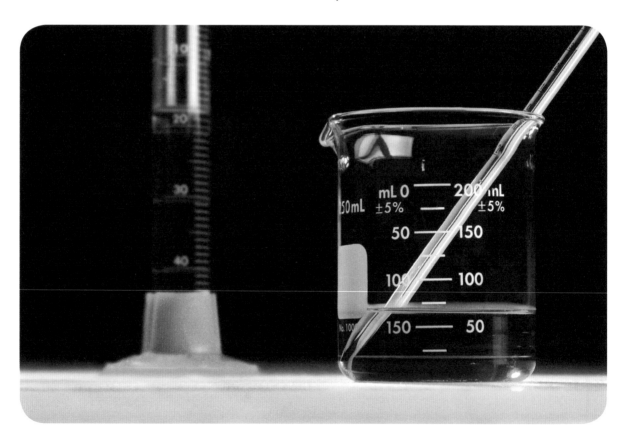

The effect of changing conditions on equilibrium

A chemical system is the reactants and products of a reversible reaction together in a closed container.

If a chemical system is at equilibrium it means that the rate of the forward reaction equals the rate of the reverse reaction.

There are three conditions that can be changed in a chemical system. These are:

- the **temperature** of the system
- the **pressure** of a system if any of the reactants or products are gases
- the **concentration** of any of the reactants or products.

A change in position of equilibrium means that there is a shift towards either the reactants or products.

If a chemical system is at equilibrium and one or more of the three conditions is changed then the position of equilibrium will **shift** to cancel out the change and we get either more reactants or more products. This is called Le Chatelier's Principle.

For example:

- If the temperature is increased, the equilibrium will shift in favour of the reaction which is accompanied by a decrease in temperature. This means that if the reverse reaction is endothermic then more reactants are formed.

- If the **pressure** is increased, the position of equilibrium will shift in favour of the reaction that would lead to a decrease in pressure. This means that the side with **fewer gas molecules** will be favoured. If the numbers of gas molecules on both sides are equal then pressure will have no effect.

- If you increase the **concentration** of one of the products, then the system will try and lower its concentration by forming more reactants.

A catalyst has no effect on the position of equilibrium. It speeds up how quickly equilibrium is reached.

A catalyst speeds up both the **forward** and **reverse** reactions **equally**.

DO IT!

Consider the reversible reaction shown below:

$2C_2H_4(g) + O_2(g) \rightarrow 2CH_3CHO(g)$

CH_3CHO is a substance called ethanal. The forward reaction is exothermic.

Explain to someone the best conditions for getting as much ethanal as possible.

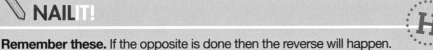

NAIL IT!

Remember these. If the opposite is done then the reverse will happen.

- Increasing the concentration of one of the reactants shifts the equilibrium to the right and favours the forward reaction to make more products.

- Increasing the pressure means that the equilibrium will shift to lower the pressure and it does this by making fewer gas molecules. Important – pressure has no effect if the number of gas molecules does not change in the reaction.

- Increasing the temperature always favours the endothermic reaction. This is because the endothermic reaction lowers the temperature and this counteracts the change.

SNAPIT!

Consider the industrial manufacture of methanol from carbon monoxide gas (CO) and hydrogen gas (H_2). A copper catalyst is used in this process.

$$CO(g) + 2H_2(g) \underset{\text{endothermic}}{\overset{\text{exothermic}}{\rightleftharpoons}} CH_3OH(g)$$

Consider the effects of some changes on the position of this equilibrium.

H

Change	What happens	Explanation
Increase the concentration of CO gas ⟶	Equilibrium shifts to right-hand side	The equilibrium moves in order to lower the concentration of CO gas by reacting it with hydrogen to make more methanol.
Increase the pressure ⟶	Equilibrium shifts to right-hand side	If you increase the pressure, the system tries to lower it by making fewer gas molecules and this means more methanol is produced.
Decrease the temperature ⟶	Equilibrium shifts to right-hand side	If you decrease the temperature, the system tries to raise it, this favours the exothermic reaction so produces more methanol.

CHECKIT! ✓

H 1 What is meant by the term chemical system?

H 2 Describe what the symbol ⇌ means?

H 3 Describe what is meant by a reverse reaction.

H 4 State Le Chatelier's Principle.

H 5 When ethene (C_2H_4) gas reacts with steam (H_2O) at 300°C, ethanol (C_2H_5OH) is formed.

$$C_2H_4(g) + H_2O(g) \underset{\text{endothermic}}{\overset{\text{exothermic}}{\rightleftharpoons}} C_2H_5OH(g)$$

   a Explain the effect on the position of equilibrium when pressure is increased.

   b Explain the effect on the position of equilibrium when the temperature is increased.

   c Explain the effect on equilibrium when a catalyst is added.

1 When calcium carbonate reacts with dilute hydrochloric acid, the gas carbon dioxide is given off as one of the products.

Suggest two ways you could follow the reaction and measure the rate of this reaction.

2 The curve below shows the line-graph obtained when a gas is given off during a reaction.

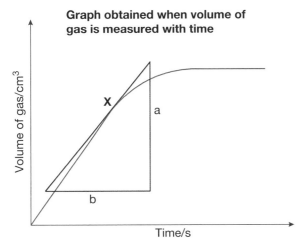

Graph obtained when volume of gas is measured with time

a Describe how you can calculate the rate of the reaction at any time using the graph.

b The value of **a** in the graph is $25\,cm^3$ and the value of **b** is $15\,s$. Calculate the rate of reaction at **X**.

3 The graph below shows two identical reactions but at different temperatures.

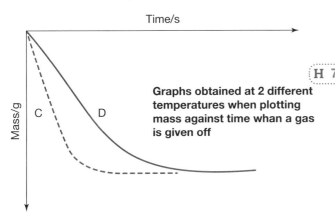

Graphs obtained at 2 different temperatures when plotting mass against time whan a gas is given off

a Describe and explain which graph (C or D) has the fastest rate of reaction.

b Explain which graph is the one given at the higher temperature.

c Using collision theory explain the effect of temperature on reaction rate.

4 When a catalyst is added to hydrogen peroxide solution the following reaction takes place.

$$2H_2O_2(aq) \rightarrow 2H_2O(l) + O_2(g)$$

a A catalyst for this reaction is manganese(IV) oxide.

  i Define the term catalyst.

  ii If 0.10 g of manganese(IV) oxide is added at the beginning, what is its mass at the end?

b Draw a reaction profile to show how a catalyst works.

c Using collision theory explain how a catalyst works

5 You are asked to investigate the effect of concentration on the rate of reaction.

What variables should be kept constant to ensure valid results.

6 When blue copper(II) sulfate crystals (formula = $CuSO_4.5H_2O$) are heated they form water and a white-grey anhydrous copper(II) sulfate powder (formula = $CuSO_4$) is formed.

The equation for the reaction is:

$$CuSO_4.5H_2O(l) \rightleftharpoons CuSO_4(s) + 5H_2O(l)$$

a Explain how you can tell the reaction is reversible.

b The forward reaction is endothermic. What can you say about the reverse reaction?

c List two observations you would make when the water is added back to the white-grey anhydrous copper(II) sulfate.

H 7 When ammonium chloride is heated the following reaction takes place:

$$NH_4Cl(s) \xrightleftharpoons[\text{exothermic}]{\text{endothermic}} NH_3(g) + HCl(g)$$

a Explain the effect on the position of equilibrium if temperature is increased.

b Explain the effect on equilibrium if pressure is increased.

Organic chemistry

Carbon compounds, hydrocarbons and alkanes

A hydrocarbon is a compound made up of hydrogen and carbon **only**.

Alkanes are hydrocarbons which have the maximum number of hydrogens and no carbon-carbon double bonds. This is why we call them saturated hydrocarbons.

The general formula of the alkanes is C_nH_{2n+2}.

The alkanes are a homologous series. This means they are a group of compounds with similar chemical properties, the same general formula and differ by a CH_2 each time.

They also show a gradation in physical properties as the molecules get bigger.

Whatever the homologous series, the way each member of the series is named (prefix) is the same.

The start of each name depends on the number of carbons in the molecule.

Number of carbons in chain	Name starts with
1	Meth–
2	Eth–
3	Prop–
4	But–

Most of the hydrocarbons in crude oil are alkanes.

DO IT!

If they are available make ball and stick models of the first four alkanes using the displayed formulae shown in the Snap It! box below.

SNAP IT!

The first four alkanes are listed below along with their molecular formulae and their displayed formulae.

Name	Molecular formula	Displayed formula
methane	CH_4	
ethane	C_2H_6	
propane	C_3H_8	
butane	C_4H_{10}	

WORKIT!

What are the molecular and displayed formulae for the alkane with 5 carbons? (3 marks)

The molecular formula can be worked out from the general formula — C_nH_{2n+2}

Here $n = 5$ so $2n+2 = 2 \times 5 + 2 = 12$

Therefore the molecular formula $= C_5H_{12}$ (1)

When you draw the displayed formula the first thing you do is put the 5 carbons in a row.

—C—C—C—C—C—

Then give each carbon four bonds.

$$\begin{matrix} | & | & | & | & | \\ -C- & C- & C- & C- & C- \\ | & | & | & | & | \end{matrix}$$

Now add the hydrogen atoms to each end of the bonds.

$$\begin{matrix} H & H & H & H & H \\ | & | & | & | & | \\ H-C- & C- & C- & C- & C-H \\ | & | & | & | & | \\ H & H & H & H & H \end{matrix} \quad (2)$$

NAILIT!

The simplest way to get displayed formulae correct is to think that each carbon has got four strong covalent bonds. Also make sure that the lines that represent the covalent bonds go direct from one atom symbol to the other atom symbol.

One of the most common errors is that candidates write down all the carbons and the bonds coming off from them but forget to draw the hydrogens.

Most of the hydrocarbons in crude oil are alkanes

✓ CHECKIT!

1 Explain what is meant by the term hydrocarbon.

2 Explain what is meant by the term homologous series.

3 Give the general formula of the alkanes. Write the molecular formula of the alkane with 6 carbons.

Crude oil, fractionation and petrochemicals

Crude oil is a mixture of many hydrocarbons and this means that it can be separated into its components using a physical method.

These hydrocarbons are miscible and have similar boiling points and therefore they can be separated by fractional distillation.

Most of the hydrocarbons are alkanes.

The separation of crude oil into its constituents is called fractionation and the different parts are called fractions.

Before the crude oil enters the fractionating tower it is heated and evaporates to form a vapour.

As it enters the tower the fractions with the higher boiling points condense lower down the tower to form liquids.

The fractions with lower boiling points continue to rise up the tower until the temperature falls below their boiling point and they also condense.

The fractions that are gases at room temperature leave the tower as gases.

See Snap It! box on page 207 for details about the fractionation.

All alkanes have simple molecular structures. The boiling points of the fractions depend on the size of the molecules in that fraction. The larger the molecules the greater the intermolecular forces and the higher the boiling points.

Hydrocarbons are good fuels. They undergo complete combustion in air (oxygen) to give carbon dioxide and water as products and this combustion gives out lots of heat energy in an exothermic reaction.

If there is not enough oxygen then incomplete combustion takes place and poisonous carbon monoxide is produced. (This is covered in more detail in the subtopic Atmospheric pollutants.)

The apparatus, shown below, is used to test for the products of complete combustion:

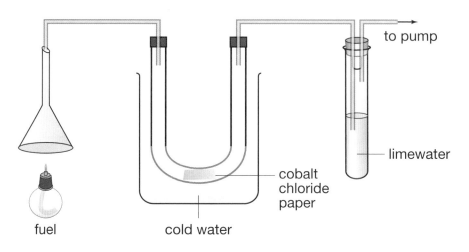

fuel   cold water   cobalt chloride paper   limewater   to pump

DO IT!

Describe how the apparatus shown can be used to identify the products of combustion. Note that cobalt chloride paper changes from blue to pink in the presence of water.

Your description should explain the use of a pump, the upturned filter funnel, the cold water and why the limewater is placed after the cobalt chloride paper.

WORKIT!

Write out a balanced equation for the combustion of propane (C_2H_3) in oxygen.   (2 marks)

The reactants are C_3H_8 and O_2. The products are H_2O and CO_2.

The best way of balancing these equations is to balance the carbons, then the hydrogens and then count up the number of oxygen atoms needed. For the oxygen you can have $\frac{1}{2}O_2$ if odd numbers are needed.

For propane you have 3 carbons and therefore 3 carbon dioxide molecules; you have 8 hydrogens and therefore 4 water molecules. This means that the total number of oxygen atoms is 10 and this means that we need 5 oxygen molecules on the reactant side.

$$C_3H_8(g) + 5O_2(g) \rightarrow 3CO_2(g) + 4H_2O(l)$$

The combustion of ethane (C_2H_6) illustrates the use of $\frac{1}{2}O_2$ because after going through the procedure we find that 7 oxygen atoms are needed on the reactant side.

$$C_2H_6(g) + 3\frac{1}{2}O_2(g) \rightarrow 2CO_2(g) + 3H_2O(l)   (2)$$

SNAPIT!

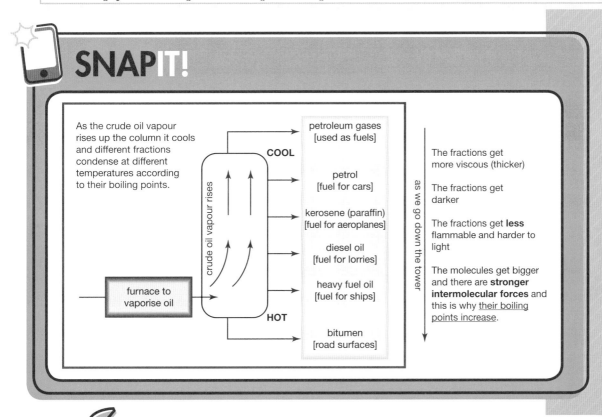

As the crude oil vapour rises up the column it cools and different fractions condense at different temperatures according to their boiling points.

crude oil vapour rises

furnace to vaporise oil

COOL

HOT

petroleum gases [used as fuels]

petrol [fuel for cars]

kerosene (paraffin) [fuel for aeroplanes]

diesel oil [fuel for lorries]

heavy fuel oil [fuel for ships]

bitumen [road surfaces]

as we go down the tower

The fractions get more viscous (thicker)

The fractions get darker

The fractions get **less** flammable and harder to light

The molecules get bigger and there are **stronger intermolecular forces** and this is why their boiling points increase.

CHECKIT!

1 List the uses of the main fractions coming from the fractionating tower.

2 Explain why it is important to separate crude oil into its fractions.

3 Explain how the boiling points of the fractions change as they go down the column.

4 One fraction X comes off the tower above another fraction Y. Compare the thickness, appearance, ease of lighting and boiling points of X and Y.

5 In the investigation of the products of combustion what are the changes observed in the cobalt chloride paper and the limewater?

6 Complete and balance the following equations for the **complete combustion** of methane (CH_4) and butane (C_4H_{10}).

   a $CH_4(g) + \_\_O_2(g) \rightarrow$ b $C_4H_{10}(g) + \_\_O_2(g) \rightarrow$

Cracking and alkenes

Cracking is the breaking down of large **alkane** molecules into **smaller** alkanes and **alkenes**.

The smaller alkanes make good fuels and the alkenes are used to make polymers.

Cracking is an important process for two main reasons:

1. **It converts fractions which have a low demand into higher demand fractions.** For example, after fractional distillation there is not enough of the petrol fraction for use as a fuel but there is more than required of the kerosene fraction. This means that some of the kerosene can be cracked to give alkanes that make up the petrol fraction.

2. **It makes useful hydrocarbons not naturally found in crude oil.** For example, cracking also gives alkenes which are not found in crude oil but are very important in the manufacture of polymers.

The general formula of alkenes is C_nH_{2n}. Remember the general formula for alkanes is C_nH_{2n+2} and the two fewer hydrogen atoms is because of the C=C double bond in the alkene molecule.

Cracking improves the economic value of the fractions coming off the tower.

In cracking the alkane vapours are passed over a heated catalyst or mixed with steam and heated up to a high temperature.

DO IT!

Refer to your notes and textbook to analyse the results of the experiment in the Snap It! box below.

SNAP IT!

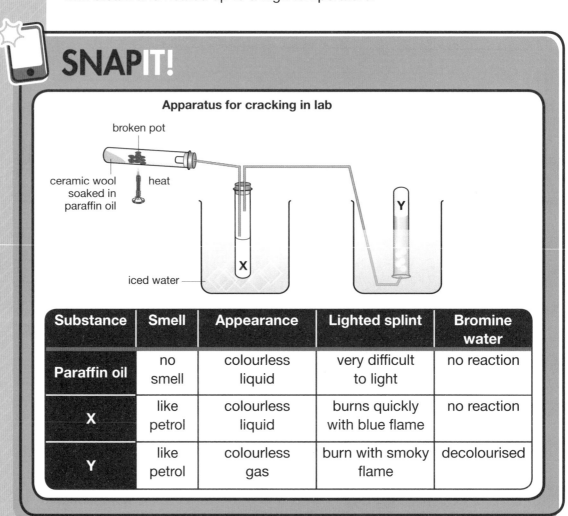

Apparatus for cracking in lab

Substance	Smell	Appearance	Lighted splint	Bromine water
Paraffin oil	no smell	colourless liquid	very difficult to light	no reaction
X	like petrol	colourless liquid	burns quickly with blue flame	no reaction
Y	like petrol	colourless gas	burn with smoky flame	decolourised

MATHS SKILLS

Balancing equations: You may have to use general formulae to identify the products in a cracking reaction.

Remember that the general formula of the alkanes is C_nH_{2n+2} and for the alkenes it is C_nH_{2n}.

WORKIT!

In the following reactions, identify the alkanes and alkenes: (2 marks)

$$C_8H_{18} \quad \rightarrow \quad C_4H_8 \quad + \quad C_4H_{10}$$
Alkane   Alkene   Alkane   (1)

$$C_{10}H_{22} \quad \rightarrow \quad C_2H_4 \quad + \quad C_4H_8 \quad + \quad C_4H_{10}$$
Alkane   Alkene   Alkene   Alkane (1)

An oil refinery

NAILIT!

The important thing about cracking is that it turns chemicals that are not very useful into ones that are very useful and economically important.

The amounts of short chain alkanes which are produced by fractionation of crude oil are not enough to meet demands. Cracking overcomes this shortfall.

CHECKIT!

1 Define cracking.

2 Describe how cracking is carried out in industry.

3 a Give the general formula for alkenes.

  b Describe the chemical test for alkenes.

  c Give a use for alkenes.

4 Describe the economic importance of cracking.

5 Complete the following equations and identify the alkanes and alkenes in the reactions.

  a $C_6H_{14} \rightarrow C_2H_6 +$ _____

  b $C_8H_{18} \rightarrow C_2H_4 + C_3H_6 +$ _____

  c $C_{12}H_{26} \rightarrow 3C_2H_4 +$ _____

For additional questions, visit:
www.scholastic.co.uk/gcse

1   a   Identify the physical property which allows us to separate crude oil into its fractions using fractional distillation.

   b   As we go up the fractionating tower what happens to the following properties:
      i   viscosity
      ii   boiling point
      iii ease of lighting?

   c   Explain the trend in boiling point as you go up the tower.

2   Alkanes are the main components of crude oil.

   a   What is the general formula of the alkanes?

   b   Complete and balance the following equations for the complete combustion of the two hydrocarbons methane and ethane:
      i $CH_4(g) + O_2(g) \rightarrow$
      ii $C_2H_6(l) + O_2(g) \rightarrow$

   c   The diagram below shows the apparatus used to investigate what is formed when an alkane burns in air.

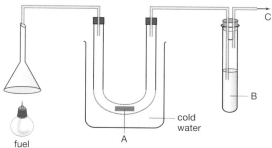

   i   Give the correct labels for A, B and C.
   ii   Describe what happens to A. What does the change show?
   iii Describe what happens to the liquid in B. What does the change show?

3   a   Why is it necessary to carry out cracking on alkanes which have large molecules?

   b   Complete the following equations:
      i $C_{10}H_{22} \rightarrow C_4H_8 +$
      ii $\rightarrow C_3H_6 + C_6H_{14}$

Pure substances and formulations

Pure substances are either single elements or single compounds.

Everyday descriptions of pure substances are inaccurate. For example, pure spring water is a solution of various minerals and gases.

Pure substances melt and boil at specific temperatures.

Mixtures melt and boil over a range of temperatures.

A formulation is a mixture that is designed as an improvement upon a pure substance on its own.

Examples of formulations are metal alloys, drugs and paints.

For example, a drug has in its formulation the active chemical and other substances that stop the drug from going off and make it easy to swallow.

DO IT!

Look up the ingredients of a simple over-the-counter drug. A painkiller like Nurofen is a good one to do. How do the ingredients of the formulation improve the drug? Research 'The ingredients of Nurofen'.

MATHS SKILLS

Using formulae and inserting values into them using percentages.

The formulae used are:

$$\text{Percentage} = \frac{mass\ of\ component}{total\ mass} \times 100\%$$

$$\text{Number of moles} = \frac{mass}{M_r}$$

WORKIT!

A drug formulation in tablet form weighs 500 mg. The two components are 350 mg of a stabiliser (X) with a relative formula mass 70, and the 150 mg of the active drug itself (Y) which has a relative formula mass of 125.

What is the percentage composition of the tablet in terms of:

a mass (2 marks)

The percentage by mass of X = 350/500 = 70% = 70% (1)

Therefore the percentage by mass of Y = 100 - 70 = 30% (1)

H b moles? (4 marks)

The number of moles of X = mass/M_r = 350 × 10^{-3}/70 = 5 × 10^{-3} mol

The number of moles of Y = mass/M_r = 150 × 0^{-3}/125 = 1.2 × 10^{-3} mol (1)

Therefore the total number of moles = 5 × 10^{-3} mol + 1.2 × 10^{-3} mol = 6.2 × 10^{-3} mol (1)

Mole percentage for X = (5 × 10^{-3}/6.2 × 10^{-3}) × 100% = 80.6% (1)

This means that the mole percentage for Y = 100 − 80.6% = 19.4% (1)

SNAPIT!

A distinctive melting point is a criterion for purity. The apparatus used is shown below along with a typical heating curve for a **pure** substance.

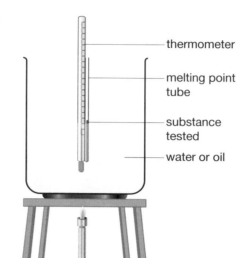

- thermometer
- melting point tube
- substance tested
- water or oil

melting point

temperature °C

time

NAILIT!

A possible question on formulations is finding the percentage composition of a formulation, and at the Higher Tier, converting the masses present to moles.

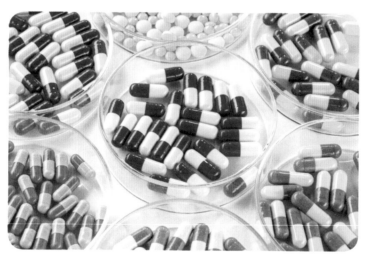

Drugs are formulations of chemicals

CHECKIT!

1 What is a pure substance?

2 Why isn't 'pure orange juice' really pure?

3 What is a formulation?

4 Give one example of a product which is a formulation and give brief details of its composition.

Chromatography

This is one of the required practicals and you could be questioned on it in the exam. The practical emphasises that you carry out a practical safely and accurately, make measurements and make conclusions using these measurements.

Chromatography is a technique that can be used to separate mixtures into their components and identify these components.

Chromatography involves two phases – a stationary phase and a mobile phase.

In paper chromatography, paper is the stationary phase and a liquid solvent is the mobile phase.

When substances are added to the paper, the mobile phase carries them through the paper. The distance a compound moves on the paper depends on its relative attraction for the paper and the solvent. The R_f value for a substance is equal to the distance moved by its spot divided by the distance moved by the solvent.

Compounds that have a higher attraction for the paper and a low attraction for the solvent spend a lot of time on the paper and move up the paper slowly.

Compounds that have a higher attraction for the solvent and have less of an attraction for the paper move quickly up the paper.

At the end of the experiment the spots obtained after the solvent is run up the paper is called a chromatogram.

Mixtures give more than one spot and pure compounds give only one spot. If the substances being separated are colourless then a locating agent is needed to show how far they have moved. Sometimes a UV lamp can be used.

 ## Practical Skills

Paper chromatography is often used to separate a mixture of coloured substances such as those found in inks and food colourings.

Chromatography requires a container, a supply of solvent(s), a ruler for measuring R_f values, chromatography paper and very thin capillary tubes for adding the test substances to the paper.

A pencil line is drawn a short distance from the bottom of the paper and this is called the baseline. The substances to be investigated are added at regular intervals along this line.

If the starting spots are too large then it will be very difficult to separate the mixture into easily identified substances. This is because the spots get larger as they rise up the paper.

The choice of solvent is not limited to water and should be chosen on the basis that it gives a good separation.

When measuring the R_f value of a substance the measurement is taken from the line where the substances are added to the paper and the middle of the spot obtained after the solvent has run up the paper.

SNAP IT!

A typical chromatogram using paper chromatography

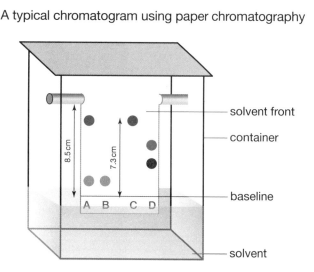

- solvent front
- container
- baseline
- solvent

8.5 cm   7.3 cm

A B   C D

MATHS SKILLS

Calculating the R_f value using experimental results.

When you express your answers for the R_f value you should give the answer to 3 significant figures unless asked to do otherwise. For example, if your calculation gives the answer as 0.65723 you should write down 0.657.

Using and rearranging the formula

$$R_f = \frac{\text{distance moved by spot}}{\text{distance moved by solvent}}$$

NAIL IT!

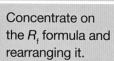

Concentrate on the R_f formula and rearranging it.

WORK IT!

In a chromatography experiment the spot for a substance X moved 12 cm and the solvent front moved 15 cm.

a What is the R_f value for X using this solvent? (1 mark)

The R_f value = 12/15 = 0.8   (1)

b In a second experiment under the same conditions, the solvent front moves 25 cm. What distance would the spot move in this second experiment? (2 marks)

$$R_f = \frac{\text{distance moved by spot}}{\text{distance moved by solvent}} \quad (1)$$

Rearrange this formula so that distance moved by spot = R_f × distance moved by solvent.

This means that the distance moved by spot = 0.8 × 25 cm = 20 cm.   (1)

CHECK IT!

1 In paper chromatography what is the stationary phase and what is the mobile phase?

2 The baseline is drawn in pencil. Why do you not draw the baseline in pen?

3 This question concerns the chromatogram shown in the Snap It! box above.

   a Which of the three substances B, C and D is/are a pure substance? Explain your answer.

   b i How can you tell that A is a mixture?

      ii What substances make up the mixture A?

   c Calculate the R_f value for substance C.

Testing for gases

Several chemical reactions produce a gas or gases. In solutions this produces effervescence (fizzing).

Hydrogen is a flammable gas. The test for hydrogen is a burning splint which is extinguished (put out) with a 'pop'.

The equation for this combustion reaction is: $2H_2(g) + O_2(g) \rightarrow 2H_2O(l)$

Oxygen (O_2) supports combustion. The test for oxygen is a glowing splint which relights in oxygen.

Carbon dioxide reacts with a solution of limewater to give solid calcium carbonate. When carbon dioxide is passed into limewater, the limewater turns cloudy or milky.

The equation for this reaction is:

$CO_2(g) + Ca(OH)_2(aq) \rightarrow CaCO_3(s) + H_2O(l)$
       colourless   white solid
          liquid

Chlorine (Cl_2) bleaches blue litmus paper (or UI paper). This is used as the test for chlorine.

DO IT!

Print out the diagrams of the tests in the Snap It! box below and then write down a list of comments to accompany them.

SNAP IT!

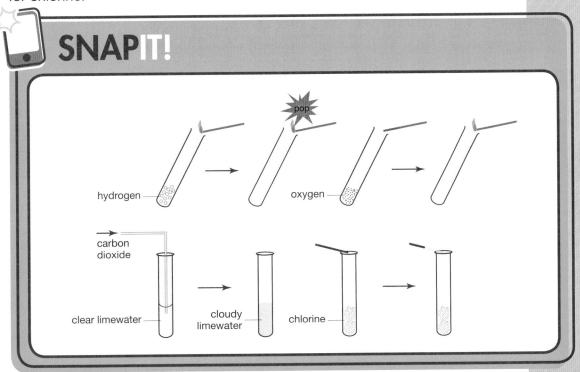

CHECK IT!

1 Describe how you would test for the gases produced in the following reactions:

   a $2H_2O_2(aq) \rightarrow 2H_2O(l) + O_2(g)$

   b $CaCO_3(s) + 2HCl(aq) \rightarrow CaCl_2(aq) + CO_2(g) + H_2O(l)$

   c $Mg(s) + 2HCl(aq) \rightarrow MgCl_2(aq) + H_2(g)$

   d $2NaCl(aq) + 2H_2O(l) \rightarrow 2NaOH(aq) + H_2(g) + Cl_2(g)$

Chemical analysis

1 Describe your observations when the following actions are carried out.

   a Hydrochloric acid is added to solid calcium carbonate.

   b Nitric acid and silver nitrate solution is added to:

      i   sodium iodide solution

      ii   potassium chloride solution

      iii potassium bromide solution

      iv   sodium chloride solution.

   c Hydrochloric acid and barium chloride solution are added to:

      i   sodium sulfate solution

      ii   sodium chloride solution.

Chemistry of the atmosphere

The composition and evolution of the Earth's atmosphere

The present composition of the Earth's atmosphere is 78% (about four-fifths) nitrogen, 21% (about one-fifth) oxygen and 1% of other gases like argon and the other noble gases, carbon dioxide, and water vapour.

This composition has evolved over billions of years.

In the early days of the Earth's existence there was a lot of volcanic activity and this gave rise to an atmosphere containing lots of water vapour and carbon dioxide along with methane, nitrogen and ammonia.

As the Earth cooled, the water vapour condensed and formed the oceans, rivers and lakes.

At some stage in the Earth's history, life began and algae and photosynthetic bacteria carried out photosynthesis.

In photosynthesis, glucose is made from the reaction between carbon dioxide and water. Oxygen is a by-product of this reaction.

$$\text{carbon dioxide(g)} + \text{water(l)} \xrightarrow{\text{light}} \text{glucose(s)} + \text{oxygen(g)}$$

$$6CO_2(g) + 6H_2O(l) \xrightarrow{\text{light}} C_6H_{12}O_6(s) + 6O_2(g)$$

As the oxygen in the atmosphere increased because of photosynthesis, animals evolved.

As the oxygen increased, the amount of carbon dioxide was reduced by various processes. These processes were: dissolving in water; limestone formation; photosynthesis; crude oil and natural gas formation; coal formation.

Once nitrogen was formed it remained in the atmosphere because it is very unreactive. This explains its high concentration in the atmosphere.

> ## MATHS SKILLS
>
> The composition of the atmosphere is usually expressed in percentages and the percentage of a gas can be found using the equation
>
> $$\text{percentage of gas} = \frac{\text{volume of gas}}{\text{total volume}} \times 100\%$$

DO IT!

Look at the apparatus in the Snap It! box. Use your notes or textbook to find out how it works.

SNAP IT!

The apparatus below is used to find the composition of air.

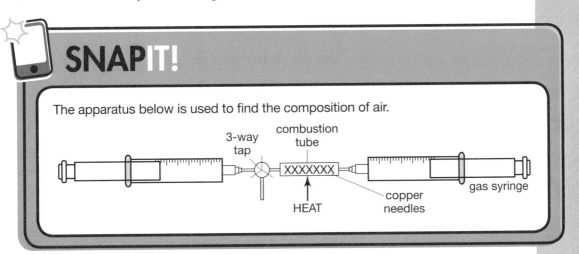

NAIL IT!

This table shows the ways that carbon dioxide has been removed during the lifetime of the Earth.

Removal process	Result
As the oceans formed carbon dioxide dissolved in the water to form insoluble solid metal carbonates.	These carbonates formed sediments so the carbon dioxide was locked in solids.
Photosynthesis uses carbon dioxide and water to make sugars such as glucose.	Carbon dioxide removed from the air.
The carbon dioxide in the oceans was converted into calcium carbonate by plankton and other marine organisms.	Their remains were compressed to form limestone.
Sometimes these dead marine organisms were covered with mud and compressed by other layers.	These remains formed crude oil and natural gas.
A similar process happened with dead land plants.	These remains formed coal as a sedimentary rock.

WORKIT!

In an experiment to work out the composition of air, the apparatus was the same as that shown in the Snap It! box on page 217.

$100\,cm^3$ of air was drawn into the apparatus through the 3-way tap. After heating, the volume of gas present was reduced to $79\,cm^3$.

a What is the percentage of oxygen in air? (2 marks)

The volume of gas has gone down because the oxygen reacted with the copper needles.

This means that the percentage of oxygen $= \dfrac{\text{volume of oxygen}}{\text{Total volume}} \times 100\%$

$\qquad\qquad\qquad\qquad = \dfrac{21}{100} \times 100\%$ (1) $= 21\%$ (1)

b What is the equation for the reaction taking place in the apparatus? (1 mark)

$2Cu + O_2 \longrightarrow 2CuO$ (1)

CHECKIT! ✓

1 List the main gases found in the early atmosphere of the Earth.

2 Give four ways that carbon dioxide has been removed from the atmosphere during the Earth's lifetime.

3 Write the word and symbol equations for photosynthesis.

Climate change

A key thing to remember in this topic is that all evidence collected by scientists has to be peer reviewed. This means that research has to be examined and tested by other scientists in the same field.

The media are also important because the interpretation they put on scientific evidence can influence public opinion and behaviour.

Most of the solar radiation reaching the Earth is shortwave UV and visible light. The energy reflected back by the Earth is longer wavelength infrared radiation.

This infrared radiation is absorbed by gases in the atmosphere which then re-emit this energy in all directions but a lot goes back to the Earth which then warms it up. This natural increase in temperature is known as the greenhouse effect.

The two main greenhouse gases are carbon dioxide (CO_2) and methane (CH_4). Scientists have discovered a correlation between the rise and fall of carbon dioxide in the atmosphere and the average global temperature.

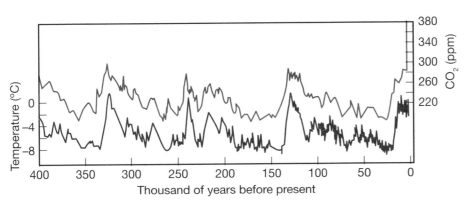

The rapid rise in carbon dioxide concentration has been linked to the burning of fossil fuels such as crude oil and coal to generate energy.

Deforestation can also have an effect on carbon dioxide in the atmosphere. Deforestation's effect is twofold because it reduces the number of trees that take up carbon dioxide by photosynthesis, and then carbon dioxide is emitted as the felled trees are burned to clear the land.

Two big sources of methane are:

• livestock farming (from animal digestion and waste decomposition)

• rubbish decay in landfill sites.

To stop the rise in carbon dioxide concentration humans need to consider how to decrease the burning of fossil fuels.

This has proved difficult for various reasons:

• a lack of affordable alternative energy sources

• economic growth relies on cheap energy

• objections to the idea that global warming is caused by humans

• lack of international co-operation.

DOIT!

Research one of the **possible consequences** shown in the Snap It! box, to see if it is actually taking place. What evidence is there?

SNAPIT!

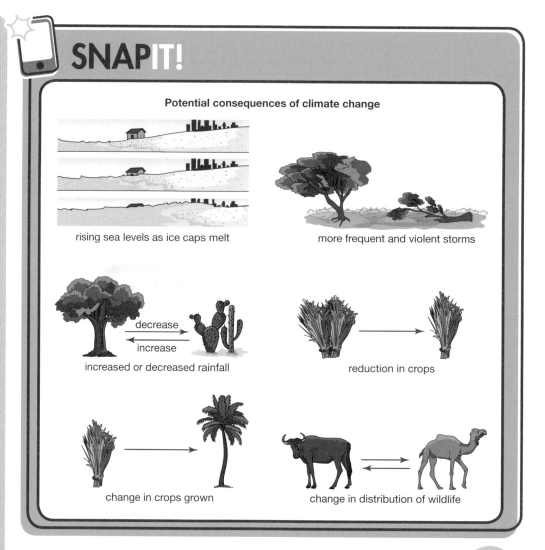

Potential consequences of climate change

rising sea levels as ice caps melt

more frequent and violent storms

decrease
increase
increased or decreased rainfall

reduction in crops

change in crops grown

change in distribution of wildlife

NAILIT!

One question that you might have to answer is 'Why have some countries been slow to reduce their burning of fossil fuels?' Possible answers are as follows:

- Some people think that climate change is just one part of a natural cycle.
- Some people believe that the evidence is still not 100% reliable.
- There is a difficulty with predicting the possible effects of climate change because different scientific models have different limitations and therefore make different predictions.
- In the short term, the alternatives to burning fossil fuels are more expensive.
- Lobbying by the petroleum and coal industries.
- Governments are slow to replace fossil fuels with other forms of energy.
- Inability of international bodies like the UN to get all nations to agree to reduce their use of fossil fuels.

CHECKIT!

1 Name the **two** main greenhouse gases.

2 Explain why the rise in carbon dioxide in the atmosphere is often blamed for global warming.

3 Name the type of radiation absorbed and re-emitted by greenhouse gases.

4 Describe the causes of the rise in carbon dioxide concentration in the atmosphere.

The carbon footprint and its reduction

The carbon footprint of a product or activity is the total amount of carbon dioxide and greenhouse gases emitted over the lifetime of that product or activity.

There are a number of ways to remove or reduce the carbon footprint.

- Increase the use of alternative energy supplies. For example, solar cells, wind power and wave power.
- Energy conservation involves reducing the amount of energy used by using energy-saving measures such as house insulation or using devices that use less energy.
- In Carbon Capture and Storage (CCS) the carbon dioxide given out by power stations can be removed by reacting it with other chemicals and then stored deep under the sea in porous sedimentary rocks, especially those that used to be part of oilfields.
- Carbon taxes and licences. Penalising companies and individuals who use too much energy by increasing their taxes.
- By removing carbon dioxide from the air using natural biological processes, especially photosynthesis. This is done by planting trees or trying to increase marine algae by adding chemicals to the sea. This is called carbon offsetting.
- Using plants as biofuels, because plants take in carbon dioxide, when they are burned they only release the same amount of carbon dioxide, which is zero net release. This makes them carbon-neutral.

NAILIT!

You may be asked why people, companies or countries do not reduce their carbon footprints. Possible answers are as follows:

- People are reluctant to change their lifestyle. For example, still using large cars instead of smaller more fuel-efficient ones.
- Even energy-saving devices have a carbon footprint because of the processes used in the extraction of materials to make them and the energy used in their manufacture and in their disposal.
- Countries and companies still find it more economic to use lots of energy.
- Countries may not co-operate with each other.
- There is still disagreement that climate change is man-made and about its causes.
- People are still not sure of the facts and do not know about the possible consequences of climate change.

DOIT!

Look at your notes on aluminium extraction by electrolysis. Why does this process have a large carbon footprint?

SNAPIT!

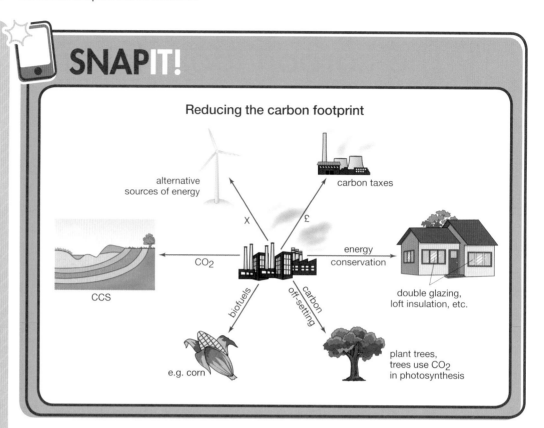

Reducing the carbon footprint

CHECKIT! ✓

1 Describe what is meant by the term carbon footprint.

2 Explain the following terms:

   a Carbon offsetting

   b Carbon capture and storage.

3 List two reasons why people are not reducing their carbon footprint.

4 Photovoltaic cells and other alternative sources of energy reduce the need for burning fossil fuels but they do have a carbon footprint. Explain why.

[H 5] The data below shows the savings in carbon dioxide emissions by two different methods of home insulation when a **detached house** is insulated.

Method	Reduction in carbon dioxide emissions/kg
Loft insulation	990
Cavity wall insulation	1100

   a Which of these reductions would be less for a terraced house? Explain your answer.

   b Calculate how many moles of carbon dioxide would have been emitted without insulating the loft of the detached house. ($M_r(CO_2) = 44$)

   c Calculate the volume of carbon dioxide that would have been emitted.

Atmospheric pollutants

An atmospheric pollutant is something that is introduced into the atmosphere and has undesired or unwanted effects.

When hydrocarbons undergo complete combustion in air they produce carbon dioxide and water along with energy.

If there is insufficient (not enough) air then incomplete combustion occurs and carbon monoxide (CO) and particulates (small soot particles) are formed.

Carbon monoxide (CO) is a toxic gas. It combines with haemoglobin in the blood and reduces the capacity of the red blood cells to carry oxygen. This means the cells around the body do not receive sufficient oxygen. This can cause death.

The reaction is a reversible one:

$CO(g) + HbO_2 \rightleftharpoons O_2(g) + HbCO$  Hb = haemoglobin

Particulates cause global dimming (reflect sunlight back out to space before it reaches the atmosphere) so that less sunlight gets through to the Earth's surface. They are also irritants and cause damage to the lungs.

At the high temperatures caused by combustion in car engines, nitrogen and oxygen can combine to form oxides of nitrogen. These cause respiratory problems. Eventually these oxides dissolve in water to form acid rain. Acid rain causes weathering of buildings and damages plants and aquatic life.

Fossil fuels contain sulfur and when this burns in air it forms sulfur dioxide (SO_2). This is a very acidic gas and when it dissolves in clouds of water vapour it causes acid rain. It also causes respiratory problems.

To reduce the formation of sulfur dioxide many petrochemicals are desulfurised.

DO IT!

Construct a mind map based on climate change, greenhouse gases and pollution.

STRETCH IT!

The equations for the formation of carbon monoxide and particulates follow the same rules as for the balancing of any other equation. You should also remember that you are allowed to use ½ molecules of oxygen in these equations and water is always formed as the other product.

The burning of methane is the simplest example of the combustion of a hydrocarbon. Just for comparison, the equation for the complete combustion of methane is as follows:

$CH_4(g) + 2O_2(g) \rightarrow CO_2(g) + 2H_2O(l)$

The equation for carbon monoxide formation is:

$CH_4(g) + 1½O_2(g) \rightarrow CO(g) + 2H_2O(l)$

For particulates (which are carbon):

$CH_4(g) + O_2(g) \rightarrow C(g) + 2H_2O(l)$

Notice for both of these less oxygen is used up and the methane undergoes incomplete combustion.

NAIL IT!

As a rule, the oxides of all non-metals (except hydrogen) are acidic.

Make sure you are able to write the balanced equations for forming sulfur dioxide and oxides of nitrogen.

For sulfur dioxide:

$S(s) + O_2(g) \rightarrow SO_2(g)$

For oxides of nitrogen, the first oxide is nitrogen monoxide (NO):

$N_2(g) + O_2(g) \rightarrow 2NO(g)$

Another oxide of nitrogen is nitrogen dioxide (NO_2). This is very dangerous as it damages the lungs and produces acid rain:

$N_2(g) + 2O_2(g) \rightarrow 2NO_2(g)$

SNAP IT!

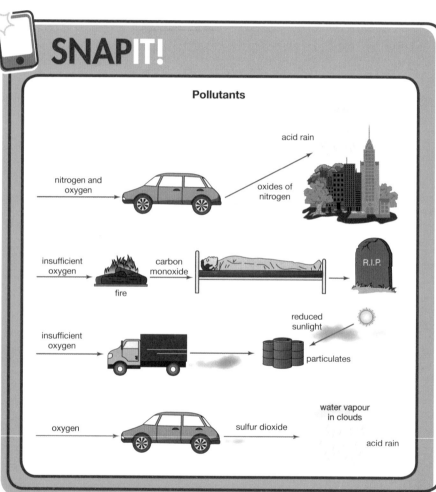

Acid rain causes damage to materials like this stone wall

CHECK IT!

1 List the four pollutants formed by burning fossil fuels and for each one give its effects.

2 Explain why a coal or wood fire should always be in a well-ventilated room.

H 3 a i Write the balanced symbol equation for the complete combustion of methane to give carbon dioxide.

ii Write the balanced symbol equation for the incomplete combustion of methane to give carbon monoxide.

1   Name the main gases in the atmosphere of the early Earth.

2   List five ways by which the amount of carbon dioxide in the atmosphere has been reduced.

3   Describe the composition of the present atmosphere of the Earth.

4   Name two greenhouse gases and explain how they cause the greenhouse effect.

5   Describe the evidence that rising levels of carbon dioxide are causing climate change.

6   Why might some areas expect flooding as a result of climate change?

7   Explain what is meant by the term 'carbon footprint'.

8   Describe five ways by which the carbon footprint of an organisation or individual person can be reduced.

9   Explain why some countries might be reluctant or slow to reduce their carbon footprint.

10 What is a pollutant?

11 Carbon monoxide is a pollutant.

    a i   What is the formula of carbon monoxide?

       ii How is it formed?

       iii What are its undesired effects?

    b Oxides of nitrogen are pollutants.

       i   How are they formed?

       ii What are their undesired effects?

12 Explain why sulfur should be removed from petrol.

Using resources

Finite and renewable resources, sustainable development

The natural resources used by chemists to make new materials can be divided into two categories – finite and renewable. Finite resources will eventually run out. Examples are fossil fuels and various metals even though we are able to get some valuable materials from places like the oceans.

Renewable resources are ones that can be replaced at the same rate as they are used up. They are derived from plant materials. An example is ethanol, which is made from sugar from fermentation. Ethanol can be used as fuel for cars instead of petrol which is extracted from the finite resource, crude oil.

Many of the Earth's natural resources are running out and if they are used at the current high rates they will be depleted (used up) very soon. In order to increase the lifetime of these finite resources, the industry has to develop processes that increase the lifetime of natural resources.

Sustainable development meets the needs of present development without depleting natural resources. In a sustainable process:

- there is a high yield
- there are few waste products
- there is very little impact on the environment and the products should not harm the environment.

SNAP IT!

Sustainable processes should do the following:

- have reactions with high atom economy
- use renewable resources
- have as few steps as possible
- use catalysts.

MATHS SKILLS

One of the main concerns related to sustainability is the number of years left before certain finite resources are exhausted.

The remaining reserves are usually in very large amounts and they are usually expressed in millions (10^6), billions (10^9) and trillions (10^{12}).

In order to answer these questions you should be able to manipulate numbers by expressing them in standard form.

WORKIT!

In 2013, the known USA reserves of natural gas were 646 trillion m^3.

At the present rate, the consumption is 765 billion m^3 per annum.

Calculate how many years the USA has left before its natural gas is depleted. (3 marks)

The reserves $= 646 \times 10^{12} m^3 = 6.46 \times 10^{14} m^3$

The consumption per annum $= 765 \times 10^9 m^3 = 7.65 \times 10^{11} m^3$ (1)

Therefore the number of years left $=$ reserves/consumption per annum

$= 6.46 \times 10^{14} / 7.65 \times 10^{11}$ (1)

$= 844$ years   (1)

An oil pump in a field of rapeseed

✓ CHECKIT!

1 What is meant by the following terms:

  a finite resource

  b renewable resource

  c sustainable development?

2 Describe four characteristics of a sustainable process and for each one explain why it increases the sustainability of the process.

3 The USA has 17.7 billion metric tonnes of coal reserves remaining. The annual consumption of coal in the USA is 175 million metric tonnes per annum. How long will it be before the USA runs out of its own coal?

What could be done to extend this time?

Life cycle assessments (LCAs)

DOIT!

Draw a table that summarises the advantages and disadvantages to the environment of using paper and plastic bags.

A life cycle assessment is an analysis of the environmental impact of a product at each stage of its lifetime, from its production all the way to its disposal.

The stages in a life cycle assessment are as follows:

- The extraction/production of raw materials.
- The production process – making the product, the packaging and any labelling.
- How the product is used and how many times it is used.
- The end of the life of the product – how is it disposed of at the end of its lifetime? Is it recycled?

The diagram in the Snap It! box below shows a typical life cycle assessment.

The use of energy, resources and waste production can be calculated or measured reasonably accurately.

On the other hand, pollution effects are more difficult to measure or calculate.

LCAs can be used by companies to avoid unnecessary generation of waste and lead them to rethink their procedures.

SNAPIT!

What happens to product after its lifetime comes to an end
- Products that can be recycled easily
- Products that produce low toxic products when disposed of

Raw materials
- Either extract raw materials or
- Use recycled materials

Manufacturing
- Measures to prevent air and water pollution
- Energy conservation
- Use of recycled raw materials and components

Usage
- Products that use less power
- Products that use less water and chemicals

Distribution of product
- Simple cheap packaging
- Short delivery distances

WORKIT!

What is a more sustainable product – a paper bag or a plastic bag?

Paper bags come from trees - a renewable resource. (1) Plastic (polyethylene) bags are made from ethene which is produced during cracking of petrochemicals - a finite resource. (1)

During the production/manufacture stage, paper has a greater impact on the environment than plastic bags. Paper bag production consumes much more water, produces more acidic greenhouse gases and has a greater global warming effect. (1)

Paper bags are damaged by water and less easily reused. (1)

Paper bags are heavier and after use generate 5 times more solid waste than plastic bags. However, unlike plastic bags, paper-bag waste biodegrades easily and can be used to make compost rather than put in landfill. (1)

In conclusion, plastic bag manufacture has fewer unwanted effects than for paper bags. However, over the whole life cycle of the products, paper bags are more sustainable because they are made from sustainable raw materials and do not cause a waste/disposal problem. (1)

CHECK**IT!**

1 What is a life cycle assessment?

2 Name the four stages/parts of a life cycle assessment.

3 The table below shows the greenhouse emissions in grams of greenhouse gas per kilowatt hour of power produced for four ways of producing electricity. The measurements are made over the lifetime of the generating device.

Method of electricity generation	Greenhouse gas emissions (in g/kWh) over the lifetime of the device
Silicon photovoltaic cell	45
Coal power station	900
Natural gas power station	420
Nuclear power station	30

a Comment on, and explain the values for, greenhouse emissions of the coal and natural gas power stations.

b Comment on, and explain the values for, the photovoltaic cell and the nuclear power station.

4 The flowcharts below show the greenhouse emissions as a percentage of the total for stages in the lifetimes of a photovoltaic cell and a coal power station.

Solar photovoltaic

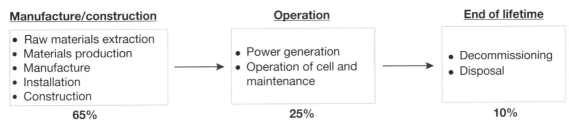

Manufacture/construction	**Operation**	**End of lifetime**
• Raw materials extraction • Materials production • Manufacture • Installation • Construction	• Power generation • Operation of cell and maintenance	• Decommissioning • Disposal
65%	**25%**	**10%**

Coal power station

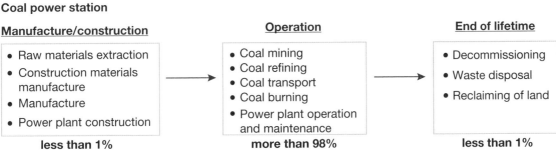

Manufacture/construction	**Operation**	**End of lifetime**
• Raw materials extraction • Construction materials manufacture • Manufacture • Power plant construction	• Coal mining • Coal refining • Coal transport • Coal burning • Power plant operation and maintenance	• Decommissioning • Waste disposal • Reclaiming of land
less than 1%	**more than 98%**	**less than 1%**

a Explain why the bulk of the emissions for the photovoltaic cell come at the beginning.

b For the coal power station only 1% of the greenhouse emissions come at the beginning of its lifetime and only 1% comes at the end of its lifetime. Explain these relatively small numbers.

Alternative methods of copper extraction

H

Copper is extracted from copper-rich ores by smelting and electrolysis.

In smelting, use is made of the fact that copper is less reactive than carbon so the ore is roasted with carbon. For example, malachite, an ore which is mostly copper(II) carbonate, is first decomposed to copper(II) oxide.

$$CuCO_3(s) \rightarrow CuO(s) + CO_2(g)$$

The copper(II) oxide is then reduced to copper by the carbon.

$$2CuO(s) + C(s) \rightarrow Cu(s) + CO_2(g)$$

If the ore contains mainly copper(II) sulfide (CuS) then reaction with oxygen produces impure copper and sulfur dioxide.

$$CuS(s) + O_2(g) \rightarrow Cu(s) + SO_2(g)$$

Some of the copper(II) sulfide produces copper(II) oxide which can then be reduced using carbon as shown above.

The other method is electrolysis. In this method sulfuric acid is added to give copper(II) sulfate solution. Electrolysis gives pure copper at the cathode.

If the copper produced by smelting is impure then this impure copper is made the anode in electrolysis. It dissolves to give copper(II) ions which are then deposited on the cathode as pure copper.

copper anode $\xrightarrow{\text{loses 2 electrons}}$ copper(II) ions in solution $\xrightarrow{\text{gains 2 electrons}}$ copper on cathode

- If the copper ores are low-grade ores (low in copper) then it is uneconomical to extract copper from them using the usual methods so alternatives are used. Bioleaching is a process where bacteria use copper sulfide as a source of energy and separate out the copper, which can be filtered off from the liquid produced. Bioleaching is slow but uses only about 30 to 40% of the energy used by copper smelting.

- Phytomining takes advantage of plants that take up copper from slag heaps that contain low-grade copper ores. The copper concentrates in the plant and is left in the ash when the plants are burned. Sulfuric acid is added to the ash to give copper(II) sulfate solution.

Phytomining is an environmentally friendly but slow method of extraction.

Copper can be extracted from this copper(II) sulfate solution by adding scrap iron. The iron displaces pure copper from the solution.

$$Fe(s) + CuSO_4(aq) \rightarrow Cu(s) + FeSO_4(aq)$$

DO IT!

Record an MP3 describing the extraction of copper by smelting, bioleaching and phytomining. Alternatively explain the processes to a revision partner.

NAIL IT!

H

The electrolysis of copper sulfate can use different electrodes. If inert electrodes (electrodes that do not take part) are used then copper ions from the solution are discharged at the cathode as copper.

If a copper anode is used then the copper at the anode dissolves to give copper(II) ions (Cu^{2+}). These copper ions gain electrons as they are discharged at the cathode to give pure copper. This means that copper can be purified by making impure copper the anode in electrolysis.

SNAPIT!

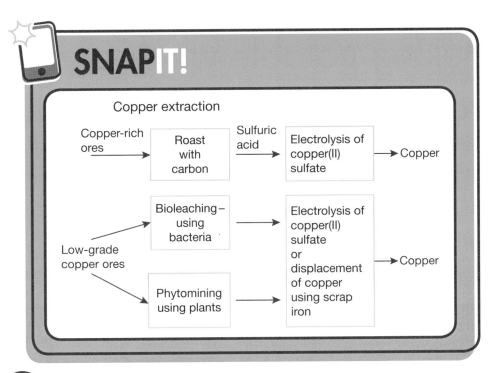

Copper extraction

Copper-rich ores → Roast with carbon → [Sulfuric acid] Electrolysis of copper(II) sulfate → Copper

Low-grade copper ores → Bioleaching – using bacteria → Electrolysis of copper(II) sulfate or displacement of copper using scrap iron → Copper

Low-grade copper ores → Phytomining using plants → Electrolysis of copper(II) sulfate or displacement of copper using scrap iron → Copper

STRETCHIT!

The half equations for the electrolysis of copper using copper electrodes are as follows:

At anode $\quad Cu(s) \rightarrow Cu^{2+}(aq) + 2e^-$

At cathode $\quad Cu^{2+}(aq) + 2e^- \rightarrow Cu(s)$

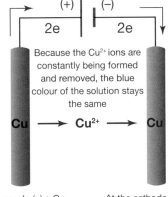

Because the Cu^{2+} ions are constantly being formed and removed, the blue colour of the solution stays the same

$Cu \longrightarrow Cu^{2+} \longrightarrow Cu$

At the anode (+) a Cu atom loses 2 electrons to form Cu^{2+} ion in solution

At the cathode (–) a Cu^{2+} ion gains 2 electrons to form Cu atom

CHECKIT!

H 1 a What is a low-grade copper ore?

b i A low-grade ore of copper contains 0.5% copper. What is the mass of copper in 1 tonne (10^6 g) of this copper ore?

ii How many moles of copper are present in this mass of copper? [A_r(Cu) = 63.5]

H 2 a Why is carbon used in copper smelting?

b Describe the two stages in the extraction of copper from copper(II) carbonate using carbon. Give the balanced symbol equations for each step.

H 3 a What organisms are used in bioleaching?

b Give the advantages and disadvantages of bioleaching.

H 4 How is impure copper purified using electrolysis?

Making potable water and waste water treatment

Potable water is water that is safe to drink.

This type of water does not have to be pure. It usually contains small concentrations of salts and no microbes. It can also have chemicals added. For example, fluoride can be added to reduce dental decay in children. In areas where water is plentiful, the impure water goes through a series of processes to make it potable (see Snap It! box below).

In arid countries, sea water is made potable by distillation or reverse osmosis. Israel, for example, gets very little rain and uses reverse osmosis to get about 70% of its water. These methods use a lot of energy.

NAIL IT!

This topic is closely aligned with that of subtopics Mixtures and compounds and Identifying ions in an ionic compound. You should revise these together as the purity of water is an important concept. You should know how to make sure that your water is pure or impure.

DO IT!

Write the different steps for water purification on same-sized cards or bits of card. Jumble or shuffle them and put them in the correct order. Compare with the diagrams in the Snap It! boxes.

Repeat for the treatment of waste.

SNAP IT!

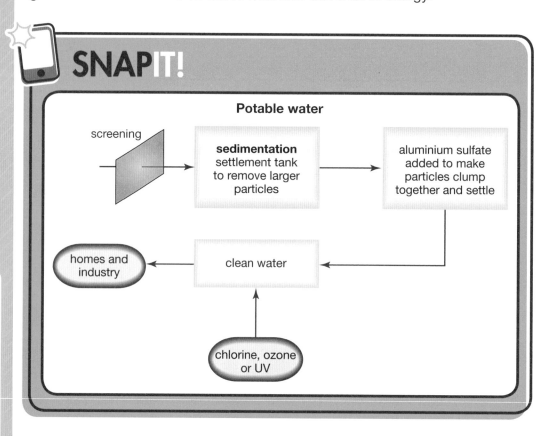

Large amounts of waste water are produced by industry and domestic consumers.

Before it is returned to the environment this water is treated using a number of processes including filtration/screening, sedimentation or settling.

Large pieces of grit and soil are separated by screening or filtration and the remaining liquid is passed into settling tanks.

After sedimentation the sludge obtained is treated using anaerobic digestion (carried out in the absence of oxygen). This produces methane which can be used as a fuel to run the sewage plant and a solid that can be used as a fertiliser or as a fuel. The liquid or effluent is then treated using aerobic digestion (in the presence of oxygen) and then returned to the environment.

SNAPIT!

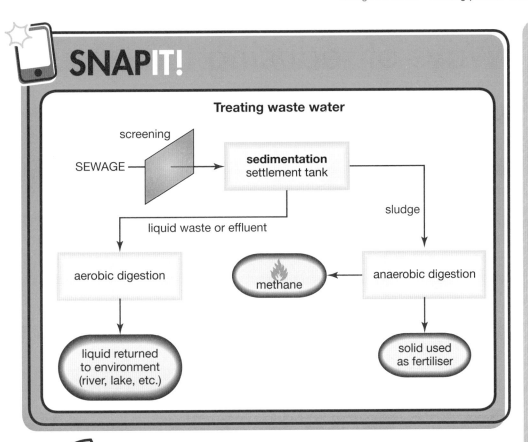

Treating waste water

CHECKIT!

1 **a** What is meant by potable water?

   b How could you show that potable water contained:

   i dissolved solids   **ii** chloride ions?

   c **i** Describe a chemical test for pure water.

   ii Describe a test for pure water based on a physical property.

2 Which chemical technique does screening resemble?

3 What do you understand by the following terms:

   a anaerobic   **b** aerobic   **c** sedimentation?

4 **a** Why would the production of methane in the anaerobic digestion of solid waste increase the sustainability of water treatment processes?

   b Give any other ways that the process is made more sustainable.

5 Some water can be extracted from underground rocks such as limestone. What stage in the purification process may be unnecessary if this is the source of water?

H 6 A sample of water was thought to contain **chloride** <u>and</u> **bromide** ions. The water would not pass for human consumption if bromide ions were present. Two students were given the task of showing that bromide ions were present.

   Student I suggested adding silver nitrate solution along with nitric acid.

   Student II suggested passing chlorine gas through the water.

   a Explain why the process suggested by student I would not give a clear-cut result.

   b Explain why the method suggested by student II would work. What observation would confirm the presence of bromide ions?

Ways of reducing the use of resources

If we do not reduce our use of different materials then the sources of these materials will run out.

Extraction of metals from their ores requires lots of energy and therefore will also further deplete the reserves of fossil fuels still available.

Extraction also leads to the creation of more waste and mining the metal ores has bad impacts on the environment.

- Glass bottles can be reused or the glass can be crushed and melted to be reformed. This saves energy and conserves resources that are used to make the glass.

- Aluminium extraction requires lots of electrical and heat energy and the ore of aluminium (bauxite) is running out. Bauxite mining and concentration has a bad impact on the environment. Recycling aluminium saves 95% of the energy used in extraction and produces 95% less greenhouse gases.

- Separating iron from other metals is relatively easy because it is magnetic. Recycling iron and steel saves a lot of energy (and because of this a lot of fossil fuels) and reduces the emission of greenhouse gases and pollutants. Scrap iron is also used to help in the production of steel.

- Plastics can be sorted and then recycled or incinerated. They can also be cracked to give hydrocarbon fuels and alkenes.

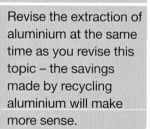

SNAPIT!

Reusing and recycling reduces

- use of fossil fuels
- greenhouse gas emissions
- mining of ores
- negative impact on environment.

CHECKIT! ✓

1 Give at least three advantages of recycling metals and glass.

2 a Why is so much aluminium recycled?

   b How could aluminium be separated from iron in a recycling plant?

3 Sometimes waste plastics can be cracked to give alkenes as one of the products. Why can this be thought of as a way of recycling?

Analysis and purification of a water sample

This is one of the required practicals and you could be questioned on it in the exam. The practical emphasises that you carry out a practical safely and accurately and safely use a range of equipment to purify and/or separate chemical mixtures including evaporation and distillation.

Practical Skills

Questions you should be able to answer:

- What apparatus can I use?
- How can you test the water sample for purity?
- What chemical test should be used?

NAILIT!

Questions you could ask yourself:

- How can you show that there is a dissolve solid in the solution you are given? What apparatus would you use?
- What tests would you use to identify any dissolved solids?
- How do the pieces of apparatus shown in the diagrams work to purify the water?
- How would you show that the purified water was indeed pure?

SNAPIT!

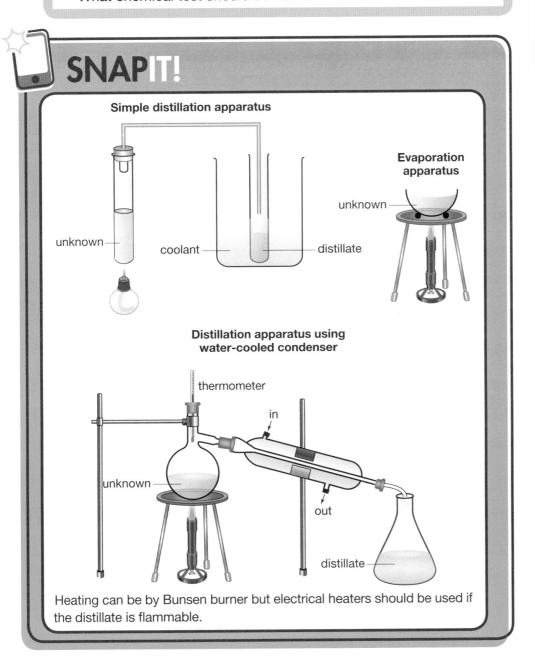

Heating can be by Bunsen burner but electrical heaters should be used if the distillate is flammable.

WORKIT!

A sample of water was evaporated to dryness and a residue was obtained which was white in colour.

When the residue was tested using the flame test no colour was obtained.

Samples of the water were added to separate test tubes and the following results were obtained:

1   On addition of sodium hydroxide the solution gave a white precipitate.

2   A separate sample gave no precipitate with silver nitrate solution but did give a dense white precipitate with barium chloride solution.

3   After using a simple distillation apparatus to purify the water, the distillate gave very faint white precipitates with both sodium hydroxide and separately with barium chloride solution.

4   When a condenser and flask were used to purify the water, the distillate gave no precipitates with the sodium hydroxide solution or barium chloride solution.

What do these results show? (6 marks)

The white residue showed that the water was impure and did contain a dissolved solid.   (1)

The white colour of the residue indicates that the original solution did not contain a transition metal, otherwise it would have been coloured.   (1)

The result with sodium hydroxide solution shows that the original solution contained magnesium ions.   (1)

The result with silver nitrate solution showed that no halide ions were present but the result with barium chloride solution did show the presence of sulfate ions.

These results show that the dissolved substance in the water was magnesium sulfate.   (1)

After using the simple distillation apparatus the fainter precipitates show that less of the dissolved solid was present but that the purification process was not completely successful.   (1)

After distillation using the condenser no precipitate with either sodium hydroxide or barium chloride solution showed there wasn't any dissolved substance in the water. This showed that purification was successful.   (1)

CHECKIT!

1   A sample of water turned universal indicator paper red. It gave no residue when it was evaporated to dryness. When tested with silver nitrate solution it gave a white precipitate.

    a i   Name the apparatus you could use instead of the UI paper.

       ii Explain the advantage of using this apparatus over the UI paper.

    b   Suggest the substance dissolved in the water.

2   A sample of water gave a white residue when it was evaporated to dryness. The water also gave a yellow flame in the flame test and when tested with silver nitrate solution gave a yellow precipitate. After distillation no coloured flame was given in the flame test and no precipitate was given in the silver nitrate test.

    What conclusions can you draw from these results?

or additional questions, visit:
www.scholastic.co.uk/gcse

1 a Explain the difference between a finite and a renewable resource.

b Ethanol can be obtained from sugar by fermentation and from the reaction between ethene and steam. Explain which of these two processes is more sustainable.

2 The diagram below shows the main steps in the treatment of water to give potable water.

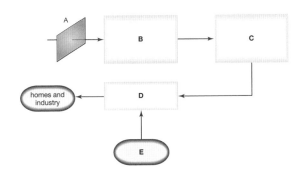

a i Describe what is happening at A.

ii Name what is added at C and explain why it is added.

iii Explain why the water at D is not fit to drink.

iv Describe the process taking place at E and explain why it is important.

b List two ways by which potable water is obtained from seawater.

c Potable water is not pure water.

i Outline how you could show that potable water contains dissolved impurities.

ii Describe a physical test which could be used to show pure water has been made.

d When waste water is treated the sludge formed after settlement is digested anaerobically.

i Define the term anaerobic.

ii List two useful products from anaerobic digestion.

H 3 a Explain what is meant by the following three terms when applied to the extraction of copper:

i smelting   ii phytomining

iii bioleaching.

b You are given a solution of copper(II) sulfate. Give two ways you would obtain pure copper from the solution.

4 a What is meant by a life cycle assessment?

b What are the four stages in the product's lifetime that are analysed for their impact on the environment?

Energy

Energy stores and systems

NAIL IT!

It is always better to say **dissipated energy** instead of wasted energy.

Energy is **not** an object. Never use the word 'heat' as a noun. For example, say 'The energy is dissipated to the surroundings by work done against friction', not 'Heat is produced by friction'.

A **system** is an object or a group of objects.

The **energy** in a system informs us whether changes in the system can or cannot happen.

No matter what changes happen in a system, the total amount of energy in the system always stays the same. However, the energy can be transferred around different parts of the system.

We can think of these different parts of a system as **energy stores** (for example, gravitational energy stores and thermal energy stores).

Energy cannot be created or destroyed – it can only be transferred to different stores within the system.

When you describe a change in a system:

- choose the start point and end point of the change
- identify the energy stores at those points
- consider which stores empty out and which stores fill up.

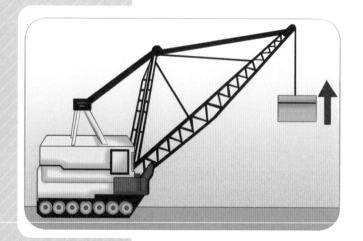

A crane lifts an object from the ground to its highest point. As it lifts the object, the chemical energy store in the fuel of the crane empties a little, and the gravitational energy store of the crane fills. Parts of the crane also heat themselves and their surroundings by work being done against **friction**, filling the thermal store and 'wasting' some energy (dissipating it to the surrounding air).

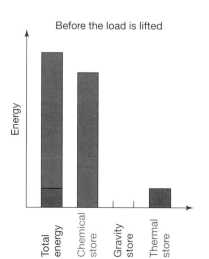

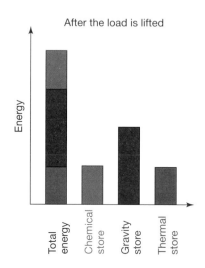

Changes in a system can happen by:

- heating
- **work done** by **forces**
- work done when an electric current flows.

SNAP IT!

Use your phone to film a ball being dropped.

DO IT!

Play your video back and pause it randomly. Starting from the drop and ending at the point you pressed pause, describe the changes in the system and how energy is transferred from one energy store to other stores.

CHECK IT!

1 You are asked to describe a change in a system as a series of energy transfers.

What are the three steps you need to do?

2 How is energy transferred when:

  a a lamp is switched on

  b an electric kettle is boiling water

  c a car is accelerated

  d chips are fried in a pan over the hob?

3 Think of two more examples of how energy is transferred.

Changes in energy stores

Kinetic energy

There is a kinetic store of **energy** associated with a moving object.

The greater the **mass** and the faster it is moving, the greater the energy in the kinetic store.

We can calculate the kinetic energy of a moving object using the equation:

$$\textbf{kinetic energy} = \textbf{0.5 mass} \times \textbf{(speed)}^2$$

$$E_k = \frac{1}{2} m v^2$$

- E_k = kinetic energy (unit: **joule**, J)
- m = mass (unit: **kilogram**, kg)
- v = **speed** (unit: **metre** per second, m/s)

WORKIT!

A car and a lorry are both travelling on the motorway at 70 mph (31.3 m/s). The car has a mass of 2650 kg, and the lorry has a mass of 5230 kg. Calculate the kinetic energy of the car and the kinetic energy of the lorry. (3 marks)

$E_k = \frac{1}{2} m v^2$

$E_k \text{ (car)} = 0.5 \times 2650 \times 31.3^2 = 1298089 \text{ J (1)}$

$E_k \text{ (lorry)} = 0.5 \times 5230 \times 31.3^2 = 2561889 \text{ J (1)}$

When they travel at the same speed, the kinetic energy store is 'fuller' for the lorry than for the car. This is because the lorry has more mass than the car. (1)

Gravitational potential energy

There is a gravitational potential store of energy associated with the height of an object.

The **higher** the object and the **greater** its **mass**, the greater the **energy** in the **gravity store**.

We can calculate the gravitational potential energy of a raised object using the equation:

$$\textbf{gravitational potential energy} = \textbf{mass} \times \textbf{gravitational field strength} \times \textbf{height}$$

$$E_p = m g h$$

- E_p = gravitational potential energy (unit: joule, J)
- m = mass (unit: kilogram, kg)
- g = **gravitational field strength** (unit: **newton** per kilogram, N/kg)
- h = height (unit: metre, m)

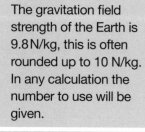

WORKIT!

A cricket ball is batted vertically upwards with an initial speed of 28.5 m/s. The mass of the ball is 155.9 g. What height will the ball reach? Assume the gravitational field strength is 10 N/kg. (3 marks)

$E_k = \frac{1}{2} m v^2$

$E_k = 0.5 \times 0.1559 \times 28.5 = 63.3 \text{ J}$ (1)

At its highest point the kinetic store is empty as the ball stops just before falling back down again, so the gravity store is full with the same amount of energy when the ball was batted. (1)

$(E_k = E_p = 63.3 \text{ J})$

$E_p = m g h$

$h = \frac{E_p}{mg} = \frac{63.3mg}{0.1559} \times 10 = 40.6 \text{ m}$ (1)

> Don't forget to convert the mass into kg first.

> Find the kinetic energy the ball started with.

> Remember the question states that g = 10 N/kg

Elastic potential energy

There is an elastic potential store of energy associated with an object that can be stretched, twisted, or compressed.

The greater the **extension** of a spring, the greater the **energy** in the **elastic store**.

We can calculate the elastic potential energy of a stretched or compressed spring using the equation:

elastic potential energy = 0.5 × spring constant × (extension)²

$$E_e = \frac{1}{2} k e^2$$

- E_e = elastic potential energy (unit: joule, J)
- k = **spring constant** (unit: newton per metre, N/m)
- e = extension (unit: metre, m)

> This equation only works if a spring is not stretched too much. If it is overstretched, it will pass its **limit of proportionality**, and this equation will not work.

DO IT!

Start making a list of the equations you will need to remember for physics.

NAILIT!

Only the elastic potential energy equation will be given to you in the exam. You will need to remember both the kinetic energy equation and the gravitational potential energy equation.

DO**IT!**

Write a short paragraph that describes the similarities and differences between the E_k and E_e equations.

WORK**IT!**

The spring of a fishing scale extends by 8 cm when a fish is hung on it. What is the elastic potential energy associated to the spring, if the spring constant is 28.9 N/m? (2 marks)

Remember to convert the units.

$e = 8 \text{ cm} = 0.080 \text{ m}$

$E_e = \frac{1}{2} k e^2$ (1)

$\quad = 0.5 \times 28.9 \times 0.080$

$\quad = 0.09 \text{ J}$ (1)

CHECK**IT!**

1 An eagle of mass 5.2 kg dives to catch a rat on the ground from a height of 58 m.

What was the gravitational potential energy associated to the eagle before diving?

2 Mo Farah can run 10 000 m in about 26 minutes and 47 seconds. His mass is 58 kg.

Calculate the average energy in the kinetic store when Mo Farah is running a 10 km race.

3 Cyclist A rides the same bike as cyclist B and they both have the same mass. Cyclist B rides at one third of the speed of cyclist A. How much greater is E_k for cyclist A compared to cyclist B? Explain how you calculated your answer.

Energy changes in systems: specific heat capacity

Specific heat capacity is the **energy** needed to increase the temperature of 1 kg of a substance by 1°C.

The **higher** the specific heat capacity of a substance, the **more** energy is needed to increase its temperature.

Materials with high specific heat capacity can store large amounts of energy in their associated **thermal** store when they are hot. This means these materials can give out more energy when they cool down.

Water has a very large specific heat capacity, therefore it is a good substance to use in central heating systems.

NAILIT!

Make sure you know how to write and read numbers in **standard form**. For example, 1 150 000 in standard form is 1.15×10^6.

Water is often used in central heating systems.

We can calculate the thermal energy put into a **system**, or the thermal energy given out from a system, when its temperature changes using the equation:

change in thermal energy = mass × specific heat capacity

× temperature changes

$$\Delta E = m \, c \, \Delta\theta$$

- ΔE = change in thermal energy (unit: **joule**, J)

- m = **mass** (unit: **kilogram**, kg)

- c = specific heat capacity (unit: joule per kilogram per degree Celsius, J/kg °C)

- $\Delta\theta$ = temperature change (unit: degree Celsius, °C)

NAILIT!

This equation will be given to you. Make sure you know how to use it.

Practical Skills

You will have investigated in class how to find out the specific heat capacity of a substance. Think about the apparatus and method you used. Depending on the substance tested, it may have looked like the diagrams below.

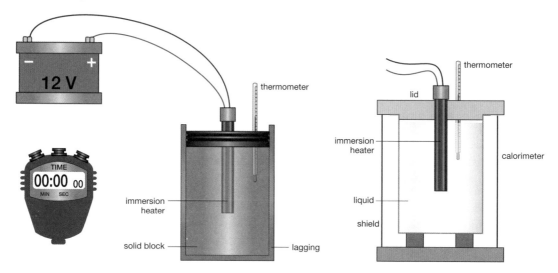

- How did you link the decrease in one energy store (or work done) with the increase in temperature?

- How did you link this with an increase of energy in the thermal store?

With the data you collected, you can find out the specific heat capacity of the substances you tested.

To calculate the specific heat capacity, rearrange the equation $\Delta E = m\,c\,\Delta\theta$:

$$c = \frac{\Delta E}{m \times \Delta\theta}$$

Substance	Energy supplied (J)	Mass (kg)	Initial temperature (°C)	Final temperature (°C)	Temperature change $\Delta\theta$ (°C)	Specific heat capacity (J/kg °C)
Aluminium	18 000	0.250	23	102	79	911
Water	18 000	0.385	22	33	11	4250
Steel	18 000	1.000	24	61	37	

- Use the rearranged equation and the data in the table to calculate the specific heat capacity of steel.

- Suggest three ways to improve the accuracy of your results.

MATHS SKILLS

We could use an energy meter to measure the energy supplied by an immersion heater. But we can find out this energy from the power rating of the heater.

$power = \dfrac{energy}{time}$

$energy = power \times time$ ◄ Rearrange the equation so energy is the subject.

$60\,W \times 300\,s = 18\,000\,J$

So each calorimeter is supplied with 18 000 J of energy in 5 minutes (300 seconds).

WORKIT!

A fish and chip shop needs to warm up 5.00 kg of oil from 22°C to 160°C to fry some cod fish. The specific heat capacity of the oil is 1670 J/kg°C.

How much energy needs to be transferred from the chemical store in the gas hob providing the flame to the thermal store associated with the oil to make this change in temperature? (2 marks)

$\Delta E = m\,c\;\Delta\theta$ (1)

$= 5.00 \times 1670 \times (160 - 22)$

$= 5.00 \times 1670 \times 138 = 1152\,300$ J ◄— Remember to put units in your answer.

$= 1.15 \times 10^{6}$ J (1)

CHECK**IT!**

1 300 g of water is brought to boiling temperature. The water is then left to cool to room temperature (25°C). The specific heat capacity of water is 4200 J/kg°C. How much energy is released by the thermal energy store associated with the water as it cools?

2 The specific heat capacity of water is 4200 J/kg °C. The chemical energy store from the fuel in the diagram will decrease by 4200 J as the temperature of 1 kg of water increases by 1°C. What will be the increase in the thermal energy store of the water?

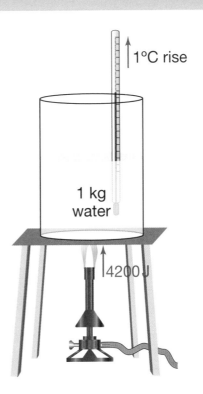

3 Material A has twice the mass of material B, but half the specific heat capacity of material B. Choose the correct statement.

   a Material A will need more energy to raise its temperature by 1°C.

   b Material B will need more energy to raise its temperature by 1°C.

   c Both materials will need the same amount of energy to raise their temperature by 1°C.

Power

Power is the amount of **energy** transferred or **work done** by, or to, a **system** every second.

One **joule** of energy transferred in one second = one **watt** of power.

We can calculate the power in a system using the equation:

- P = power (unit: watt, W)
- E = energy (unit: joule, J)
- t = time (unit: second, s)

$$power = \frac{energy\ transferred}{time}$$

$$P = \frac{E}{t}$$

$$power = \frac{work\ done}{time}$$

$$P = \frac{W}{t}$$

- P = power (unit: watt, W)
- W = work done (unit: joule, J)
- t = time (unit: second, s)

NAILIT!

Have you noticed that both energy and work done are measured in joules (J)? In physics, work done is just another way of transferring energy between the energy stores of a system. They both cause changes in the system. You can find more about work done in the Forces chapter.

MATHS SKILLS

You will need to remember and apply the equations on this page.

WORKIT!

Jasmine was late for school yesterday, so she ran upstairs to her classroom which was 15 m up from the ground floor. It took her 5 seconds. Today, Jasmine is on time, so she walks to her classroom and takes 20 seconds to climb the stairs. What was Jasmine's power output on both days, if her mass is 52 kg? Assume the gravitational field strength is 10 N/kg. (3 marks)

Use the equation for gravitational potential energy to work out the energy as

$E_p = mgh = 52\ kg \times 10\ N/kg \times 15\ m$

$= 7800\ J$ (1)

So, her power is:

When Jasmine goes upstairs the same amount of energy is transferred from the chemical store associated to her body to the gravity store when she's on top of the stairs.

g = 10 N/kg

$P\ (running) = \frac{E_p}{t} = \frac{7800}{5} = 1560\ W$ (1)

$P\ (walking) = \frac{E_p}{t} = \frac{7800}{20} = 390\ W$ (1)

CHECKIT!

1 You need to boil a glass of water using a microwave oven. Which power rating will be the quickest? Explain your answer.

   a 650 W   b 850 W

2 How much energy is transferred by an 11 W light bulb every second when it is turned on?

3 An 1100 W electric drill takes 6.3 seconds to drill a hole in a wall. How much energy has been transferred to the kinetic energy store and the thermal energy store in that time?

Energy transfers in a system

Conservation of energy and dissipated energy

The principle of **conservation of energy** says that **energy** is always conserved. Energy can be transferred usefully, stored or **dissipated**, but it cannot be created or destroyed.

In other words, the total energy in a **system** is the same before and after a change happens in the system. The energy just gets redistributed to different energy stores in the system. So, there is *no net change* in the total energy of the system.

Every time a change happens in a system, some energy is dissipated. This dissipated energy gets stored in less useful ways. For example, when water evaporates from a hot bath it increases the amount of energy in the **thermal** store associated with the bathroom.

WORKIT!

Explain why a ball never bounces higher than the height it is dropped from. (3 marks)

When the ball hits the floor part of the kinetic store (where energy is stored usefully) empties and the thermal store associated with the ball and surroundings increases. (1) The total energy in the system is still the same, but the energy in the thermal store is too spread out (dissipated) to be transferred back to the kinetic

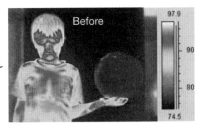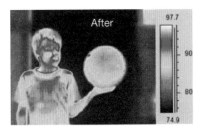

A thermal infrared image of a ball before (left) and after (right) being bounced

store (1) – this energy is 'wasted' and the ball cannot bounce back to where it was dropped from. (1)

Reducing unwanted energy transfers

It is important to reduce unwanted energy transfers in a system so that less energy is 'wasted'.

Car engines and other machines reduce the 'wasted' energy transferred by heating by using lubricants like engine oil between mechanical parts. This reduces the **friction** between moving parts of the car so less work is done against friction.

Thermal insulation is used in houses to reduce 'wasted' energy transferred by heating from the house to its surroundings.

Different materials have different thermal conductivity. The higher the thermal conductivity of a material, the quicker energy can be transferred by **conduction** through that material.

DOIT!

Take a metal tray and a plastic tray that are about the same size. Feel both trays with your hands, then place an ice cube on each tray. Which ice cube melts first? Is this what you were expecting? Can you explain what happened in terms of thermal conductivity?

The rate a house cools depends on the thickness of its walls and on the thermal conductivity of the walls. The thinner the bricks, the quicker the energy is conducted through them from the inside of the house to the outside. Cavity wall insulation is sometimes used between the inner and outer walls of houses. It reduces the amount of energy conducted through the walls because it has low thermal conductivity.

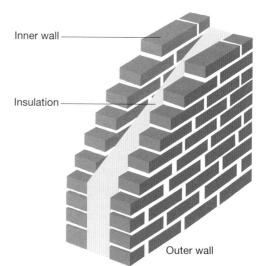

Inner wall

Insulation

Outer wall

Practical Skills

You will have investigated in class the effectiveness of different materials as thermal **insulators**.

You might have used hot water and a copper container like the one in the diagram. When the copper container is filled with water, it can be wrapped with different materials to test the temperature of the water at the start and after a set time.

Think about how you carried out your investigation.

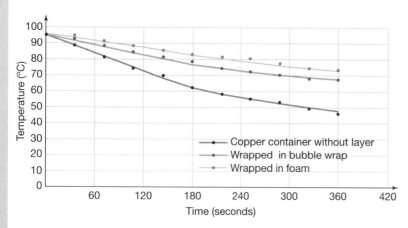

Copper container without layer
Wrapped in bubble wrap
Wrapped in foam

You might have used layers of bubble wrap, foam, and then left the copper container without any wrapping.

• Why is it important to use the same amount of water each time you test a different material?

• Which material did you conclude was the best thermal insulator? Why?

• What does this tell you about the thermal conductivity of each material?

• What do you think makes bubble wrap and foam good thermal insulators? What do you think makes copper a poor thermal insulator?

CHECK IT!

1 Describe three features Olympic cyclists use to reduce unwanted energy transfers.

2 How is energy dissipated in a toaster?

3 Describe two advantages of reducing unwanted energy transfers from a car engine.

Efficiency

Devices and machines are designed to do specific jobs. A specific amount of **energy** needs to be transferred into a device so that it can do its job. This is called the total input energy.

But only some of the total input energy is transferred into the device in a useful way that lets it do its job.

The rest of the total input energy is **dissipated**. It gets **stored** in less useful ways. This is because all changes to a **system** result in some energy being dissipated.

The energy put in (total input energy transfer) compared with the useful energy given out (useful energy output transfer) is called the efficiency of the device or system.

$$\text{efficiency} = \frac{\text{useful output energy transfer}}{\text{total input energy transfer}}$$

We can also calculate efficiency using the useful **power** output and the total power input:

$$\text{efficiency} = \frac{\text{useful power output}}{\text{total power input}}$$

NAILIT!

Efficiency can only be a number between 0 and 1. So if you get something bigger than 1, check your equation and that you have put the right numbers into it.

NAILIT!

Make sure you can remember and use these equations.

WORKIT!

In a 100 W light bulb, 93 W of its power goes to increase the thermal energy store of the room. In a 11 W LED, only 1 W of its power goes to increase the thermal energy store of the room. Calculate the efficiency of both lights, and explain which light is more efficient. (4 marks)

The useful power output of the 100 W light bulb is only
100 W − 93 W = 7 W (1)
But the useful power output of the LED is 11 W − 1 W = 10 W.

> Raising the temperature of the room is not the job a light bulb is designed to do. So increasing the thermal energy store is an unwanted energy transfer.

So, the efficiency of each light is:

$$\text{Efficiency (100 W light bulb)} = \frac{\text{useful power output}}{\text{total power output}} = \frac{7}{100} = 0.07$$

$$= 0.07 \times 100$$

$$= 7\% \text{ efficient (1)}$$

$$\text{Efficiency (11 W LED)} = \frac{\text{useful power output}}{\text{total power output}} = \frac{10}{11} = 0.91$$

$$= 0.91 \times 100$$

$$= 91\% \text{ efficient (1)}$$

The LED is much more efficient than the 100 W light bulb because most of the total input energy is transferred in a useful way. (1)

MATHS SKILLS

Efficiency can also be written as a percentage. So if a device has an efficiency of 0.23, you can also say that it is 23% efficient. To get the percentage, just multiply the efficiency by 100.

Devices and systems can be made more efficient. In Higher Tier exams you could be asked to describe ways to increase the efficiency of an intended energy transfer. So:

- Think about what the 'job' of the system is and identify the associated energy transfer, i.e. the intended energy transfer.

- Identify the unwanted energy transfers and suggest ways to reduce them.

DO IT!

Take photos of different devices in your house and make a short narrated slideshow or video where you describe how unwanted energy transfers are, or could be, reduced.

Keeping track of how efficient appliances are, can save money

CHECK IT!

1 Why is only some of the total input energy transferred to a system in useful ways?

2 Look at the equation for efficiency. Why can the efficiency of a device never be more than 1?

3 Your body is about 25% efficient at transferring energy from the chemical store of the food you eat by mechanical work when you exercise. Why is this low efficiency an advantage if you are trying to lose weight?

National and global energy resources

Earth has many **energy** resources. Different **energy stores** are associated to energy resources and this energy can be shifted (or transferred) where it is needed.

Renewable energy resources can be replaced (replenished) as they are used. Examples of renewable energy resources are biofuel, wind, hydroelectricity, geothermal, the tides, the Sun and water waves.

Non-renewable energy resources cannot be replenished. If we keep using them, they will eventually run out. Examples of non-renewable energy resources are fossil fuels and nuclear fuels.

NAILIT!

Do not confuse 'replenished' with 'recycled'. **Replenished** means that when a resource is used, more of it is replacing it. **Recycled** means that one object or resource is **reused** in a useful way again.

Energy resource	Energy store	Renewable or non-renewable	Main uses	Reliability	Environmental impact
Fossil fuels (coal, oil and gas)	Chemical energy store	Non-renewable Estimated to run out by 2080	Transport Heating Generating electricity	Very reliable and high power output	Releases carbon dioxide, which is partly responsible for global warming
Nuclear fuel (mainly uranium)	Nuclear energy store	Non-renewable Estimated to be available for more than 200 years	Generating electricity	Very reliable and very high power output	Makes dangerous radioactive waste that needs to be buried underground for thousands of years
Biofuel	Chemical energy store	Renewable	Transport Heating Generating electricity	Very reliable and medium power output	Low negative impact because it is 'carbon neutral' (it releases the same amount of carbon dioxide as it absorbed when the organisms the fuel is made of were alive)
Wind	Kinetic energy store	Renewable	Generating electricity	Unreliable because it depends on the weather, and very low power output	Windmills take up big areas that could be used for farming Birds can be killed by windmill blades

Energy resource	Energy store	Renewable or non-renewable	Main uses	Reliability	Environmental impact
Hydroelectricity	Gravitational potential energy store	Renewable	Generating electricity	Very reliable and medium power output	Big areas need to be flooded to build dams, which can have a negative effect on the local ecosystem, landscape and people
Geothermal	Thermal energy store	Renewable	Generating electricity Heating	Reliable, but only available in some areas In some geothermal sites, the thermal energy store might empty out completely Medium power output	Very low negative impact
Tides	Kinetic energy store	Renewable	Generating electricity	Very reliable and potentially very high power output, but difficult to harness	Tidal barrages can negatively affect birds that feed on mud flats that are exposed when the tide goes out, and block sewage and other waste from being carried out to sea
Sun	Nuclear energy store	Renewable	Generating electricity Heating	Very unreliable because the weather, which dictates light levels, can be highly changeable. It is only available during daylight hours Low power output	Low negative impact, but big areas are needed to harness it
Water waves	Kinetic energy store	Renewable	Generating electricity	Unreliable because it depends on the weather Low power output	Very low negative impact

SNAP IT!

Take photos of the table of energy resources on the previous pages and use them to revise wherever you are.

STRETCH IT!

Fossil fuels, as well as being non-renewable, release harmful gases into our atmosphere and the environment, like carbon dioxide and sulfur dioxide. Fossil fuels still make up about 87% of the world's energy use.

How do you think these percentages have changed in the past 30 years? How could they change in the future?

Research the environmental, political, social, ethical and economic issues of using different energy resources.

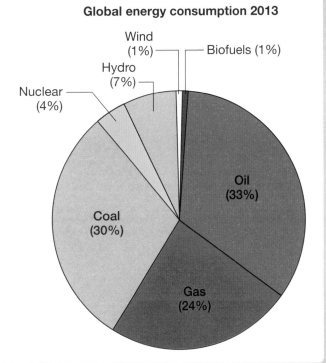

Global energy consumption 2013

Wind (1%)
Biofuels (1%)
Hydro (7%)
Nuclear (4%)
Oil (33%)
Coal (30%)
Gas (24%)

CHECK IT!

1 What is the difference between renewable and non-renewable energy resources?

2 What is one advantage of using fossil fuels to generate electricity?

3 Suggest one disadvantage of using the Sun to generate electricity in the UK.

Energy

1 Which of the following are renewable and non-renewable energy resources?

a Nuclear fuels   f The Sun

b Water waves   g Geothermal

c Coal   h Oil

d Biofuel   i The tides

e Hydroelectricity   j Natural gas

2 Rearrange the following equations to find m, e and h respectively:

a *kinetic energy = 0.5 × mass × (speed)²*

$$E_k = \frac{1}{2} mv^2$$

b *elastic potential energy = 0.5 × spring constant × (extension)²*

$$E_e = \frac{1}{2} k e^2$$

c *gravitational potential energy = mass × gravitational field strength × height*

$$E_p = m g h$$

3 In the equation *change in thermal energy = mass × specific heat capacity × temperature change* [$\Delta E = m c \Delta\theta$] the temperature is normally expressed in degrees Celsius (°C). Explain why expressing $\Delta\theta$ in degrees kelvin (K) would not affect the results of your calculations.

4 a Describe the difference between lift A which lifts a 70 kg load up three floors in 20 seconds and lift B which can lift the same load up three floors in 15 seconds.

b Assuming the lifts are operated in the same building and that the power of lift B is 343 W, what is the height of each floor in this building? ($g = 9.8$ m/s²)

5 Two ice cubes are placed on two blocks of equal size, but different materials. One block is made of aluminium and the other block is made of plastic. Which ice cube will melt first? Explain why this happens.

6 A battery powered toy car has a total power input 1.5 W. The mass of the car is 520 g and reaches a top speed of 2.2 m/s. Calculate the percentage efficiency of the toy car.

Standard circuit diagram symbols

A **switch** can break or complete an **electric circuit**. A switch allows a device to be used only when we need to, by switching it on or off.

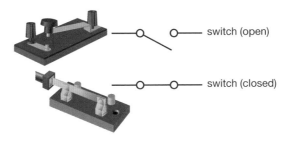

switch (open)

switch (closed)

+ ⚡ —|⊢— cell

+ ⊣|----|⊢ − battery

+ ⚡ + ⚡ + ⚡ + ⚡

Cells and **batteries** are power supplies for a circuit. They provide the **potential difference** (p.d.), which drives the **charges** around the circuit. A battery is a collection of cells connected in **series** with each other.

A **lamp** has a very thin filament (wire), usually made of tungsten. This gets very hot when a **current** flows through it, so it glows and emits **visible light**.

lamp

A **fuse** is a safety component. It is made of a wire that has lower melting point than other components. This means that if too much current is flowing through the circuit, the wire in the fuse will melt, breaking the circuit and stopping the high current from flowing and damaging other components or devices on the circuit.

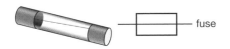

fuse

Voltmeters measure the potential difference (p.d.) across components in a circuit. They always need to be connected in series with other components. Voltmeters need to have a very big **resistance** so that hardly any current flows through them.

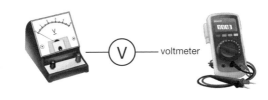

(V) voltmeter

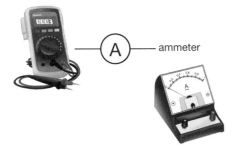

(A) ammeter

Ammeters measure the current through a circuit. They always need to be connected in series with other components. The resistance of an ammeter needs to be as close to zero as possible so that the current reading is not affected by its presence in the circuit.

A **diode** lets current flow through it in one direction only. It is useful when we want to direct the current in some parts of the circuit but not in others. The diode points from +ve to −ve.

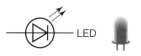

diode

 LED

A **light-emitting diode** (LED) lets current flow through it in one direction only, just like a diode, but when current flows through an LED, it emits light.

Fixed **resistors** and variable resistors oppose (resist) the flow of current with a particular resistance. This means they can be used to set the current in a circuit to the value that we want. Fixed resistors have a fixed value of resistance. However, variable resistors can be set to different values of resistance within a particular range.

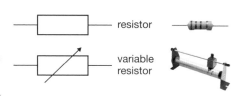

resistor

variable resistor

A **thermistor** is a resistor that changes its resistance when its temperature changes. Generally, the resistance of a thermistor decreases when its temperature increases.

 thermistor

The resistance of a **light-dependent resistor** (LDR) changes when the amount of light shone on it changes. Normally, LDRs have very high resistance; when light is shone on them, their resistance drops a lot.

 LDR

SNAP IT!

Make flash cards showing the standard circuit diagram symbols to test your knowledge of each component. Take pictures on your phone so you can revise on the go.

CHECK IT!

1 What is the main difference between a voltmeter and an ammeter?

2 Explain the main difference and similarity between LEDs and diodes.

3 What components would you use in a circuit that needs to switch on when it gets dark, but only when the current flows through it in one particular direction?

Electrical charge and current

An electric **current** is a rate of the flow of **charge**. Whenever a charge is flowing, there is a current.

So an electric current is a measure of the rate of flow of electrical charge over time.

We can calculate the electrical charge flowing through a point of a **circuit** in a given time using the equation:

charge flow = current × time $Q = I\,t$

- Q = charge flow (unit: **coulomb**, C)
- I = current (unit: **ampere**, or amp, A)
- t = time (unit: second, s)

For electrical charge to flow through a circuit, the circuit must be closed and at least one of its components must be a source of **potential difference** (p.d.).

In a single closed loop of a circuit, the current is the same at any point.

The free **electrons** in a circuit fill all wires and components of the circuit all the time, even when the circuit is broken (or open). So, when a potential difference is applied to a circuit, the charges (electrons) will all start flowing along the circuit at the same time, wherever they are in the circuit. That is why a light bulb lights up almost straight away after the switch is turned on.

The free electrons move inside the circuit at a speed of about 1 cm per minute, so if you had to wait for them to go from the **battery** to the light bulb every time you turned the switch on you would be in darkness for quite a long time.

MATHS SKILLS

You need to remember and be able to apply this equation.

SNAPIT!

Draw a flow chart to show what happens when the switch in a simple circuit containing a cell, a light bulb, a switch and wires in series is turned on. Take a photo of your flow chart to use as a revision tool.

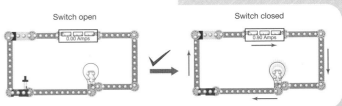

Switch open   0.00 Amps   Switch closed   0.90 Amps

WORKIT!

A charge of 0.24 C flows through a **resistor** in a circuit over a time of 14 seconds. What is the current in the circuit? Write your answer in milliamps. (2 marks)

To find out the current (I), we need to rearrange the equation $Q = I\,t$ to $I = \frac{Q}{t}$.

So, $I = \frac{0.24}{14} = 0.017$ A. (1)

$0.017 \times 1000 = 17$ mA (1) ◄────

Remember to convert your answer to milliamps
1A = 1000 mA

CHECKIT!

1 What is an electrical current?

2 Why does a light bulb light up almost straight away after a switch has been turned on?

3 How much charge flows through a light bulb in a circuit with 0.205 A current if the circuit is closed for 2 minutes and 33 seconds and then opened again?

Current, resistance and potential difference

The value of electrical current (*I*) in a circuit is determined by the value of **potential difference** (*V*) we choose for our power supply and by the value of the **resistance** (*R*) of the components we add to the circuit.

We can calculate p.d., resistance or electric current using the equation:

potential difference = current × resistance

$$V = IR$$

- *V* = potential difference, p.d. (unit: **volt**, V)
- *I* = current (unit: **ampere**, or amp, A)
- *R* = resistance (unit: **ohm**, Ω)

You cannot change electrical current (*I*) in a circuit directly, because its size depends on both the resistance (*R*) of the components and the potential difference (*V*) across those components.

NAIL IT!

You might have heard your teacher using the word 'voltage' to mean potential difference (p.d.), but only potential difference (or p.d.) will be used in questions.

MATHS SKILLS

You need to remember and be able to apply the equation above. The bigger the potential difference across the components of a circuit, the bigger the current. The bigger the resistance of the components, the smaller the current. We can get a better idea of this if we rearrange the equation to:

$$I = \frac{V}{R}$$

A good way to describe the relationship between *I*, *V* and *R* is to say that the current (*I*) is **directly proportional** to the potential difference (*V*) and **inversely proportional** to the resistance (*R*) at a constant temperature.

DO IT!

Find out about conductive ink and how this can be used in special pens and printers to make 'DIY' circuit boards on paper.

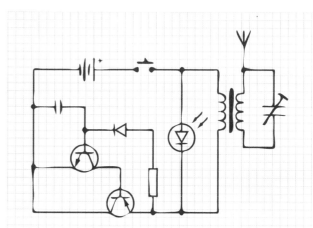

Practical Skills

You should have done practical work to investigate how different components affect the resistance of electrical circuits.

How the length of a wire changes the resistance in a circuit

The **longer** the wire, the **bigger** the resistance. This makes sense because every component has a certain resistance, so adding a longer wire in **series** with the other components increased the resistance.

In this investigation, the wire had to be at a constant temperature.

Why was this important?

It is important because, if you had changed the length of the wire and the temperature together, you could not be sure which of these two changes caused a change in resistance. So the test would not have been fair.

How different combinations of resistors in series and parallel affect the resistance in the circuit

- Adding more resistors in series **increased** the resistance in the circuit.

- Adding more resistors in **parallel** decreased the resistance in the circuit.

In both investigations you need to be able to read **circuit diagrams** and know how to set up a circuit.

Make sure you know which way around the **ammeter** and **voltmeter** go. The ammeter is always in series with the components, and the voltmeter is always in parallel with the components.

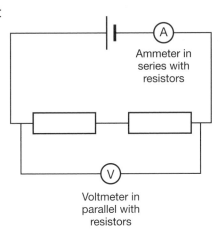

Ammeter in series with resistors

Voltmeter in parallel with resistors

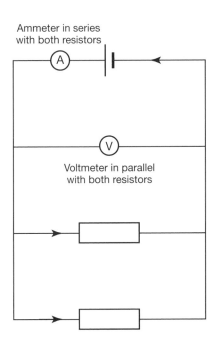

Ammeter in series with both resistors

Voltmeter in parallel with both resistors

CHECK**IT!**

1 What is the relationship between electric current, potential difference and resistance?

2 A current of 1.2 A flows through an appliance of resistance 1.3 kΩ. What is the p.d. across the appliance?

3 A torch uses two 1.5 V AA cells. When it is switched on, the current through the light bulb is 150 mA. Calculate the resistance of the light bulb in the torch.

Resistors

The electric **current** through an **ohmic conductor** (at constant temperature) is **directly proportional** to the **potential difference** across the ohmic conductor (**resistor**).

So the **resistance** stays constant as the current changes.

This means that a graph of current against potential difference (called an *I–V* graph) for an ohmic conductor is a straight line going through the origin (through zero), like the graph to below.

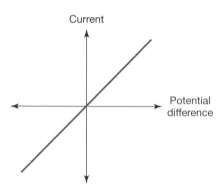

Current against potential difference for an ohmic conductor

MATHS SKILLS

You should know that the equation of a straight line is $y = mx + c$ and that $I = \dfrac{V}{R} = \dfrac{1}{R}V + 0$

So, if R is constant:

$y = I$

$m = \dfrac{1}{R} \rightarrow$ the gradient of your graph is $1/R$, and $R = \dfrac{1}{Gradient}$

$x = V$

$c = 0$ because the line goes through 0.

I–V characteristics of non-ohmic components

Non-ohmic components like **lamps**, **diodes**, **light-emitting diodes** (LEDs), **light-dependent resistors** (LDRs), and **thermistors** do not have constant resistance. Their resistance changes when the current through them changes.

Filament lamp

Filament bulbs contain a thin coil of wire, this is called the filament. When an electric current passes through the filament it heats up and emits light. When the temperature of the filament in a lamp increases due to the electric current passing through it, its resistance increases too. The graph to the right shows that the current is not directly proportional to the voltage.

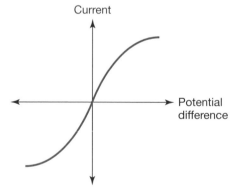

Current against potential difference for a filament lamp

STRETCH IT!

Draw the *I–V* graph for a filament lamp. Show in which parts of the graph the resistance is fairly constant, and where the temperature of the filament starts to increase.

Hint: where the resistance is constant, the graph should look like the *I–V* graph for an ohmic conductor, in other words, it should be linear.

Diode and LED

The *I–V* graph for a diode and an LED is very similar – see graph to the right. Diodes and LEDs both let current flow in only one direction. This means that the resistance of diodes and LEDs is very high in the reverse direction.

Thermistor

The resistance of a thermistor decreases when its temperature increases.

Thermistors are used in thermostats in your home to 'tell' your heating system when to turn on/off, or in your computer to turn on its fan when it is getting too hot.

LDR

The resistance of an LDR decreases when the amount (intensity) of **light** shone on it increases.

LDRs are used in street lamps to turn them on at night.

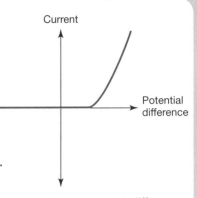

Current against potential difference for a diode or an LED

Practical Skills

You should have done practical work to investigate the *I–V* graphs of ohmic conductors (resistors), filament lamps, and diodes (the graphs on this page and on page 260).

Remember that to measure the resistance of an electrical component, you measure the current through it and the potential difference across it.

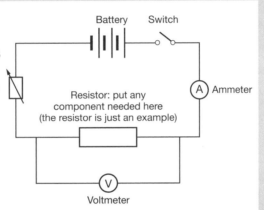

You need to be able to draw and use this **circuit diagram** when investigating how the resistance of a component changes.

You also need to be able to use *I–V* graphs to describe the properties and functions of a circuit component and to say whether each property is **linear** or **non-linear**.

Just remember that the **ohmic** component of an *I–V* graph is **linear**, and the others that are not a straight line are non-linear.

But remember that some parts of *I–V* graphs can be both linear and non-linear. This means that a component can be linear over a particular range of potential difference, but become non-linear for another range, like the filament lamp that is nearly linear for some values of potential difference.

DO IT!

Stretch your legs perpendicularly to your body and imagine your body to be the *y*-axis and the line formed by your legs as the *x*-axis. Now use your hands to make the shape of the graphs for each of the components in this topic.

CHECK IT!

1 Describe the relationship between current through and potential difference across an ohmic conductor.

2 Look at the *I–V* graph of a diode and describe how its resistance changes as the potential difference goes from negative values to positive and increasing values.

3 Describe the difference between a linear and a non-linear relationship between two variables.

Series and parallel circuits

Electrical **circuits** can be wired in **series** or **parallel**. Circuits might have some electrical components in series with each other *and* some in parallel with others, which can be very useful.

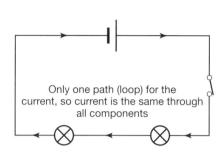

Only one path (loop) for the current, so current is the same through all components

An example of a series circuit

For components in series with each other:

- All components are in the same loop.

- The **current** is the same through each component.

- The total **potential difference** from the power supply is shared between the components.

- The total **resistance** of all components is the sum of the resistance from each component $R_{total} = R_1 + R_2$.

- The current through all components can be changed by adding or removing components (for example, adding a resistor in series with a component would decrease the current through that component).

WORKIT!

The diagram shows two 40 Ω light bulbs in series with a 1.5 V cell and two ammeters. What value of current will ammeter A_1 and ammeter A_2 measure? (2 marks)

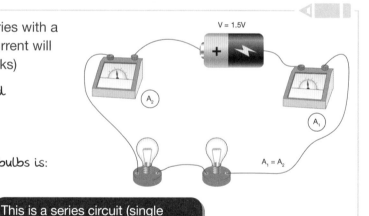

The light bulbs are in series, so their combined resistance (total resistance) is

R_{total} = 40 + 40 = 80 Ω (1)

The current through both ammeters and light bulbs is:

$$I = \frac{V}{R}$$

$$I = \frac{1.5}{80} = 0.019\,A \ (1)$$

This is a series circuit (single loop), so the current is the same in all parts of the circuit.

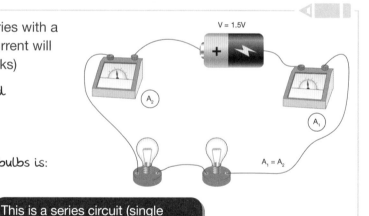

An example of a parallel circuit

components can be operated separately

For components in parallel with each other:

- Each component is individually connected to the power supply in its own loop – this means that the potential difference is the same across each component.

- The total current through the circuit is the sum of the currents through each branch/loop of the circuit.

- The total resistance of two components in parallel is less than the smallest resistance from individual components.

- Switching individual loops (circuits) in parallel on/off will not affect other loops (houses are wired with parallel circuits so that we can turn different appliances on/off separately from each other).

NAILIT!

At GCSE, you do not need to calculate the total resistance of two resistors joined in parallel. You just need to explain why adding more components in parallel will decrease the total resistance of the circuit.

The potential difference across each branch in a parallel circuit is the same. So each loop acts as an individual circuit with the power supply. The individual currents from each loop are $I = V/R$. These all sum up before the circuit splits in individual loops and where the loops join together again at the other terminal of the power supply. This means that the total current through the circuit increases, and because $V = I \times R$, the total resistance in the circuit must decrease.

DOIT!

Fill your lungs with air, and blow through two straws 'in series' (one straw stuck inside the other as in the picture).

Then, put the two straws side by side 'in parallel', and empty the air from your lungs through them.

Which setting emptied your lungs more quickly? How does this compare with electrical components in series and in parallel?

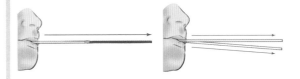

WORKIT!

In the circuit shown in the diagram, the current measured by the ammeter is 0.5 A. Resistor R_1 has a resistance of $15\,\Omega$. What is the current through R_1 and R_2, and what is the resistance of R_2? (3 marks)

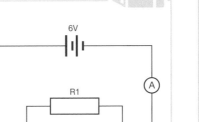

The potential difference is the same across both resistors (6 V), so we can find the current through R_1 using the relationship

$$I = \frac{V}{R} \rightarrow I_1 = \frac{V}{R_1}$$

$$I_A = \frac{6}{15} = 0.4A \quad (1)$$

$$I_{total} = I_1 + I_2 \rightarrow I_2 = I_{total} - I_1$$

The total current through the circuit is the sum of the currents through the individual resistors.

$$I_2 = 0.5 - 0.4 = 0.1\ A \quad (1)$$

Now we can calculate R_2 using $R = \frac{V}{I} \rightarrow R_2 = \frac{V}{I_2}$

$$R_2 = \frac{6}{0.1} = 60\,\Omega \quad (1)$$

STRETCHIT!

Christmas lights are wired in series with each other. Why could it be dangerous if they were wired in parallel?

CHECKIT!

1 Look at the series circuits in the Work It! example and identify what potential difference there is across each of the two light bulbs.

2 Explain why the potential difference across each component in a parallel circuit is the same.

3 A 9 V battery is connected in series with two resistors R_1 and R_2. The resistance of R_1 is twice as large as the resistance of R_2. What is the potential difference across R_1 and R_2 individually?

Mains electricity

dc and ac current

Most **circuits** you use in school are connected to a **direct current** (dc) **battery**/power supply. This means that the **potential difference** provided by the battery does not change in direction (sign) or in value.

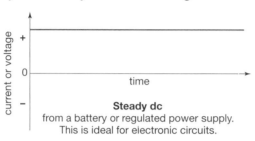

Steady dc
from a battery or regulated power supply.
This is ideal for electronic circuits.

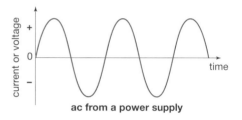

SNAP IT!

Write the difference between ac and dc current and take a photo of it to revise later.

To transfer **energy** efficiently across the country from power stations to homes, schools, and workplaces, an **alternating current** (ac) potential difference must be used. This is why the **mains electricity** in your home is ac. An ac potential difference continually alternates in direction (sign) and changes value.

The ac supply in the mains electricity in your home has a **frequency** of 50 Hz and a potential difference of 230 V.

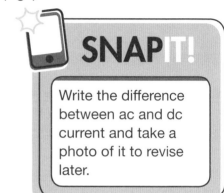

ac from a power supply

Mains electricity

Most appliances are connected to the mains electricity with a **three-core cable** plug. To make them easy to tell apart, the **insulation** around each of the three wires is colour coded.

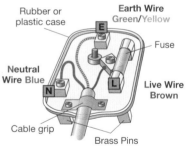

- **Live wire** – *brown*. This carries the ac potential difference from the power supply. The potential difference between the live wire and the earth wire is about 230 V.

- **Neutral wire** – *blue*. This completes the circuit. It is at, or close to, the same potential as the earth wire (0 V).

- **Earth wire** – *green and yellow stripes*. This is a safety wire. It is at 0 V potential. It carries **current** only if a **fault** happens in the circuit, to stop the appliance operating (stop it becoming 'live').

NAIL IT!

If a fault happens and the live wire becomes connected to the case around an appliance, the earth wire will let a big current flow through the live wire and the earth wire. This 'overflow' of current will heat the **fuse**, which melts away, breaking the circuit. This is why exposed metal parts of appliances should always be connected to the earth wire – to make sure the current does not flow through anyone using it, which is very dangerous.

Circuits in the mains electricity supply are wired in **parallel**. This means that the live wire of an appliance could still be connected (and dangerous) even if a **switch** in the mains circuit is open (off).

CHECK IT!

1. Explain why the live wire in the cable of a kettle could still be dangerous even if the electric kettle is turned off.

2. What is the difference between ac and dc?

3. How would the current through an appliance designed to work in the UK be affected if you used the same device in the USA, where the mains potential difference is 110 V?

Electrical power, energy transfers in appliances and the National Grid

Power

The **power** of an electrical device or appliance depends on the **potential difference** across the appliance and the current through it.

We can calculate the power transferred in an appliance using the equations:

$$\text{power} = \text{potential difference} \times \text{current}$$
$$P = VI$$

$$\text{power} = (\text{current})^2 \times \text{resistance})$$
$$P = I^2 R$$

- P = **power** (unit: **watt**, W)
- V = **potential difference** (unit: **volt**, V)
- I = **current** (unit: **ampere**, or **amp**, A)
- R = **resistance** (unit: **ohm**, Ω)

The power of an appliance is the rate of **energy** (unit: **joule**, J) transferred by the appliance. In other words, how much energy is transferred each second.

So, a 40 W light bulb transfers 40 J of energy each second.

MATHS SKILLS

You need to recall and apply these two equations. If you remember that $V = I R$, you can see how both equations are the same. In fact, we have substituted $I R$ (V) instead of V in $P = V I$ to get $P = I^2 R$.

NAILIT!

Always remember that the potential difference from your mains electricity (household sockets in the UK) is 230 V.

WORKIT!

What current will flow through an 800 W microwave oven when it is operating? (1 mark)

By rearranging the power equation $P = VI$, we get

$$I = \frac{P}{V} = \frac{800}{230}$$

Remember that the mains potential difference is 230 V.

$$= 3.5 \text{ A (1)}$$

SNAPIT!

Take photos of the power labels from at least three of your home appliances. Then try to calculate the current flowing through them. How does their stored energy change?

NAILIT!

You need to be able to describe, and give examples of, how the power ratings for some home appliances relate to changes in stored energy when you are using them.

Energy transfers in circuits

Every time a **charge** flows in a **circuit**, electrical work is done.

The amount of energy transferred by an appliance changes depending on:

- the time the appliance is operating
- the power of the appliance.

We can calculate the amount of energy transferred by electrical work using the equations:

$$\textbf{energy transferred} = \textbf{power} \times \textbf{time}$$
$$E = Pt$$

$$\textbf{energy transferred} = \textbf{charge flow} \times \textbf{potential difference}$$
$$E = QV$$

- E = energy transferred (unit: **joule**, J)
- P = power (unit: **watt**, W)
- t = time (unit: second, s)
- Q = **charge flow** (unit: **coulomb**, C)
- V = **potential difference** (unit: **volt**, V)

WORKIT!

A 9 V battery is used in series with a 1200 Ω resistor and a 340 Ω resistor. How much energy is transferred by the resistors if the switch is closed for 2 minutes? (3 marks)

The total resistance in the circuit is $R_{total} = R_1 + R_2 = 1200 + 340 = 1540\ \Omega$

$I = \dfrac{V}{R} = \dfrac{9}{1540}$ ← Rearrange $V = IR$

$\quad = 0.006A$ (1)

The total power of both the resistors is

$P = VI = 9 \times 0.006$

$\qquad = 0.053\ W$

The circuit is operating for time

$t = 2\ minutes = 60 \times 2 = 120\ s$ (1)

So, the energy transferred in that time is

$E = Pt = 0.053 \times 120$

$\qquad = 6.3\ J$ (1) ← Don't forget to put units in your answer!

Everyday appliances are designed to transfer energy in **useful** ways. For example, energy is shifted from the chemical store in a power station to the **kinetic** store in a drill connected to the **mains** via electrical working, or to the **thermal** store in the heating element of a kettle via electrical working.

The National Grid

Energy is transferred from power stations to homes, schools and workplaces through a system of **transformers**, cables and pylons called the **National Grid**.

There are two types of transformers:

- **step up**

- **step down**.

A transformer changes the value of potential difference in a circuit, but it keeps the electrical power the same.

Transformers can work only if the current is **alternating current** (ac).

Step-up transformers increase the potential difference across the circuit they are connected to.

1 This decreases the current through the circuit.

2 Fewer heating effects.

3 Less energy dissipated (wasted).

Step-down transformers decrease the potential difference across the circuit they are connected to. This increases the current through the circuit → useful for industrial and home appliances that need bigger currents.

NAILIT!

Some questions may give an amount of power in terms of megawatts (see question 3 in the Check It! below). Remember that a megawatt (MW) is a unit of power equal to one million watts.

The National Grid

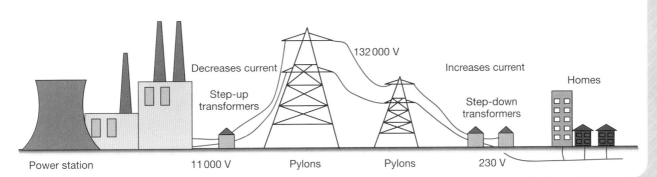

132 000 V

Decreases current

Step-up transformers

Increases current

Homes

Step-down transformers

Power station   11 000 V   Pylons   Pylons   230 V

CHECK**IT!**

1 What does the power of an appliance depend on?

2 Describe the energy transfers and the changes in **energy stores** when a car engine is working.

3 The safety guidelines for laser pointers are that their power should not be greater than 1 mW. If a safe laser pointer uses two AAA cells, which each have a voltage of 1.5, what will be the current through its laser diode?

1 Match the component with the correct symbol and the symbol with the correct description.

Light Dependent Resistor	⊗	It lets current flow through it in one direction only, but when current flows through it, it emits light.
Light Emitting Diode		It is a collection of cells connected in series with each other.
Lamp		It opposes the flow of current with a certain resistance and it can be set to different values of resistance within a specific range.
Variable Resistor		It is a resistor that changes its resistance depending on how much light is shone on it.
Battery		It is a resistor that changes its resistance depending on its temperature.
Thermistor		It has a very thin filament (wire), usually made of tungsten, which gets very hot when a current flows through it, so it glows and emits visible light.

2 a What is electric current?

b Explain why an ammeter needs to have resistance as close to zero as possible.

3 Rearrange the equation **charge flow = current × time** to the formula that would allow you to calculate the time taken by a certain amount of charge to flow through a point in a circuit.

4 a From the relationship $V = IR$ show what the gradient of a graph of Current vs Potential Difference would be.

b Using the same relationship what would be the current flowing through a 12 kΩ resistor when a p.d. of 9 V is applied across it?

c An electronic circuit connected to an LDR is set to turn a set of garden lights on when the resistance of the LDR increases beyond 40 MΩ. The graph to the right shows how the resistance of this LDR changes with light intensity. What is the minimum light intensity needed for the garden lights to automatically switch off at dawn?

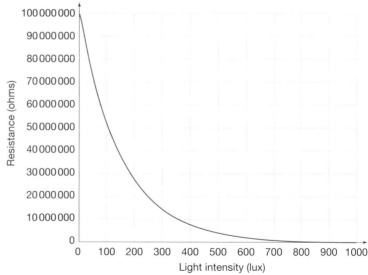

d A light bulb of resistance 2.5 kΩ is connected to three 1.5 V cells in series. What charge would flow through the bulb, if the circuit is switched on for 1.5 minutes?

Particle model and density of materials

The **density** of a material tells us how much **mass** of that material is in one unit of **volume** (1 m³).

Materials denser than water will sink. Materials less dense than water will float.

We can calculate the density of a material using the equation:

$$density = \frac{mass}{volume}$$

$$\rho = \frac{m}{V}$$

- ρ = density (unit: **kilogram** per metre cubed, kg/m³)
- m = mass (unit: kilogram, kg)
- V = volume (unit: **metre** cubed, m³)

NAILIT!

You need to recall and apply this equation. If you struggle to remember which way around the mass and volume go in the equation, think of the units of density, kg/m³, which is a mass divided by a volume.

You might be asked to calculate the volume of regular shapes. For example, the volume of a cube is
$V = l \times l \times l = l^3$,
or in other words the length of its side multiplied by itself three times.

The particle model

Everything is made of particles (atoms and molecules). Materials that have more massive particles that are closer together are more dense than other materials.

The same material can change density as it **changes state**. For example, ice is made of water molecules that are in a **solid state**. The solid state of water is less dense than its **liquid state** (ice molecules are spread out over a bigger volume than water molecules). This is why ice floats in water.

Molecules of water
(liquid state)

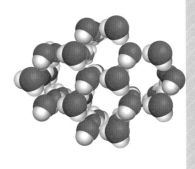

Molecules of water as ice
(solid state)

Most materials expand when the temperature increases, so become less dense. The red water in the image to the right is at much higher temperature than the cold (blue) water at the bottom, so the red dyed water floats on top of the blue dyed water.

DOIT!

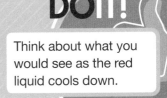

Think about what you would see as the red liquid cools down.

SNAPIT!

Find a small irregular object denser than water and drop it gently inside a measuring jug filled to the 300 ml mark with water. Take a photo of the jug before and after adding your object and calculate the density of your object in kg/m³.

WORKIT!

What is the density of the irregular rock in the diagram? (2 marks)

The mass of the rock is m = 25 g = 0.025 kg, and its volume is the difference in the volumes recorded in the measuring cylinder before and after dropping in the rock,

$V = 10 \text{ cm}^3 = 0.00001 \text{ m}^3$ (1)

So, using the density equation, we get

$P = \dfrac{m}{V} = 0.0025/0.00001$

$= 2500 \text{ kg/m}^3$ (1)

Remember to put mass into kg.

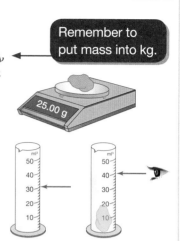

Practical Skills

You might have used micrometers and vernier callipers to measure the dimensions of regular objects and to calculate their volume in density investigations.

Reading vernier and micrometer scales can be tricky.

The Vernier scale in the diagram reads 2.6 mm.

The 0 mark on the lower scale is between 2 mm and 3 mm on the upper scale, so we know that the vernier calliper is measuring an object that is 2 mm and something wide. To find out the tenth of millimetres measured by the calliper, we need to check that the mark on the lower 0–10 scale aligns exactly with one of the marks on the top scale. In our diagram, this is mark 6, so the reading is 2.6 mm, or 0.0026 m.

Most micrometers can measure objects' thicknesses and diameters to the nearest 100th mm (0.01 mm). A full turn of the thimble measures a length of 0.5 mm.

So, you first read the numbers on the sleeve (in our example, 5.50 mm).

Then, read which mark on the thimble aligns with the middle line on the sleeve (in our example, 0.12 mm).

Finally, add the two readings together (5.50 + 0.12 = 5.62 mm).

Can you convert all these readings to cm or m?

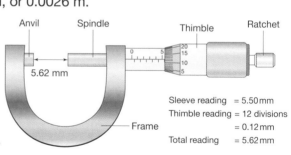

Sleeve reading = 5.50 mm
Thimble reading = 12 divisions
= 0.12 mm
Total reading = 5.62 mm

CHECKIT!

1 What does the density of a material depend on?

2 The density of the element mercury is 13 534 kg/m³. What volume would 2.3 kg of mercury have?

3 0.250 kg of fresh water has a volume of 250 cm³. The average density of the human body is 985 kg/m³. Explain (with calculations) why a person would float better in the Dead Sea than in a fresh water lake, knowing that 0.250 kg of Dead Sea water has a volume of about 202 cm³.

Changes of state and internal energy

Changes of state

The mass of a substance is always conserved when it changes state of matter, i.e. the total mass **before** the change is equal to the total mass **after** the change.

Changes of state are reversible which means that the material gets back to its original state, if the change is reversed.

When a substance changes state its internal energy changes too.

Internal energy

The total energy in the kinetic and potential stores associated with the particles that make up a system (their movement and interactions) is the internal energy of that system.

Heating a system changes the energy stored within the system by increasing the energy associated with the particles that make up the system.

This increase in internal energy either raises the temperature of the system or causes a change of state.

Specific latent heat

The **latent heat** is the **energy** needed for a substance to change state.

When a substance changes state, the energy supplied changes the internal energy of the system (the **energy stored**), but not its temperature.

This is why heating and cooling curves for different substances show that the temperature remains **constant** during a change of state.

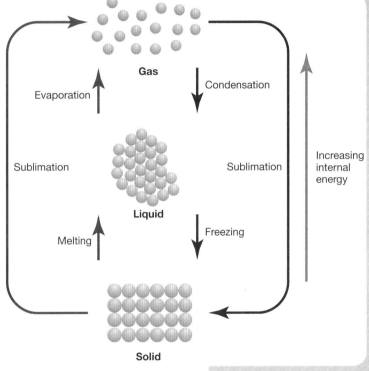

MATHS SKILLS

In this topic you will need the equation to calculate the change in thermal **energy** of a **system** again. This depends on the **specific heat capacity** of a substance (page 243).

change in = mass × specific × temperature
thermal energy   heat capacity   change

$$\Delta E = mc\,\Delta\theta$$

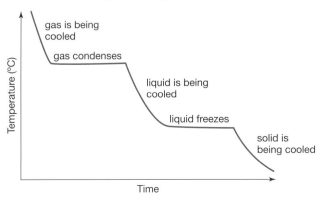

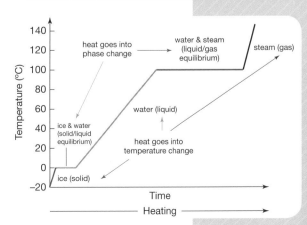

These graphs show state change compared to temperature change

The **specific latent heat** of a **material** is the amount of **energy** needed for 1 kg of that substance to **change state** (without a change in temperature).

SNAPIT!

Take a photo of the cooling and heating graphs on the previous page and keep it with you to revise their features wherever you are.

The equation to find the energy needed to change the state of a substance is:

energy for a change of state = mass × specific latent heat

$$E = mL$$

- E = energy (unit: **joules**, J)
- m = mass (unit: **kilogram**, kg)
- L = specific latent heat (unit: **joules per kilogram**, J/kg)

Specific latent heat of fusion – change of state from solid to liquid.

Specific latent heat of vaporisation – change of state from liquid to vapour.

WORKIT!

108 800 J are supplied to 320 g of ice at 0°C for it to melt completely. Calculate the specific latent heat of fusion of water.

The mass of the ice is

$m = 320$ g $= 0.320$ kg.

Use the equation $E = mL$ and rearrange

$L = \dfrac{E}{m} = \dfrac{108\,800}{0.320}$

$\qquad = 340\,000$ J/kg

CHECKIT!

1 What is the difference between specific heat capacity and specific latent heat?

2 How would the sections where the ice is melting and the water is boiling in the heating curve above change if a larger amount of water were used, but the same power were provided to the system over time?

3 The specific latent heat of vaporisation of alcohol is 896 kJ/kg. How much energy needs to be supplied to 2.3 kg of alcohol for it to completely evaporate?

Particle model and pressure

Particle motion in a gas

The particles of a **gas** always move with **random** motion. This means they move in all directions and at different speeds (random **velocity**).

The temperature of a gas depends on the average **kinetic energy** of its particles.

For a gas at constant **volume**, increasing the temperature will increase the **pressure** exerted by the gas. This is because the gas particles collide against (hit) each other and the walls of their container with a bigger average **force**.

So, the higher the temperature of a gas (at constant volume), the higher its pressure.

A pressure cooker works by keeping most of the steam from boiling water inside the fixed volume of the container. So, as the temperature increases, the pressure increases too. Temperatures higher than 100° C can be reached, and so the food cooks faster.

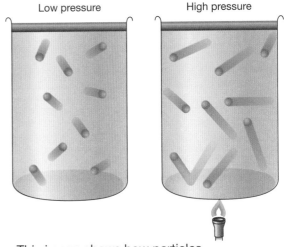
Low pressure   High pressure

This image shows how particles move at different pressures

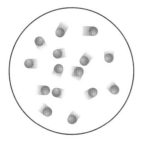

The gas particles on the right have more kinetic energy than those on the left. So, the gas on the right is at a higher temperature.

Pressure cookers reach higher temperatures so food cooks faster

Pressure in gases

Gas particles keep colliding against (hitting) the walls of their container. Each collision exerts a tiny force on the walls of the container, but there are lots and lots of particles colliding at high velocities. So, if we add together all the collisions, we can measure a net force produced by the pressure perpendicular (at right angles, 90°) to the wall of the container, or to any other surface that the gas is in contact with. The image on the right shows how gas particles behave within a container.

If the temperature of a gas stays constant, but its volume changes, its pressure changes too.

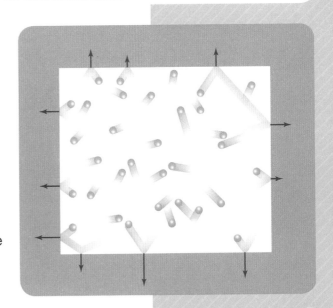

The bigger the volume of a gas (at constant temperature), the lower its pressure.

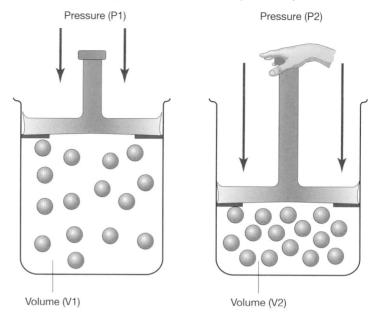

Pressure (P1)   Pressure (P2)

Volume (V1)   Volume (V2)

For a gas at the same temperature (same average kinetic energy):

- In a bigger volume, fewer collisions will happen per second. This is because there is more space for the gas particles to move around. So the pressure of the gas decreases. We say that the gas 'expanded'.

- In a smaller volume, more collisions will happen per second. This is because each particle has less space to move before it hits a wall. So the pressure of the gas increases. We say that the gas is 'compressed'.

The equation linking the volume and pressure of a gas kept at constant mass and temperature is:

$$\textbf{pressure} \times \textbf{volume} = \textbf{constant}$$
$$[\textbf{\textit{pV}} = \textbf{constant}]$$

- p = pressure (unit: **pascal**, Pa)
- V = volume (unit: **metre** cubed, m³)

WORKIT!

A diver is exploring the sea bed at a depth where the water pressure is 3.5×10^5 Pa. At that depth, the air bubbles that come out of her mouth have a volume of 1.7 cm³. The pressure at the surface of the water is about 1.0×10^5 Pa. What will be the volume of the bubbles at the surface (assuming the temperature of the air inside the bubbles stays the same)? (3 marks)

First convert the volume to m³.

$1.7 \text{ cm}^3 = 1.7 \times 10^{-6} \text{ m}^3$ (1)

Now use the equation pV = constant

$P_1 V_1 = P_2 V_2 \rightarrow V_1 = \dfrac{P_1 V_1}{P_2}$ (1)

$= \dfrac{3.5 \times 10^5 \times 1.7 \times 10^{-6}}{1.0 \times 10^5}$

$= 6.0 \times 10^{-6} \text{ m}^3$ (1)

Increasing the pressure of a gas

Work done is the transfer of energy by a force.

Applying a force on the particles of a gas increases the **internal energy** of the gas.

For example, if you use a pump to force air into a car's tyre, you increase the number of particles in the tyre and the average velocity of the air particles. So you increase the average kinetic energy of the air particles, or in other words, you increase the temperature of the gas.

> Remember that everything is made of particles, and particles are atoms and molecules (see page 269).

Summarise the content in this subtopic using no more than 50 words.

STRETCH IT!

Look at this thermal image. Notice how just a few pushes with the foot pump increases the temperature of the air inside the tube.

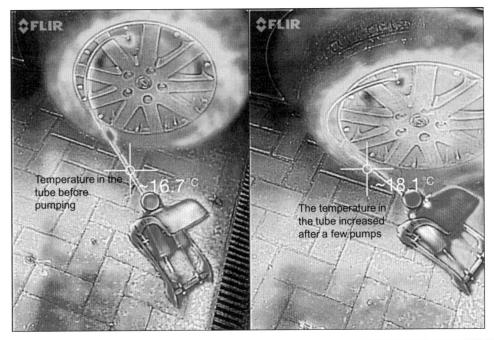

CHECK IT!

1 Describe how the particles of a gas move.

2 Explain why pV might not be constant when the volume of a bicycle tyre is increased by pumping air into it with a foot pump.

3 A gas is contained inside a sealed syringe with a movable piston at an initial volume V_1 and initial pressure p_1. If the piston of the syringe is moved to a final volume $V_2 = 3 V_1$, what will be the difference between p_1 and the final pressure p_2?

1 a What are the units of density?

b The density of olive oil is $\rho_o = 812$ kg/m³ and the density of ethanol is $\rho_o = 785$ kg/m³.

   i Which fluid will float on top of the other, if they are both poured in a glass cylinder?

   ii What is the ratio of the mass of ethanol and the mass of oil ($m_e : m_o$) for an equal volume of ethanol and oil?

c The equation for the volume of a cylinder is $\pi r^2 h$. What is the density of a cylinder of steel of height 12 cm, diameter 24 mm and mass 0.437 kg?

2 a What changes of state are associated with latent heat of fusion and latent heat of vaporisation?

b A toy soldier made of tin 185 g, is melted in a fire. The specific latent heat of fusion for tin is 59 000 J/kg. Calculate the energy transferred to increase the internal energy of the tin when the toy soldier melts.

3 Complete the boxes in the graph below to label the different areas of the cooling curve with reference to change of state.

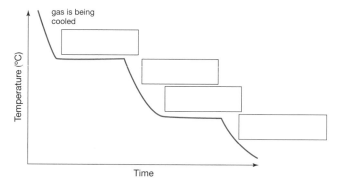

4 Explain why increasing the temperature of a gas in a gas cylinder causes the pressure of the gas to increase as well. Use the particle model of matter in your answer.

5 The image below shows the same syringe with trapped air inside it in two different situations – in hot water and in icy water. Explain why the volume of air inside the syringe is different in the two situations.

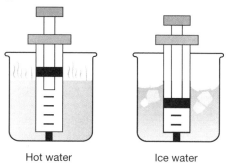

Hot water   Ice water

6 The volume of air trapped in a syringe at room temperature is measured to be 20 ml. The piston is very slowly pulled to allow the air inside the syringe to expand to 55 ml. The piston is then held in that position.

a Calculate the pressure of air inside the syringe when the volume is 55 ml, knowing that atmospheric air pressure is 101 325 Pa.

b Which assumption did you have to make to complete your calculation in part a?

The structure of the atom

Atoms are the building blocks of matter. They have a very small radius (1×10^{-10} m).

Atoms have a positively charged nucleus, which has a tiny radius – smaller than 1/10 000 the radius of an atom.

The nucleus is made of neutrons (neutral particles) and protons (positive particles).

Electrons are negatively charged particles that orbit the nucleus. As the radius of these orbits (or electron clouds) is much bigger than the radius of the nucleus, most of the volume of an atom is empty space.

Electrons are arranged at different energy levels (different distances) from the nucleus. Electrons can move further from or closer to the nucleus by taking in (absorbing) or giving out (emitting) electromagnetic radiation.

- When an electron **absorbs** electromagnetic radiation, it will move to a **higher** energy level (further from the nucleus).

- When an electron **emits** electromagnetic radiation, it will move to a **lower** energy level (closer to the nucleus).

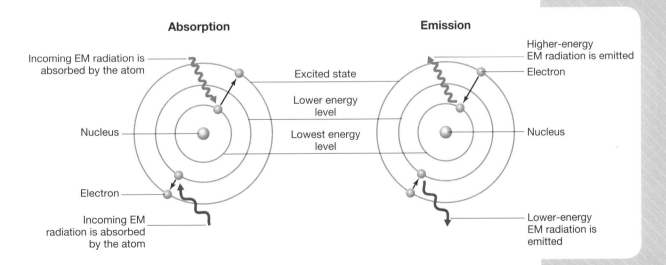

Mass number, atomic number and isotopes

The number of electrons and protons in an atom is the same, so the overall **electrical charge** of an atom is zero.

Atoms of the same element have the same number of protons. This is the atomic number of the element. For example, the atomic number of carbon is six because all carbon atoms have six protons in their nucleus.

The sum of the number of protons and neutrons in an atom is the **mass number** of the atom.

The atomic and mass numbers can be represented like this:

(Mass number)   23
(Atomic number) 11 **Na**

The same element can have different isotopes. These are atoms that have the same atomic number, but different mass numbers. This means that they have the same number of protons, but different numbers of neutrons. An example is below showing the three isotopes of hydrogen:

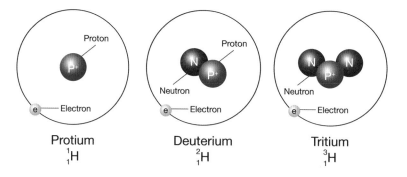

Protium
^1_1H

Deuterium
^2_1H

Tritium
^3_1H

Sometimes atoms can lose some outer electrons, so the atoms become positively charged ions.

SNAP IT!

Take a picture on your phone of the ways to find the atomic number, mass number and number of neutrons.

Number of protons = atomic number = mass number − number of neutrons

Number of neutrons = mass number − atomic number

Mass number = number of neutrons + atomic number

CHECK IT!

1 What is the overall charge of an atom that has lost two outer electrons?

2 Uranium-238 has a mass number of 238 and atomic number of 92. What is the number of neutrons and protons in its nuclei?

3 The radius of an atom is bigger than the radius of its nucleus by how many orders of magnitude?

Developing a model of the atom

The flow chart below shows some of the different discoveries that scientists have made to create the model of the atom that we use today.

As new discoveries are made and checked using experiments, scientists might suggest even more accurate models of the atom in the future.

SNAPIT!

Make a copy of this flowchart and take a picture with your phone, so that you can revise the development of the model of the atom wherever you are.

Before the discovery of the electron — Greek philosophers believed that the atom was the most basic building block of all things, and that it could not be divided into smaller components.

Thompson's atomic model (1897) — the electron was discovered and the plum pudding model was developed. This suggested that the atom is a sphere of positive charge with negative electrons embedded in it, like raisins in a plum pudding.

Electron

Sphere of positive charge

Rutherford's model (1912) — alpha particles (positive nuclei of helium) were shot at high speed at gold atoms. Some went through undisturbed, some were deflected slightly, but some were bounced back at very large angles. This suggested that most of the mass of an atom is concentrated in a very small, positively charged nucleus, with electrons orbiting around it.

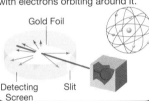

Gold Foil

Detecting   Slit
Screen

Chadwick proves the existence of neutrons (1932) — the mass of nuclei was too large to be made only of protons, James Chadwick proved another neutral particle called a neutron is inside the nucleus of atoms with the protons.

Discovery of the proton (1919) — more experiments showed that the charge of a nucleus could be divided into a whole number of smaller charges called protons, all with the same positive charge.

Bohr's energy levels (1913) — Rutherford's model was adapted by showing that electrons could only orbit the nucleus of an atom at specific distances (radii). This was important because this new model agreed with experimental measurements.

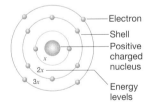

Electron

Shell

Positive charged nucleus

Energy levels

NAILIT!

Remember that the development of a model of the atom is a good example of how scientific methods and theories change over time.

CHECKIT!

1 Describe the plum pudding model of the atom.

2 What is the main difference between Rutherford's nuclear model of the atom and Bohr's model?

3 Explain why the large deflections of very few alpha particles hitting gold atoms is evidence that most of the mass of an atom is concentrated in the centre of the atom (the nucleus).

Radioactive decay and nuclear radiation

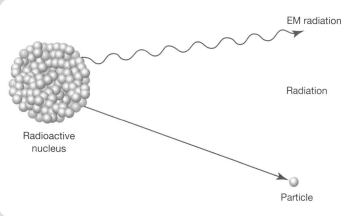

EM radiation

Radiation

Particle

Radioactive nucleus

Some elements are unstable. To become more stable, the nuclei of their atoms give out (emit) **radiation**.

By emitting radiation, the nucleus changes. This is a random process called radioactive decay.

To become more stable, a nucleus can:

- emit some of its mass (particle emissions) as nuclear radiation, or
- decrease its internal energy by emitting high-frequency electromagnetic radiation.

Each type of nuclear radiation emitted by unstable nuclei is **ionising**. This means that it can remove electrons from the outer orbits of other atoms, leaving them positively **charged**. If atoms in your cells are ionised, it can be very damaging to your body.

The types of nuclear radiation emitted by unstable nuclei are shown in the table below:

	Alpha (α)	Beta (β)	Gamma (γ)	Neutron (n)
Nature	A nucleus of helium $^4_2$He. Two protons and two neutrons	An electron e^-	An electromagnetic wave	A neutral particle inside the nucleus of an atom
Charge	+2	−1	0	0
Mass	Relatively large	Very small	No mass	Large ($\frac{1}{4}$ of mass of α)
Range in air	3–5 cm	17 m	100s of metres	100s or 1000s of metres
Ionising effect	Very strong	Weak	Very weak	Very strong

Both alpha particles (α) and neutrons are heavy particles and have very high ionising power. As alpha particles are blocked by skin, they are most dangerous when the radioactive source is inside you. Neutrons are more dangerous when the source is outside your body, because neutrons have very high penetration power.

Gamma rays have no mass or charge, so do not interact with other particles as strongly as alpha particles. This means that gamma rays are not very ionising, but they are also not easily decelerated, so they are more penetrating than alpha particles. Neutrons are not typically able to ionize an atom directly because of their lack of charge, but they are indirectly ionizing, as they are absorbed by stable nuclei. These nuclei will become unstable by the absorption of a neutron, and more likely to emit ionizing radiation of another type. Neutrons are, in fact, the only type of radiation that is able to turn other materials radioactive.

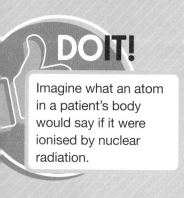

DO IT!

Imagine what an atom in a patient's body would say if it were ionised by nuclear radiation.

Different nuclear radiation has different penetration power.

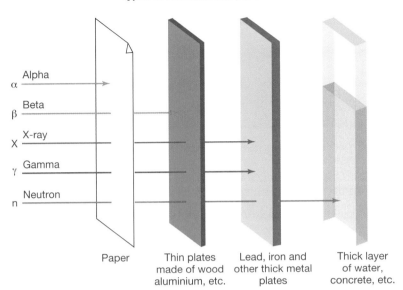

Types of radiation and penetration

α Alpha

β Beta

X X-ray

γ Gamma

n Neutron

Paper | Thin plates made of wood aluminium, etc. | Lead, iron and other thick metal plates | Thick layer of water, concrete, etc.

A Geiger counter can detect and measure ionising radiation - alpha, beta and gamma

Neutrons can easily pass through (penetrate) most materials, even iron. But they are slowed and blocked by materials that are rich in hydrogen, such as hydrocarbons and water. Thick concrete walls are good radiation shields because they can block both neutrons and gamma rays (γ).

WORKIT!

Why are thick concrete walls a good radiation shield? (3 marks)

Alpha particles (α) are blocked by paper, and beta particles (β) are blocked by aluminium, so a layer of concrete will be good at shielding people from them. (1)

Fewer gamma rays can pass through a heavy material like lead or concrete, so a concrete wall would give good protection from gamma rays. (1)

Neutrons are the most penetrating type of radiation, but they can be absorbed by hydrogen-rich materials. This is because concrete is made by mixing cement and water (which contains a lot of hydrogen atoms), thick concrete will give good protection from neutron radiation too. (1)

NAILIT!

For this question you need to compare the penetration power of the different types of radiation.

CHECKIT!

1 What are the main types of nuclear radiation?

2 Why are neutrons the most dangerous type of radiation?

3 Why is gamma radiation more penetrating than alpha particles?

Nuclear equations

As chemical equations represent chemical reactions, nuclear equations represent radioactive decay.

The symbols for representing alpha particles and beta particles are:

- $_2^4$He or α particle is a nucleus of helium, so it has an atomic number of 2 (2 protons) and mass number of 4 (2 protons and 2 **neutrons**).

- $_{-1}^{0}$e or β particle is a fast-moving electron from the nucleus, so it has an atomic number of -1 to represent its negative **charge**, but mass number of 0 because it does not have a nucleus.

When nuclear radiation is emitted from a nucleus, it can change the **mass** and/or the charge of the nucleus.

This nuclear equation shows how **alpha decay** changes both the charge and the mass of a nucleus:

$$_{86}^{219}\text{radon} \longrightarrow {}_{84}^{215}\text{polonium} + {}_2^4\text{He}$$

The α emission takes away two protons and two neutrons from the nucleus of the radon atom. This means that its atomic number falls by 2 and its mass number falls by 4. Both charge and mass in its nucleus fall, and so the radon changes into a different element.

This nuclear equation shows how beta decay increases the charge, but does not change the mass of a nucleus:

$$_{6}^{14}\text{carbon} \longrightarrow {}_{7}^{14}\text{nitrogen} + {}_{-1}^{0}\text{e}$$

NAILIT!

At GCSE, you don't need to remember these two examples. You only need to know what happens in alpha decay and beta decay and how to balance atomic numbers and mass numbers.

DOIT!

Draw a table showing the general rules of how the atomic and mass numbers change when a nucleus emits alpha or beta particle, or gamma radiation.

NAILIT!

Make sure you know the Greek letters used for the different types of radioactive decay:

α alpha

β beta

γ gamma

WORKIT!

Complete the following nuclear equations. (2 marks)

$$^{60}_{27}\text{Co} \rightarrow {\color{gray}\underline{}}\text{Co} + \gamma \qquad ^{42}_{19}\text{K} \rightarrow {\color{gray}\underline{}}\text{Ca} + {\color{gray}\underline{}}\text{e}$$

γ-rays have no atomic number and no mass, so there is no change in the atomic number and mass number of cobalt in the equation.

$$^{60}_{27}\text{Co} \rightarrow {}^{60}_{27}\text{Co} + \gamma \quad (1)$$

β particles have atomic number = −1 and mass number = 0, so the atomic number of potassium (K) will increase by 1 (and become a calcium nucleus), and the mass number will not change.

$$^{42}_{19}\text{K} \rightarrow {}^{42}_{20}\text{Ca} + {}^{0}_{-1}\text{e} \quad (1)$$

NAILIT!

The proton/atomic number of a nucleus tells you what the element is in the periodic table. So, a change in atomic number in nuclear equations always means that the nucleus has become a different element, even if its mass number has not changed.

Periodic table of the elements

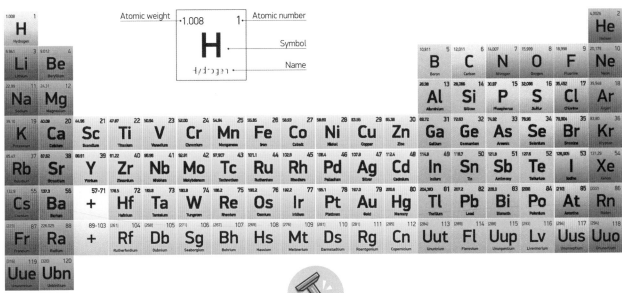

NAILIT!

Make sure you can find information from the periodic table, it will be useful when completing nuclear equations.

CHECKIT!

1 What tells you which element in the periodic table a nucleus belongs to?

2 How would you represent a neutron in a nuclear equation?

3 What changes would a neutron emission cause in a nucleus?

Half-life of radioactive elements

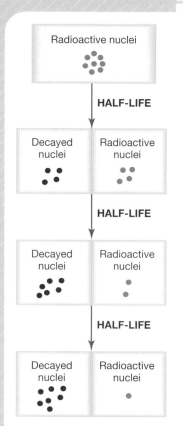

Radioactive decay is a **random** process. This means that it is impossible to predict which of the nuclei in a sample of a radioactive isotope will decay next, or when a nucleus will decay. However, we can make predictions about decay when there are lots of unstable nuclei.

The decay rate of a radioactive isotope depends on:

- The element in the sample.

- How many undecayed nuclei are in the sample.

The *more* undecayed nuclei there are, the bigger the decay rate. This means that, on average, more nuclei decay per second when we start measuring the radiation of an isotope sample than when we *stop* measuring the radiation.

The decay rate of a radioactive isotope does not depend on the temperature and pressure of the sample. Even at very high temperatures and pressures the radioactive decay rate of an element is not affected.

We can define the half-life of a radioactive isotope in two ways:

- The time taken for the number of nuclei in a sample of a radioactive isotope to halve.

- The time taken for the count rate (or **activity**) from a sample of a radioactive isotope to fall to half its original level.

Half-life measured in terms of radioactive nuclei, look at how the numbers of radioactive and decayed nuclei change every half-life

Half-life in terms of count rate halving

The random nature of radioactive decay is linked to the half-life of different isotopes. So, other random events also behave like a sample of decaying nuclei.

For example, the level of foam on top of a fizzy drink will fall in a random way. We can find its 'half-life' by plotting a graph of the height of the foam against time.

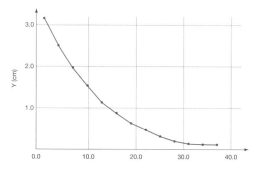

DO IT!

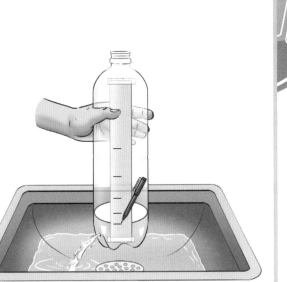

Take a big plastic bottle and pierce a hole at the bottom. Then fill the bottle with water, covering the hole with your finger. Let the water out into your sink and make a mark on the bottle every 5 seconds. Use your marks to draw a graph of the level of water left against time. You should get a curve that looks like the one for the fizzy drink foam.

NAIL IT!

You can find the half-life of a sample of radioactive nuclei from a graph which shows the number of undecayed nuclei against time. Just find the point on the activity axis (y-axis) where the activity/number of undecayed nuclei has dropped to half the original value. Then, go across the x-axis to see at what time that happened. That time is the half-life of the radioactive isotope.

SNAP IT!

Take a video of the foam from a fizzy drink and count how many seconds it takes for the level of foam to halve each time. You should notice a similar pattern to the graph on page 284.

WORK IT!

The remains of an animal found frozen in a glacier on the Alps contains $\frac{1}{32}$ the amount of carbon-14 found in living animals today. Carbon-14 is a radioactive isotope of carbon and has a half-life of 5730 years. How old might this animal be? (3 marks)

The net decline of the radioactive emission is:

1 half-life → $\frac{1}{2}$ original amount

2 half-lives → $\frac{1}{4}$ original amount

3 half-lives → $\frac{1}{8}$ original amount

4 half-lives → $\frac{1}{16}$ original amount

5 half-lives → $\frac{1}{32}$ original amount (1) ◄

If the amount of carbon-14 is $\frac{1}{32}$ of the original amount when the animal was alive, and because the amount of carbon-14 halves every half-life, this means that 5 half-lives have gone by since it died.

Each half-life is 5730 years long, so 5730 × 5 = 28 650 years. (1)

This means that the animal was probably alive 28 650 years ago. (1)

Radioactive contamination and irradiation

Soil, food and other materials can become contaminated by unwanted materials that have radioactive isotopes.

Dust and gases that have radioactive isotopes are the most dangerous, because they are very difficult to remove from the environment. Dust and gases can also be breathed into your body.

The hazard from radioactive contamination comes from the decay of the contaminating atoms.

The level of hazard is affected by how many contaminating atoms there are and the type of radiation emitted.

People working with radiation or in hazardous areas wear protective suits.

An object can be **irradiated** if it is exposed to nuclear radiation, but the irradiated object does **not** become radioactive itself.

Irradiation can be useful, for example, to make medical equipment and food safe by sterilising them, or to destroy cancer cells in patients' bodies. But the amount of irradiation a human body is exposed to should be as small as possible. That is why precautions, like lead screening, must be taken against the hazards from the radioactive sources that are used in irradiation. People who are often around radiation because of their jobs, for example, people who operate x-ray machines, may wear a radiation monitoring badge to make sure they stay within the recommended dose limits.

CHECK**IT!**

1 What does the rate of radioactive decay depend on?

2 Define the half-life of a radioactive isotope.

3 The activity of a sample drops to a quarter of its original activity after 146 days. What is the half-life of the radioactive isotope in the sample?

1 Write the number of electrons, protons and neutrons for the following elements.

a $^{23}_{11}Na$

b $^{14}_{7}N$

c $^{235}_{92}U$

d $^{208}_{84}Po$

e $^{9}_{4}Be$

f $^{14}_{6}C$

2 Explain how Bohr's model of the atom is different from Rutherford's model.

3 The two graphs below show the activity of elements A and B changes over time. Look at the graphs and answer the questions.

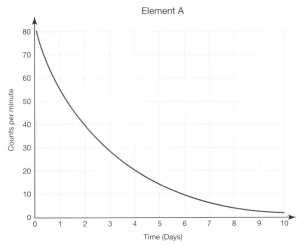

Element A

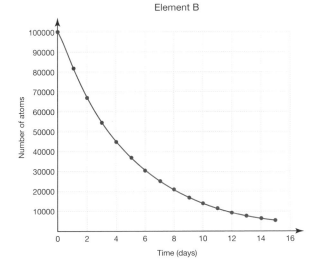

Element B

a What are the half-lives of element A and element B?

b After how many half-lives does the count per minute reading drop to 6 counts per minute in element A?

c How many days does it take for the number of radioactive atoms in element B to drop to a quarter of their original value?

4 Explain the difference between radioactive contamination and irradiation.

Forces

Forces and their interactions

Scalar and vector quantities

Scalar quantities have only magnitude. Some examples of scalar quantities are **speed**, **mass** and **energy**.

Vector quantities have magnitude and direction. Some examples of vector quantities are **velocity**, **pressure** and **forces**.

An arrow is a useful way to represent a vector quantity. The length of the arrow shows the magnitude (the size) of the vector quantity. The direction of the arrow shows the direction of the vector quantity.

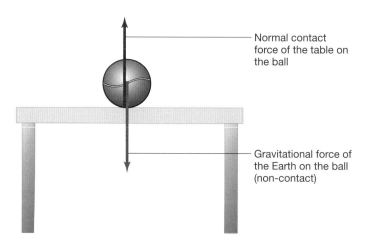

Contact and non-contact forces

When two objects interact with each other, they experience a force caused by this interaction.

Forces can be described as pushes or pulls acting on an object. All forces between objects are either:

- **contact forces** – when the objects are touching (for example, **friction**, air resistance, tension and the 'normal' contact force)

- **non-contact forces** – when the objects are not touching (for example, gravitational force, electrostatic force (from static electricity) and magnetic force).

Normal contact force of the table on the ball

Gravitational force of the Earth on the ball (non-contact)

NAILIT!

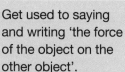

Get used to saying and writing 'the force of the object on the other object'.

For example, 'the pull of the rope on the climber'. This will help you to identify all the forces applied to an object.

Gravity

The mass of an object is a scalar quantity measured in **kilograms** (kg). The mass of an object stays the same wherever it is in the universe.

The weight of an object on the Earth is the gravitational force caused by the pull of the Earth on the object.

The direction of the pull (weight) is always towards the centre of the Earth. We can think of the weight of an object, which is a vector quantity, as the gravitational force acting at a single point called the centre of **mass** of the object.

The magnitude of weight on the Earth is caused by the **gravitational field** around the Earth. The magnitude depends on the gravitational field strength at the point where the object is. The **further** away an object is from the Earth, the **smaller** its weight.

We can calculate the weight of an object using the equation:

weight = mass × gravitational field strength

$$W = m\,g$$

- W = weight (unit: **newton**, N)
- m = mass (unit: kilogram, kg)
- g = gravitational field strength (unit: newton per kilogram, N/kg)

Weight is measured using weighing scales made of a calibrated spring-balance (a **newtonmeter**).

NAILIT!

Make sure you can remember and use this equation.

DOIT!

Weighing scales show your mass not your weight. So, measure your mass in kg with your scales. What is your weight in newtons?

WORKIT!

The spacecraft *Juno* arrived in Jupiter's orbit on 4 July 2016. Its mass was 3625 kg when it was launched from Earth. What is its weight on Earth and on the surface of Jupiter? (2 marks)
(g = 9.8 N/kg on Earth, and g = 23 N/kg on Jupiter.)

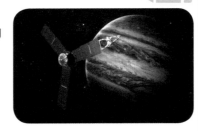

The weight of Juno on Earth is $W = m\,g = 3625 \times 9.8$

$= 35\,525$ N (1)

The weight of Juno on Jupiter is $W = m\,g = 3625 \times 23$

$= 83\,375$ N (1)

The mass of *Juno* will still be 3625 kg on Jupiter, because the mass of an object stays the same wherever it is in the universe.

MATHS SKILLS

Weight and mass are **directly proportional** ($W \propto m$). Saying 'weight is directly proportional to mass' means that, if we double the mass, the weight also doubles.

Make sure you can recognise and use the symbol for direct proportionality ($\propto$).

CHECK**IT!**

1 Name one scalar quantity and one vector quantity not listed above.

2 What is the difference between contact and non-contact forces?

3 Explain the difference between mass and weight.

Resultant forces

When more than one **force** is acting on an object, we can replace all these forces with a single force that has the same effect as all of the forces acting together.

This single force is called the **resultant force**. It is the sum of all of the forces acting on an object.

When the resultant force is zero (0 N), all of the forces acting on an object are **balanced**.

WORKIT!

1 Look at the images. Which diagram best describes the forces acting on the tennis ball? Explain your choice. (3 marks)

(Note that the yellow arrows are only showing the direction the ball is moving and are **not** a force.)

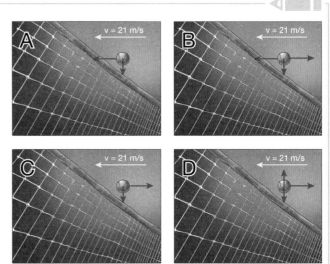

Diagram C best describes the forces on the tennis ball because the ball is no longer in contact with the racket, so there is no forward force acting on the ball. (1) The only horizontal force on the ball is the air resistance in the opposite direction to the motion of the ball. (1) So the ball is decelerating. The force pointing downward is the weight of the ball. (1)

2 The truck in the diagram is travelling east. What is the resultant force F on the truck? (2 marks)

The thrust from the engine, F_t, and the drag, F_d, are in opposite directions.

$F = F_t - F_d = 3250 - 1050 = 2200$ N (1)

The truck is accelerating eastward with a force of 2200 N. (1)

1050 N

3250 N

STRETCH IT!

The resultant force on a ping-pong ball at rest on a table and the resultant force on the same ping-pong ball falling freely (when it has reached **terminal velocity**) is the same. In both situations, the forces on the ball are equal and in opposite directions. So the sum of the forces is zero, so the resultant force is 0 N.

In physics, the effect is the same because the velocity of the ball in both situations does not change. The ball when it is on the table stays at rest, and the ball when it is falling freely carries on falling at a constant velocity (at the same speed and in the same direction). So in both situations, all the forces on the ball are balanced.

Ping-pong ball at rest on the table — Normal contact force of the table on the ball — Gravitational force of the Earth on the ball

Ping-pong ball in free fall — Air resistance of the air molecules on the ball — Gravitational force of the Earth on the ball

 ## Components of forces

A single force can be broken down ('resolved') into two components acting at right angles (90°) to each other.

Just like for resultant forces, these two components together have the same effect as the single force.

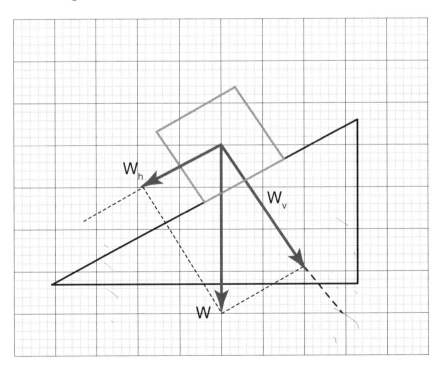

DO IT!

Draw each of these **vectors** on graph paper, and find their horizontal and vertical components:

- A force arrow of 6.5 N pointing right 30° above the horizontal line.
- A **velocity** arrow of 2.8 m/s pointing left 55° below the horizontal line.

Look at the diagram above. We can resolve the weight of the box on the ramp into two components:

- a component acting along the surface of the ramp (W_h)
- a component acting at 90° to the surface of the ramp (W_v).

We can draw a 'free body diagram' on graph paper to show all the forces acting on the box. If the weight of the box is 4 N, we can draw W 4 cm long. The **magnitude** of W_h = 3.5 N, so we can draw its arrow 3.5 cm long. And the magnitude of W_v = 2.2 N, so we can draw its arrow 2.2 cm long.

WORKIT!

A passenger plane is applying a thrust T = 63 000 N in a direction 70° north-west (70° to the left of the vertical line). The horizontal component of the drag is D = 28 000 N, and the weight of the plane is W = 14 000 N. Calculate the resultant horizontal force and the resultant vertical force on the plane. (2 marks)

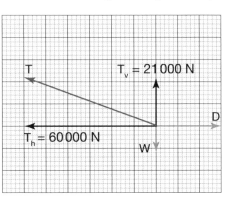

> First, we need to find the horizontal and vertical components of the thrust, T_h and T_v.

Redraw the force arrows on graph paper (making sure we keep the same proportions). We will use 1 cm on the graph paper to represent 10 000 N. (1)

Our vector diagram shows that T_h = 60 000 N (because the length of this arrow is 6 cm) and T_v = 21 000 N (because the length of this arrow is 2.1 cm). (1)

CHECKIT!

1 What is meant by a resultant force?

2 A skydiver of mass 75 kg opens his parachute. The air resistance from it is initially 1542 N. What is the magnitude and direction of the resultant force on the skydiver?

(H) 3 What are the horizontal and vertical components of a force of 5 N pulling an object to the left 45° above the horizontal line?

Work done and energy transfer

A **force** does work on an object when it causes that object to move a distance. In other words, when a force causes the **displacement** of an object, this means that the force has done work on the object.

We can calculate the work done by a force on an object using the equation:

work done = force × distance (moved along the line of action of the force)

$$W = F\,s$$

- W = work done (unit: **joule**, J)
- F = force (unit: **newton**, N)
- s = distance, or displacement (unit: **metre**, m)

One joule of work is done when a force of one newton causes a displacement of one metre. This means that:

one joule = one newton-metre → 1 J = 1 Nm

NAILIT!

Make sure you can remember and use this equation.

You also need to be able to convert joules to newton-metres, and newton-metres to joules.

Whenever work is done by a force, **energy** is transferred. For example, when an apple is dropped, the **weight** of the apple does work as it pulls the apple to the ground. The gravitational potential energy store associated with the apple empties as the **kinetic energy** store of the apple fills.

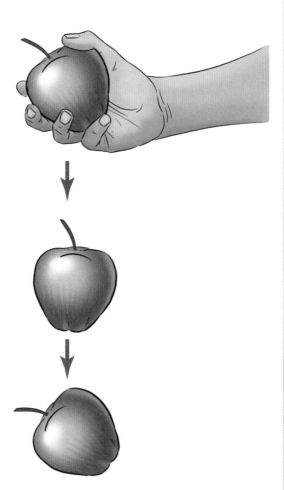

Work is also being done against another force that is acting on the falling apple. This is the frictional force of air resistance. Air resistance is a frictional force because the **particles** of air cause **friction** on the falling apple. The work done against air resistance increases the **thermal** energy store of the apple and increases the thermal energy store of the air around it.

So, the work done against a frictional force causes an increase in the temperature of an object.

WORKIT!

A ball of mass 10 kg is dropped from a height of 20 m. The same ball is then left free to roll down a ramp that is 20 m high and 50 m long. Calculate the work done by the weight of the ball in these two situations. (4 marks) (g = 9.8 N/kg)

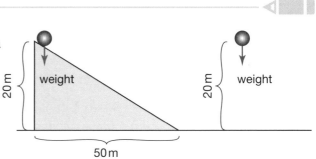

Weight of the ball = $m \, g$ = 10 × 9.8 = 98 N (1)

When the ball is dropped from a height s = 20 m, the work done by the weight is:

W = F × s = weight × s = 98 × 20

= 1960 J (1)

When the ball is free to roll down the ramp, the work done by the weight is the same as when the ball is dropped (W = 1960 J). This is because the height of the ramp is 20 m (the same as the height of the drop), and the line of action of the weight is along the vertical. (1) So any horizontal movement does not count toward the work done by the weight. (1)

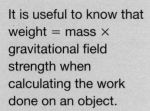

NAILIT!

It is useful to know that weight = mass × gravitational field strength when calculating the work done on an object.

DOIT!

Choose an object you can lift without too much effort and find its weight. Then, lift it to a height you have previously measured. Now, calculate how much work your pulling force did in lifting the object to that height.

CHECKIT!

1 What is the work done by a force?

2 How many joules are in 530 newton-metres?

3 A rocket has a mass of 22 000 kg. What is the minimum work that needs to be done by the thrust to lift the rocket 1.54 km? Assume that the mass of the rocket stays constant. In this question the gravitational field strength = 9.8 N/kg.

Forces and elasticity

To change the shape of a stationary object (stretch, compress or bend it), you always need to apply two **forces** on the object. If you applied only one force, you would only turn (rotate), pull or push the object.

When stretching a rubber band, one finger is pulling in one direction, while the other finger is pulling in the opposite direction.

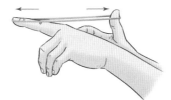

When compressing Blu-Tack against a wall, the finger is pushing on the Blu-Tack, while the wall is exerting a force on the Blu-Tack in the opposite direction.

When bending a piece of Blu-Tack, one hand is twisting the block of Blu-Tack clockwise, while the other hand is twisting it anti-clockwise.

Objects that are stretched, compressed or bent can be warped from their original shape (deformed) either elastically or inelastically.

- **Elastic deformation** → the object returns to its original shape after being stretched.
- **Inelastic deformation** → the object does **not** return to its original shape after being stretched (its shape is changed permanently).

For an elastic object, we can see the relationship between the force on the object and how the object changes shape using the equation:

$$\textit{force} = \textit{spring constant} \times \textit{extension}$$

$$\textit{F} = \textit{k e}$$

- F = force (unit: **newton**, N)
- k = spring constant (unit: newton per metre, N/m)
- e = extension, or compression (unit: metre, m)

DO IT!

Write a social media post describing the difference between elastic and inelastic deformation.

Practical Skills

In class, you will have investigated the relationship between force and extension for a spring. Think about how you calculated the spring constant of the spring from the data you collected. You could have drawn the graph linking the force and the extension as either a force–extension graph or an extension–force graph.

Force–extension graph:

If the graph is plotted with force on the y-axis and extension on the x-axis, the equation is:

$$F = k\,e$$

So, the spring constant k is the gradient of the line.

The gradient is always calculated as gradient $= \dfrac{\Delta y}{\Delta x} \rightarrow$ make sure you draw a big right-angled triangle on the graph to find Δy and Δx.

- $\Delta y \rightarrow$ highest y − lowest y (the height of your triangle)

- $\Delta x \rightarrow$ highest x − lowest x (the base of your triangle)

So, in our graph:

$$k = \text{gradient} = \frac{\Delta y}{\Delta x} = \frac{100 - 10}{1.20 - 0.12} = \frac{90}{1.08} = 83.3 \text{ N/m}$$

The spring constant of our spring is 83.3 N/m.

Extension–force graph:

If the graph is plotted with extension on the y-axis and force on the x-axis, the equation is:

$$e = \frac{F}{k}$$

So, the gradient of the line is 1/k.

In the graph:

$$\text{gradient} = \frac{\Delta y}{\Delta x}$$
$$= \frac{10.80 - 0.100}{90 - 8}$$
$$= \frac{0.980}{82}$$
$$= 0.112$$

But in this graph:

$$\text{gradient} = \frac{1}{k} \Rightarrow k = \frac{1}{\text{gradient}} = \frac{1}{0.012} = 83.3 \text{ N/m}$$

The spring constant of our spring is again 83.3 N/m.

The symbol Δ is the Greek letter Delta, which can be used as an abbreviation for 'change in' or 'difference between'.

MATHS SKILLS

The equation shows that the force on an elastic object, such as a spring, is **directly proportional** to the extension (or to the compression) of the object. However, the equation does not work any more if the spring goes above its **limit of proportionality** (because then it becomes permanently deformed).

The equation is a straight line (**linear**) relationship. So, you might be asked to find the spring constant (k) using experimental data plotted on a F-e graph and by calculating the gradient of the line.

If the spring extends above its limit of proportionality, its f-e relationship becomes **non-linear**. That is, its f to e relationship is no longer directly proportional.

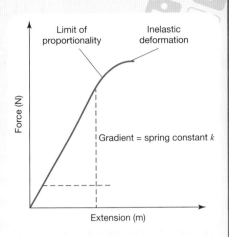

Work done and energy transfer

A force stretching or compressing a spring does work on the spring. So **energy** is transferred to the elastic potential energy store associated with the spring. If the spring has not exceeded its limit of proportionality (if it is not inelastically deformed):

- the work done on the spring = the elastic potential energy stored

We can calculate the work done on a spring when stretching or compressing it using the equation:

elastic potential energy = 0.5 × spring constant × (extension)²

$$E_e = \frac{1}{2} ke^2$$

This equation will be given to you, but make sure you know how to use it. It is also used in the Energy chapter (page 241).

CHECK**IT!**

1 What is the difference between elastic and inelastic deformation?

2 A rubber band extends by 5 cm when pulled with a force of 3.5 N. What is the spring constant of the rubber band?

3 What is the elastic potential energy of the rubber band in Question 2? What is the work done on it?

Distance, displacement, speed and velocity

Distance and displacement

Distance is a **scalar** quantity. This means that it does not have a direction.

For example, if you walked along a straight line of 100 m and then walked back to the start, you will have walked a distance of 200 m.

Displacement is a **vector** quantity. This means that it has both magnitude and direction.

Therefore, if you walked along the straight line of 100 m and then back to the start, your total displacement will have been 0 m. This is because the displacement in one direction cancels out the displacement in the other direction.

Speed and velocity

Speed is also a scalar quantity, so it does not have a direction.

Moving objects and people often change speed. The speed of a person walking, running or cycling depends on lots of factors, like the person's age, how hilly/flat the land is, how fit the person is and the distance travelled.

We can calculate the distance travelled by an object moving at a constant speed using the equation:

$$\textbf{distance travelled} = \textbf{speed} \times \textbf{time}$$

$$s = vt$$

s = distance (unit: metre, m)

v = speed (unit: metre per second, m/s)

t = time (unit: second, s)

NAIL IT!

Make sure you can remember typical values of speed for walking (about 1.5 m/s), running (about 3 m/s), cycling (about 6 m/s) and different types of transportation (like cars).

The speed of sound and the speed of the wind also change in different conditions. A typical value for the speed of sound in air is 330 m/s.

NAIL IT!

Make sure you can remember and use this equation.

MATHS SKILLS

The average speed of an object travelling with non-uniform motion (changing speed) can be thought as the constant speed the object would need to travel at to cover the distance travelled in the time taken to travel that distance.

To calculate average speed, you can rearrange the equation to:

$$v = \frac{s}{t}$$

WORKIT!

A girl usually takes 20 minutes to walk to school. Use suitable values and calculations to estimate the distance of the school from the girl's house. (1 mark)

Time taken in seconds is:

$t = 20 \times 60 = 1200$ s

If the girl is travelling at a constant speed, the distance travelled can be calculated as:

$s = vt = 1.5 \times 1200 = 1800$ m

A typical value for the speed of a person.

Distance $= 1.8$ km (approximately) (1)

The **velocity** of an object is a vector quantity, so it has both magnitude and direction. It is the speed of the object in the direction it is moving.

Circular motion

An object moving at a constant speed in a circle has a changing velocity. This is because the direction of the object keeps changing.

When a sparkler is moved in a circle at high speed, it is constantly changing direction. However, the sparks that detach from the sparkler move in a straight line in the direction the sparkler was moving when these sparks were formed.

SNAPIT!

Push a tennis ball around the inside of a hula hoop. The contact force between the ring and the ball keeps the ball moving in a circle. After one or two circles, lift the hula hoop. Take a video with your phone, and explain why the ball moves in a straight line when the hoop is lifted.

CHECKIT!

1 What is the difference between displacement and distance, and the difference between velocity and speed?

2 A builder takes 12 minutes to cycle to work. What is the estimated distance from his house to his place of work?

3 Mo Farah's time for the World Half Marathon in Cardiff 2016 was 59 minutes and 59 seconds. The distance of a half marathon is 21.098 km. What was Mo Farah's average speed in this competition?

Distance–time relationship

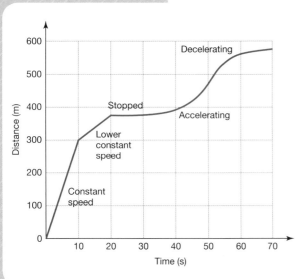

The distance travelled by an object moving in a straight line can be represented by a distance–time graph.

The gradient of a distance–time graph (how 'steep' the line is) represents the **speed** of the object. The greater the value of the gradient (the steeper the line), the faster the object.

In a distance–time graph, we can identify important features:

- A straight line means constant speed (constant gradient)

- A horizontal (flat) line means stationary (stopped)

- A steepening curve means **accelerating** object (increasing gradient)

- A flattening curve means **decelerating** object (decreasing gradient).

NAILIT!

Make sure you know how to find out the speed of an object from the gradient of a graph.

SNAPIT!

Push a toy car, or an object that can slide along a table, for 10 cm at steady speed in 2 seconds. Stop the object at 10 cm for 4 seconds and then move it further on the same line for 30 cm in 3 seconds. Take a video of the object moving and plot a distance–time graph of the object's motion.

WORKIT!

The graph shows the distance travelled by a cyclist over 70 seconds. Calculate the speed of the cyclist between 50 and 70 seconds. (1 mark)

$$\text{Gradient} = \frac{\Delta y}{\Delta x} = \frac{\text{change in distance}}{\text{change in time}}$$

$$= \frac{400 - 100}{70 - 50} = \frac{300}{20} = 15\,\text{m/s} \ (1)$$

Remember that the symbol Δ is just a short way of saying 'change in' or 'difference between two values'.

To calculate the speed of the cyclist between 50 and 70 seconds, we need to calculate the gradient of the line for that time interval.

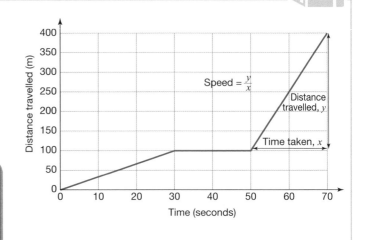

MATHS SKILLS

You might be asked to calculate the speed of an object that is accelerating or decelerating. To do this, you need to draw the **tangent** to the graph at the point needed. The tangent is a straight line that is at a right angle (90°) to the curve at a specific point.

For example, look at the distance–time graph of Usain Bolt breaking the 100 m world record in Beijing in 2008. If you want to calculate Bolt's speed 2 seconds into the race, you need to calculate the gradient of the curve at 2.00 seconds. So, first draw a long straight line that is 90° to the curve at 2.00 seconds (the red line in the diagram). Then work out the gradient of the straight line:

$$\text{Gradient} = \frac{\Delta y}{\Delta x} = \frac{52 - 0}{7.00 - 0.75} = \frac{52}{6.25} = 8.32 \text{ m/s}$$

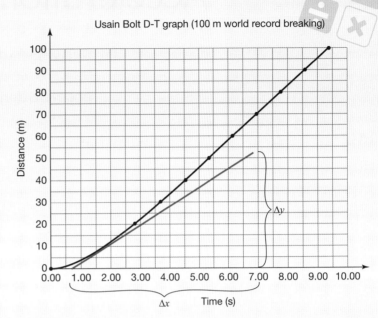

Usain Bolt D-T graph (100 m world record breaking)

H

CHECK**IT!**

1 What does the gradient of a distance–time graph tell you?

2 Why does a flattening curve show a decelerating object?

3 A distance–time graph for a racing car shows a horizontal line along the time axis between 0.00 and 0.62 seconds, then a sharp curve of increasing gradient. What might the horizontal line mean?

Acceleration

We can calculate the average **acceleration** of an object using the equation:

$$acceleration = \frac{change\ in\ velocity}{time\ taken}$$

$$a = \frac{\Delta v}{t}$$

- a = acceleration (unit: **metre** per **second** squared, m/s²)
- Δv = change in **velocity** (unit: metre per second, m/s)
- t = time (unit: second, s)

In a velocity–time graph, the gradient represents the acceleration of the object.

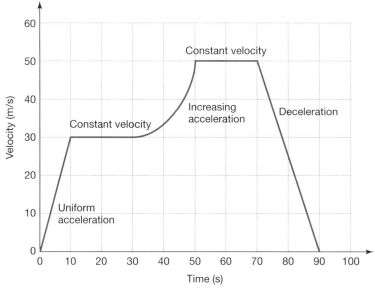

An example of a velocity-time graph

In a velocity–time graph, we can identify important features:

- A straight line means constant acceleration (constant gradient).
- A horizontal (flat) line means constant **velocity** (no acceleration).
- A steepening curve means increasing acceleration (increasing gradient).
- A straight line pointing down means slowing down (**deceleration**) (negative gradient).

NAIL IT!

When you are given a graph, make sure you always check the axes first before you do anything else:

- On *distance–time graphs*, the gradient represents *velocity*.
- On *velocity–time graphs*, the gradient represents *acceleration*.

MATHS SKILLS

You might be asked to use a velocity–time graph to calculate the distance travelled (the **displacement**) by an object. You can do this by calculating the area under the velocity–time graph.

In some velocity–time graphs, you might need to count the squares under the graph to estimate the displacement.

For example, to calculate the distance travelled in the first 30 seconds by the object in the graph, you can calculate the area under the line between 0 and 30 seconds (in the graph, the red triangle between 0 and 10 seconds and the red rectangle between 10 and 30 seconds).

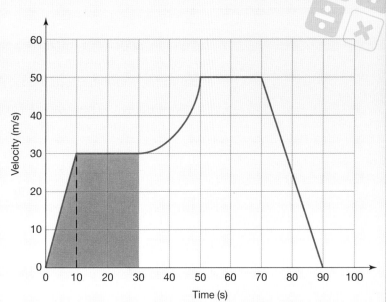

Using a velocity–time graph to calculate displacement

Area of triangle $= \frac{1}{2} \times$ base $\times$ height

$$= 0.5 \times (10 - 0) \times (30 - 0) = 0.5 \times 10 \times 30 = 150\,m$$

area of rectangle = base $\times$ height $= (30 - 10) \times (30 - 0) = 20 \times 30 = 600\,m$

So, the total distance travelled is $150 + 600 = 750$ m.

When an object is moving with uniform acceleration, we can use the equation:

(final velocity)² − (initial velocity)² = 2 × acceleration × distance

$$v^2 - u^2 = 2\,a\,s$$

- v = final velocity (unit: metre per second, m/s)
- u = initial velocity (unit: metre per second, m/s)
- a = acceleration (unit: metre per second squared, m/s²)
- s = distance (unit: metre, m)

Near the surface of the Earth, all objects that fall freely because of **gravity** fall with a constant acceleration of about 9.8 m/s².

WORKIT!

A car accelerates with constant acceleration from 12 m/s to 17 m/s. If it took 13 m to reach 17 m/s, what is the acceleration of the car? (2 marks)

$a = \dfrac{v^2 - u^2}{2s} = \dfrac{17^2 - 12^2}{2 \times 13}$ (1) ◄

Rearrange the equation $v^2 - u^2 = 2\,a\,s$

$= \dfrac{289 - 144}{26} = \dfrac{145}{26}$

$= 5.6\ m/s^2$ (1)

NAILIT!

This equation will be given to you, but make sure you know how to use it.

DOIT!

Search on the internet the typical acceleration of different objects, people and animals. For example, a fighter plane, a family car, a racing car, Usain Bolt, a cheetah, a bicycle, and so on.

303

Terminal velocity

An object falling through a fluid, like a ping-pong ball falling in air, will at first accelerate. This is because the gravitational pull of the Earth is greater than the air resistance on the object.

But as the velocity of the falling object increases, so does the upward **force** on the object (the air resistance), until the magnitude of the upward force becomes equal to the magnitude of the downward force on the object (its weight).

So, eventually, the **resultant force** on the object will be 0 N, and it will now fall at a constant speed. This is called the object's terminal velocity.

The velocity–time graph shows how the velocity of a skydiver changes during her fall.

H

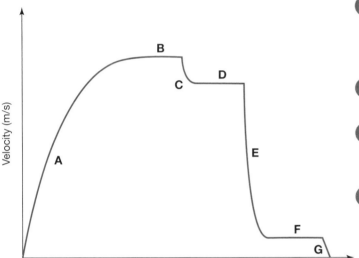

A The skydiver jumped and is accelerating. As her velocity increases, the air resistance on her body also increases, so the resultant force on her body decreases.

B The skydiver is moving at her terminal velocity.

C The skydiver spreads her arms and legs, increasing her surface area. This increases the air resistance, causing her to decelerate.

D The skydiver is moving at the terminal velocity for her spread-out body. (Her terminal velocity changed at C because the air resistance decreased when her velocity decreased).

E The skydiver opens her parachute. This increases her surface area dramatically, so the air resistance on her parachute also increases. She decelerates very quickly.

F The skydiver is moving at the terminal velocity for her increased surface area from her opened parachute.

G The skydiver decelerates as she lands on the ground.

CHECK IT!

1 Is acceleration a **scalar** quantity or a **vector** quantity?

2 A cheetah changes its speed from 18 m/s to 27 m/s in 8.2 seconds. What is its acceleration?

3 A professional racing car can go from 0 to 100 km/h with an acceleration of 2 m/s^2. Assume the acceleration is uniform. What distance will the car travel in performing this change in velocity?

Newton's laws of motion

Newton's first law

When the **resultant force** acting on an object is zero, the object will either:

- Stay stationary if the object was already stationary ('at rest').

- Continue moving with a constant **velocity** if the object was already moving. This means that it will continue moving both at the same **speed** and in the same direction.

So, the velocity will not change unless a resultant force (different from zero) is acting on the object.

Driving force

Resistive forces

For example, when a car is travelling at constant speed along a straight line, resistive forces (like air resistance) are balancing the driving force.

So, if an object is at rest it will stay still unless acted upon by an unbalanced force. If an object is moving it will continue to move at the same speed with no acceleration or deceleration until acted upon by a resultant force.

 STRETCHIT!

When you are on a bus and the bus starts moving, your body feels pulled backward. In the same way, when the bus stops, your body feels pushed forward. This happens because all objects have a tendency to stay in their state of rest or at constant motion. This tendency is called **inertia**.

 SNAPIT!

Put a tennis ball on top of a skateboard and push the skateboard forward gently, but allow it to stop against a wall. Make a video with a smartphone and comment on how this explains the inertia of the tennis ball.

NAILIT!

Make sure you can remember and use this equation.

Newton's second law

Newton's second law can be shown as an equation:

$$resultant\ force = mass \times acceleration$$

$$F = m\,a$$

- F = resultant force (unit: **newton**, N)
- m = **mass** (unit: **kilogram**, kg)
- a = **acceleration** (unit: **metre** per second squared, m/s^2)

This equation can be rearranged to:

$$a = \frac{F}{m}$$

This shows that the acceleration of an object is **directly proportional** to the resultant force acting on it, and that the acceleration is inversely proportional to the mass of the object.

In other words, the effect of the **resultant force** (the **acceleration**) is the cause (resultant force) divided by the **resistance** to the cause (mass or **inertial** mass).

MATHS SKILLS

Remember that the symbol for proportionality is ∝. Therefore:

- $a \propto F$
- $a \propto \dfrac{1}{m}$

STRETCH IT!

Inertial mass is a measure of how difficult it is to change the velocity of an object. You can see what this is when we rearrange the equation for Newton's second law to:

H

$$m = \frac{F}{a}$$

Inertial mass is defined as the ratio of force over acceleration.

Practical Skills

You will have investigated in class how changing the force affects the acceleration of an object of constant mass, and how changing the mass of an object affects its acceleration caused by a constant force. Think about the method you used.

You might have used a data logging system to measure the velocity of a vehicle on an air track at two different points, while you changed the mass or the force on the vehicle. Whatever method you used, to increase the accuracy of your results, it is important to repeat the measurements of acceleration for different masses (or forces), and then to calculate a mean value from your measurements.

You might have calculated the mass (or force) from the gradient of a graph of your results. For example:

- For constant force and changing mass, if you plotted a graph of acceleration against $\frac{1}{mass}$, you could calculate the force as the gradient of your graph.

- For constant mass and changing force, if you plotted a graph of acceleration against force, you could calculate the mass as $\frac{1}{gradient}$ of the straight line. This is because gradient $= \frac{1}{m}$.

Newton's third law

When two objects interact, they exert forces on each other that are equal in magnitude and opposite in direction.

The weight of the book is balanced by the normal contact force of the table on the book. These two forces are equal and opposite, but they are **NOT** an example of **Newton's third law**.

Remember that **interaction pairs** (Newton's third law forces) must always:

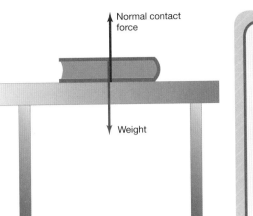
Normal contact force

Weight

- Act between two objects → the **gravitational pull** of the Earth on the book (**weight**) and the **gravitational pull** of the book on the Earth.

- Have the same nature → the **electrostatic repulsion** of the atoms of the table on the atoms of the book and the **electrostatic repulsion** of the atoms of the book on the atoms of the table.

In the example, both forces act on a single object (the book) and are different in nature – the **weight** is a **non-contact force** (**gravity**) and the **normal contact force** is an electrostatic force.

WORKIT!

A child of mass 35 kg jumps to the ground from a small wall. The child is pulled by the Earth with a force of 343 N. Use Newton's second and third laws to explain why the child moves toward the Earth, but the Earth does not seem to move toward the child at all. The mass of the Earth is 5.972×10^{24} kg. (3 marks)

Due to Newton's third law the Earth and the child attract each other with equal and opposite forces $F = 343$ N. (1)

If we use Newton's second law to calculate the acceleration of the child and the acceleration of the Earth caused by this force, we can see that the effect of the force (the acceleration) will be much bigger for the child than for the Earth, because of the much greater mass (inertia) of the Earth. (1)

a_c = child's acceleration m_c = child's mass

a_e = Earth's acceleration m_e = Earth's mass

> Due to Newton's third law the Earth and the child attract each other with equal and opposite forces F = 343 N.

$$a_c = \frac{F}{m_c} = \frac{343}{35} = 9.8 \text{ m/s}^2$$

$$a_e = \frac{F}{m_e} = \frac{343}{5.972 \times 10^{24}} = 5.7 \times 10^{-23} \text{ m/s}^2 \quad (1)$$

CHECKIT!

1 What needs to happen to change the velocity of an object?

2 What does Newton's second law tell you?

H 3 How are inertia and inertial mass related?

Stopping distance

The stopping distance of a vehicle is:

stopping distance = thinking distance + braking distance

- **Thinking distance** is the distance travelled by the vehicle during the driver's reaction time.

- **Braking distance** is the distance travelled by the vehicle because of the braking force.

The **speed** of the vehicle affects both the braking and stopping distance. In fact, the faster the vehicle, the longer it will travel during the reaction time of the driver, and the longer the distance needed to break with the same braking **force**. So, for a given braking force:

- the *faster* the vehicle, the *greater* the stopping distance.

The mind map below shows some factors that affect thinking distance and braking distance. Adverse road conditions include wet or icy roads. Vehicles might also have poor brakes and/or poor tyres.

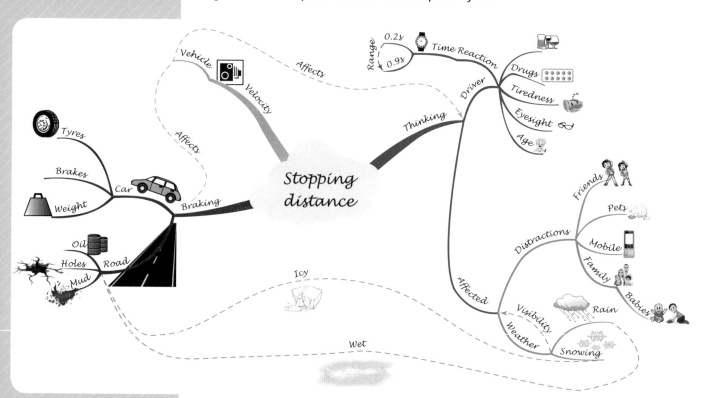

To reduce the **kinetic energy** of a vehicle, work is done by the **friction** force between the brakes and the wheels when the braking force is applied to the brakes. This causes the temperature of the brakes to increase.

For *higher* speeds, *greater* braking forces are needed to stop the vehicle within a given distance. *Greater* braking forces cause *greater* decelerations. These might cause brakes to overheat, due to increased friction, and/or the driver to skid and lose control of the vehicle.

WORKIT!

Distances

A car is travelling at 22 m/s on the motorway when the driver notices an accident ahead. Estimate the thinking distance travelled by the car after the driver notices the danger. (2 marks)

Typical reaction times are between 0.2 and 0.9 seconds. This means that the average reaction time of a person is ~0.6 s. (1) So:

$$s = vt = 22 \times 0.6$$
$$= 13.2 \text{ m (1)}$$

> The thinking distance is the distance travelled by the vehicle during the reaction time of the driver.

Forces and deceleration

A lorry travelling with kinetic energy 700 kJ performs an emergency stop in 124 m. Estimate the average braking force needed to stop the lorry. (2 marks)

The equation for work done is $W = Fs$. So:

$$F = \frac{W}{s} \text{ (1)}$$
$$= \frac{700\,000}{124}$$
$$= 5645 \text{N (1)}$$

> The work done by the braking force to stop the lorry needs to be at least equal to the kinetic energy of the lorry.

CHECK IT!

1 What factors affect the stopping distance of a vehicle?

2 Explain the potential dangers of rapid decelerations of vehicles.

3 A young and healthy driver has a reaction time of 0.3 seconds. If the thinking distance after the driver noticed a danger was 6.5 m, at what speed was the driver's car travelling?

Momentum

H

Momentum is a property of moving objects. It can be defined using the equation:

$$Momentum = mass \times velocity$$

$$p = mv$$

- p = momentum (unit: **kilogram metre** per second, kg m/s)
- m = **mass** (unit: kilogram, kg)
- v = **velocity** (unit: metre per second, m/s)

In a system that is not affected by any external forces (closed system), momentum is always conserved.

This means that the total momentum before an event (like a collision) is always the same as the total momentum after the event.

This is called conservation of momentum.

MATHS SKILLS

The equation shows that:

- the *greater* the mass, the *greater* the momentum
- the *greater* the velocity, the *greater* the momentum

NAILIT!

Momentum is a **vector quantity** because it depends on the object's velocity.

This explains why two identical balls that travel at the same **speed** in opposite directions (*v* and –*v*) and then hit each other head-on will stop after they collide. The momentum of each ball cancels out the momentum of the other ball. Therefore, the total momentum is zero before *and* after the collision.

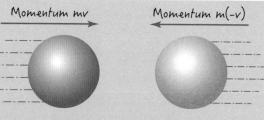

Momentum *mv*   Momentum *m*(–*v*)

MATHS SKILLS

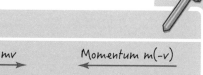

You can show this change in momentum by combining the equations for force (*F*) and acceleration (*a*)

$$F = m \times a \quad \text{and} \quad a = \frac{\Delta v}{t}$$

to get: $F = \frac{m\Delta v}{\Delta t}$

- F = force
- Δv = change in velocity
- m = mass
- Δt = change in time, or the time taken

In our new equation, $m\Delta v$ = change in momentum.

So, the force applied on an object = the rate of change in momentum.

Changes in momentum

When a **resultant force** different from zero acts on an object that can move, or that is already moving, the velocity of the object changes.

When the velocity changes, the momentum changes too.

We can improve the safety of lots of things by reducing the rate of change in momentum. For example:

- Soft surfaces like crash mats in gyms and cushioned surfaces in play areas cause the change in velocity to happen over a longer time. If a fall or collision happens, the surface reduces the rate of change in momentum and therefore, the **force** applied on the body.

- Air bags and seat belts help to increase the time taken by the driver/passengers to stop moving. This again results in a reduced rate of change in momentum.

- Cycle and motorcycle helmets reduce the rate of change in momentum of your head colliding with an obstacle, so the force on it is also reduced.

DO IT!

Find at least one object/tool used as a safety feature and explain how it works using rate of change of moment in your explanation.

WORKIT!

1 A toy car of mass $m_1 = 50$ g hits a stationary toy car of mass $m_2 = 78$ g at a velocity of 5 m/s. The stationary car has a bit of Blu-Tack on it, so the two cars stay attached to each other after they have collided. Calculate the total momentum of the two cars before and after the collision, and calculate the velocity of the two cars after the collision. (3 marks)

The total momentum is the same before and after the collision. The total momentum before the collision is:

$$P_{tot} = m_1 v_1 + m_2 v_2 = 0.050 \times 5 + 0.078 \times 0$$
$$= 0.25 + 0 = 0.25 \text{ kg m/s (1)}$$

So, the momentum of the two attached toy cars after the collision must also be 0.25 kg m/s, and their combined mass is now:

$$m_{tot} = m_1 + m_2 = 0.050 + 0.078$$
$$= 0.128 \text{ kg (1)}$$

and rearranging the momentum equation:

$$v = \frac{P}{m} = \frac{0.25}{0.128}$$
$$= 1.95 \text{m/s (1)}$$

2 A bullet is fired at 420 m/s at a wooden target and it lodges itself in the wood in 0.12 seconds. If the mass of the bullet is 4.2 g, what is the force applied by the bullet on the target? (3 marks)

Change in speed $\Delta v = 0 - 420 = -420$ m/s. (1) (The negative sign just shows that the bullet has slowed down. It is not important here, so we will carry on without it).

The mass of the bullet in kg is $m = 0.0042$ kg. Therefore:

$$F = \frac{m\Delta v}{\Delta t}$$
$$= \frac{0.0042 \times 420}{0.12} \text{ (1)}$$
$$= 14.7 \text{ N (1)}$$

CHECKIT!

1 What is the momentum of an object?

2 What is meant by conservation of momentum?

3 A tennis ball of mass 58.5 g moves at 17 m/s. What is its momentum?

1 a Draw the resultant force from the force arrows below.

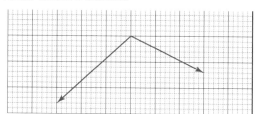

b Resolve the force in its vertical and horizontal components, and label each component F_y and F_x, respectively.

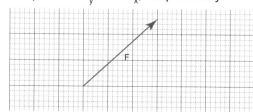

2 Copy the diagram below and draw all the gravitational forces applied in this system, and explain why you drew those forces and with that size.

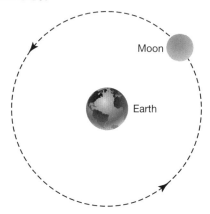

3 Draw and label the resultant force in this situation, making sure you draw the resultant force to the correct scale and label it with the correct value of force in newtons.

4 Two horses pull the Queen's carriage with a force of 523 N each. If the total work done by their forces is 378 652 J, calculate the distance they have travelled.

5 Usain Bolt takes 3.78 s to run the first 30 m from stationary, but he takes 3.95 s from the moment the starting gun is fired. Calculate the two average speeds to run 30 m from these two times, and suggest why there is a difference.

6 If the combined mass of this cyclist and bicycle is 81.5 kg, calculate the acceleration of the cyclist.

Transverse and longitudinal waves

Waves are **oscillations/perturbations** that travel through a **medium** (material).

Waves carry **energy** and can also carry information between different places.

It is the wave that travels, not the particles of the medium. We can see this because the water particles in a ripple tank only move up and down as the waves pass through a point, but they are not carried forward by the wave.

Transverse waves move up and down at right angles (90°) to the direction of travel of the wave. Examples of transverse waves are ripples in water and **electromagnetic waves**.

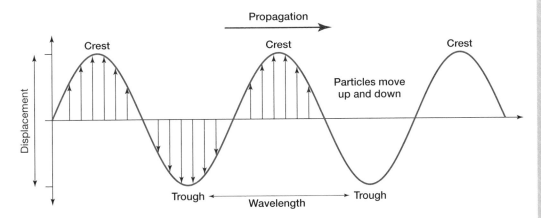

Longitudinal waves move along (parallel to) the direction of travel of the wave. Longitudinal waves have areas of compression and rarefaction. An example of longitudinal waves is **sound waves**.

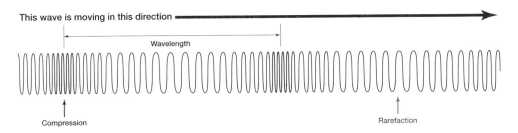

Properties of waves

Waves have four main properties that help us to describe how they move.

Amplitude is the maximum displacement of a point on a wave from its undisturbed position (from the horizontal middle line in the diagram).

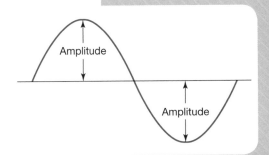

Wavelength (λ) is the distance between a point on a wave and the equivalent point on the adjacent wave (where the wave repeats itself). This is shown in the graph to the right.

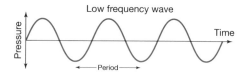

Frequency is the number of waves passing a point each second.

The wave speed is the speed the wave is moving through the medium or the speed the energy is being transferred in that medium.

The wave equation applies to all waves. It shows how wave speed, wavelength and frequency are related:

$$\textbf{wave speed = frequency} \times \textbf{wavelength}$$

$$\textbf{\textit{v} = \textit{f} } \boldsymbol{\lambda}$$

- v = wave speed (unit: metre per second, m/s)
- λ = wavelength (unit: metre, m)
- f = frequency (unit: hertz, Hz)

The period of a waves is the time taken for a full wave to pass through a point. The relationship between the period of a wave and its frequency is show in the equation:

$$\textbf{\textit{Frequency}} = \frac{\textbf{1}}{\textbf{\textit{period}}}$$

$$\textit{f} = \frac{1}{\textit{T}}$$

- T = period (unit: second, s)
- f = frequency (unit: hertz, Hz)

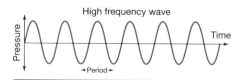

MATHS SKILLS

When waves go from a less dense medium to a denser medium, they slow down and their wavelength decreases. However, their frequency stays the same.

The diagram shows that the wavelength decreases as the ripples go into shallower water (denser medium). As the wave equation informs us, due to the frequency staying the same, the wavelength decreases and therefore the wave speed (the velocity) must decrease too.

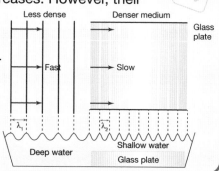

One way to measure the speed of sound in air is to put two microphones at a known distance between each other (the further, the better). Then, strike together two metal rods above one of the microphones. If we connect both microphones to an oscilloscope, or to a computer and use sound recording software, we can then calculate the time it took for the sound to travel from the microphone below the rods to the microphone further away.

We can calculate the speed of sound in air using the speed equation $v = \frac{s}{t}$ (from page 298 in the Forces chapter).

Practical Skills

You will have investigated in class how to find the speed of water waves in a ripple tank and the speed of waves in a solid medium. Think about what apparatus you needed and why it was suitable for measuring the frequency and wavelength of the waves.

For example, using the set up in the diagram to the right, putting a ruler on the white screen below the tank and using the stroboscope should let us take fairly accurate readings of the wavelength (the distance between the wave fronts).

Measuring the time taken for 10 wave fronts to pass a point, and dividing by 10, should give a fairly accurate measure of the period. For even better accuracy, we could take repeated readings and then calculate a mean value from these. We can then calculate the frequency as $f = 1/T$.

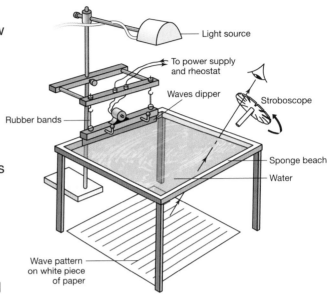

To calculate the speed of a wave in a solid, one end of a long copper rod/tube could be hit by a hammer and the signal recorded by two microphones placed next to each end of the rod could be displayed on an oscilloscope.

The time difference between the signals is the time taken for the sound wave to travel along the rod, and by measuring the distance of the rod, you can calculate the speed of the wave.

For example, a 4 m copper rod shows two peaks on the oscilloscope 8.7×10^{-4} seconds apart. Therefore the speed of the sound wave in copper is:

$$\text{speed} = \frac{\text{distance}}{\text{time}}$$

$$v = \frac{s}{t} = \frac{4}{8.7 \times 10^{-4}} = 4598 \ m/s$$

Remember that in the equation above, s is the distance travelled.

WORKIT!

A guitarist plays an 'A' note with frequency of 440 Hz. Calculate the wavelength of this sound wave. (2 marks)

The speed of sound in air is usually about 330 m/s (1), so rearranging the wave equation $v = f\lambda$:

$$\lambda = \frac{v}{f} = \frac{330}{440} = 0.75m = 75cm \ (1)$$

CHECKIT!

1 How can you work out the amplitude of a wave from a graph?

2 The frequency of the musical note 'middle C' is 261.6 Hz. Calculate the period of this wave.

3 A series of wave fronts in a ripple tank are generated with a period of 0.25 seconds, and the distance between the ripples is 1.2 cm. Calculate the speed of these water waves.

Reflection and refraction

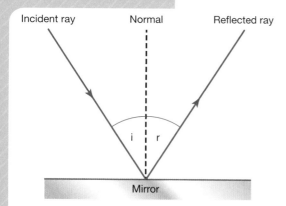

Incident ray   Normal   Reflected ray

i   r

Mirror

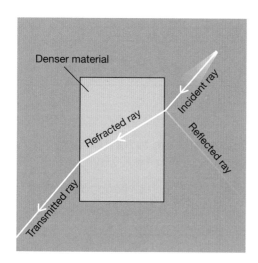

Denser material

Refracted ray

Incident ray

Reflected ray

Transmitted ray

A wave can be reflected at the boundary between two different materials. This means the wave bounces off the boundary at the same angle from the 'normal'.

The 'normal' is an imaginary line at a right angle (90°) to the boundary.

- angle of incidence (i) = angle of reflection (r)

A wave can be transmitted by a material. This means the material lets the wave through, and the wave (or part of it) comes out of the other side of the material.

A wave can be absorbed by a material. This means the material does not transmit or reflect the wave.

A wave can be **refracted** at the boundary between two different materials of different density. This means the wave changes direction at the boundary at a different angle from the 'normal'.

- If the transmitting material is denser than the first **medium**, the wave will slow down and bend towards the normal to the boundary.

- If the transmitting material is less dense than the first medium, the wave will speed up and bend away from the normal to the boundary.

- The angle of **refraction** of a wave transmitted through a material also depends on the **frequency** of the wave.

SNAPIT!

Pour some water in a glass and put the glass over a lamp until you see a reflection of the lamp on the surface of the water. Take a photo of the glass and the water from above. What is happening to the light from the lamp at the boundary between the air and the water?

 Practical Skills

You will have investigated in class the reflection of light by different surfaces. Think about the apparatus you used. You might have used a light sensor to measure the light intensity before and after the reflection. This method would let us calculate the percentage of light reflected by each material. What were the possible sources of error? For example, some light might have dissipated as the travelling light ray spread out. This can be reduced by using lenses to focus the light rays more effectively.

Practical Skills

You might also have investigated in class the refraction of light by different substances. Think about how you could have increased the accuracy of your results. For example, by repeating the measurements of the angles of refraction for the same angles of incidence and then using these measurements to calculate the mean values.

STRETCH IT!

When a wave goes from a denser to less dense material, it refracts (bends) away from the 'normal'. When the refracted wave is at a right angle (90°) from the normal, the angle of incidence is called the critical angle. For any angle bigger than the critical angle, the wave will be reflected back into the denser material. This is called total internal reflection.

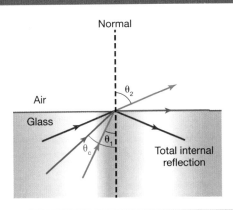

DO IT!

Put a pencil in a glass of water. Using the term refraction explain what you see.

CHECK IT!

1 What happens to a light ray incident on a mirror at 32° from the 'normal' to the mirror?

2 A wave travels from a less dense material to a more dense material, hitting the denser material at an angle more than 0° from the 'normal' to the boundary. What happens when the wave is transmitted by the denser material?

3 Jade looks at white light from a lamp through a red filter. The lamp appears to emit only red light to Jade. What has happened to the other frequencies of the light coming from the lamp?

Electromagnetic waves

Electromagnetic waves transfer **energy** from a source to an absorber. For example, the infrared signal sent by the transmitter in a remote control sent to the receiver on your TV, or the ultraviolet radiation emitted by the Sun and absorbed by the skin, which gets tanned.

Electromagnetic waves are **transverse** waves. All electromagnetic waves travel at the same **velocity** in a vacuum (empty space) or in air.

Electromagnetic waves make up a continuous spectrum called the **electromagnetic spectrum**, where they are grouped according to their **wavelength** and **frequency**.

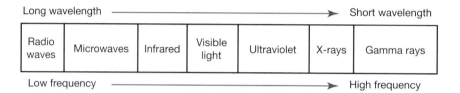

Long wavelength ⟶ Short wavelength

Radio waves	Microwaves	Infrared	Visible light	Ultraviolet	X-rays	Gamma rays

Low frequency ⟶ High frequency

The human eye can detect electromagnetic waves only within a limited range of frequencies, called **visible light**. However, we can detect the other waves of the electromagnetic spectrum indirectly using special instruments.

Properties of electromagnetic waves

Electromagnetic waves of different wavelengths are absorbed, transmitted, refracted or reflected in different ways depending on the materials they are travelling in. This affects things like the colour of objects.

Some effects happen because the waves have different velocities in different materials. For example, electromagnetic waves can be refracted at the boundary of two different materials. We can show how this happens using wave front diagrams, as shown below.

STRETCH IT!

How the wave bends depends on the densities of the different materials. If the wave starts at an angle more than 0° from the 'normal' (the bottom diagram), the wave bends toward the 'normal' (in a denser material, shown in the diagram) or away from the 'normal' (in a less dense material).

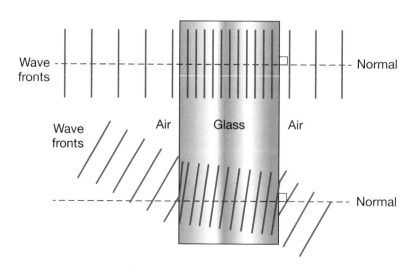

Refraction through a parallel plate

The wave slows down in the denser material (the glass in the diagram). So, the wave fronts get closer together in the denser material. The wave speeds up again in the less dense material (the air in the diagram). So, the wave fronts spread out again.

Oscillations in electrical circuits can generate radio waves, as shown in this diagram.

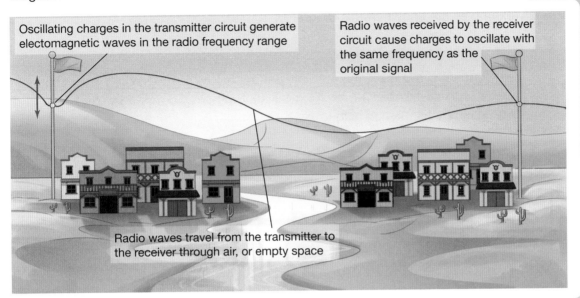

Oscillating charges in the transmitter circuit generate electomagnetic waves in the radio frequency range

Radio waves received by the receiver circuit cause charges to oscillate with the same frequency as the original signal

Radio waves travel from the transmitter to the receiver through air, or empty space

Dangers of absorbing electromagnetic radiation

The range of frequencies of electromagnetic waves in the electromagnetic spectrum can be generated by changes in atoms and in their nuclei. For example, γ-rays are generated by changes in the nuclei of atoms.

Absorbing electromagnetic waves can also result in changes in atoms and their nuclei.

The absorption of ultraviolet waves, X-rays and γ-rays are ionising and can damage human body tissue. These higher frequency EM waves carry more energy and can therefore penetrate the body further than lower frequency EM waves.

The effect of these hazards on body tissue depends on the type of radiation and the size of the dose. We measure radiation dose in sieverts (Sv), which is a measure of the risk of harm resulting from an exposure of the body to the radiation.

1000 millisieverts (mSv) = 1 sievert (Sv)

WORKIT!

It is recommended that people working in radiation environments should not be exposed to more than 20 mSv per year. A medical physicist who works with γ-rays in a hospital is exposed to 0.08 mSv each day. Is the environment in which she works following the recommendations? The employee works 5 days a week. (2 marks)

There are 52 weeks in a year, so for a 5-day working week the medical physicist is exposed to
0.08 × 5 = 0.4 mSv each week
and 0.4 × 52 = 20.8 mSv per year. (1)

The radiation dose the physicist is exposed to each year is more than the recommended dose. This shows that the hospital is not following the recommendations. (1)

NAILIT!

Some of the hazards to human tissue can be:

• Ultraviolet waves → increase risk of skin cancer and cause skin to age prematurely.

• X-ray and γ-rays → **ionising** radiation that can cause mutation of genes and cancer.

SNAPIT!

Take a photo of the applications of electromagnetic waves diagram on the next page. Explain why each of the electromagnetic waves is suitable for the practical application. Use the photo to revise on the go.

Uses and applications of electromagnetic waves

Electromagnetic waves are used for a wide range of applications. Some examples are shown in this diagram.

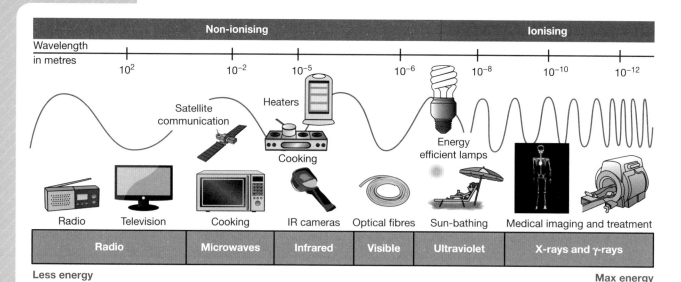

Practical Skills

You will have investigated in class how different surfaces absorb or radiate (emit or give out) infrared radiation. Think about the skills you needed. You probably focused on the method you used to collect accurate measurements (or data). For example, if you covered the sides of different beakers with sheets of different colours and/or materials, and then measured the temperature of hot water poured inside the beakers over time, a few things would have affected your results. Here are some ideas:

• Water can evaporate from the top of the beakers, so it is important to cover them with insulating material to reduce this effect.

• The thermometer/temperature probe should be completely immersed in the water and not touching the glass of the beaker, because you want to measure the temperature of the water.

• The beakers should be placed on insulating mats to reduce the effects of energy transfer by heating.

CHECK**IT!**

1 What do electromagnetic waves carry?

2 How are electromagnetic waves different from other waves, like sound or waves in water?

3 Why are some electromagnetic waves more dangerous than others?

1 **a** Use the two graphs to find the amplitude, period and wavelength of the wave. Draw and label on the graph these three properties of the wave.

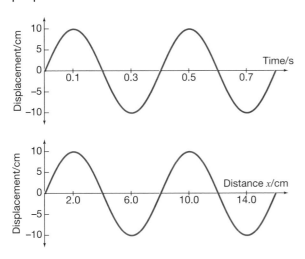

b Use the information you found in part **a** to calculate the speed of the wave.

2 Redraw the diagram below and write the parts of the electromagnetic spectrum in the correct order inside the grid.

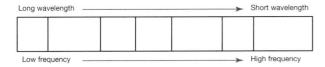

3 Match the types of electromagnetic waves with their applications.

Infrared	Television
X-rays	Satellite communication
Visible light	Electrical heaters
Radio waves	Fibre optics communication
Microwaves	Energy efficient lamps
Gamma-rays	Destroying cancer cells
Ultraviolet	Human internal hard tissue photography

4 Complete the ray diagram to show how the red laser beam enters and exits the glass block.

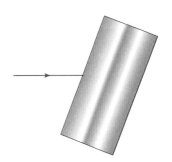

Electromagnetism

Magnetism

Magnetic poles are the places on a magnet where magnetic forces are the strongest.

- Opposite poles attract each other.
- Same poles repel each other.

These forces of attraction and repulsion are examples of **non-contact forces**.

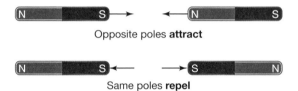

Opposite poles **attract**

Same poles **repel**

Magnetic fields

Permanent magnets generate their own **magnetic fields**.
For example, a bar magnet as shown below.

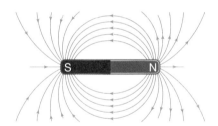

You can see the magnetic fields around this bar magnet

Induced magnets become magnets when they are placed in a magnetic field. For example, when an iron nail touches the pole of a bar magnet. When an induced magnet is removed from the magnetic field, it quickly loses most or all of its magnetism. Induced magnetism always causes a force of attraction.

A magnetic field is the area around a magnet where a magnetic force acts on another magnetic material.

Examples of magnetic materials are iron, steel, cobalt and nickel. The force between a magnet and a magnetic material is always a force of attraction.

To understand how a magnetic field affects magnetic materials, we can draw magnetic field lines around a magnet.

The strength of a magnetic field depends on the distance from the magnet. Magnetic field lines are concentrated at the poles and spread out the further they are from the magnet. This is why the magnetic field is at its strongest at the poles of the magnet, and weaker the further away from the magnet.

Magnetic field lines always point from the north pole to the south pole of a magnet. This is because, if you placed another north pole in the magnetic field, the direction of the force acting on it would be the same as the direction of the magnetic field at that point.

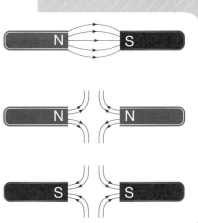

This also explains why like poles attract each other and unlike poles repel each other.

We can plot the pattern of magnetic field lines using small magnetic compasses, which contain a small bar magnet that is free to spin over a needle.

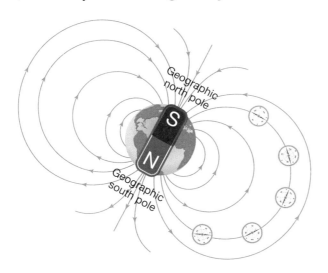

You can see how the compass needle change depending on their position in the magnetic field

The behaviour of the magnetic needle of a magnetic compass shows evidence that the Earth's core must be magnet. The Earth's north geographic pole, is actually a south **magnetic pole**.

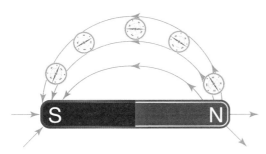

DO IT!

Write a short paragraph explaining why magnetic fields are stronger near the poles. Use the phrase 'numbers of magnetic field lines' at least once in your paragraph.

CHECKIT!

1 What is the difference between a permanent and an induced magnet?

2 How could you find the direction of the magnetic field lines of a bar magnet?

3 Explain why like poles repel each other, using the idea of magnetic field lines.

Motor effect

When a **current** flows through a **conducting** wire, a **magnetic field** is generated around the wire. Imagine gripping the wire in your right hand and pointing your thumb along the wire in the direction of the current. Your fingers curl in the same direction as the magnetic field lines.

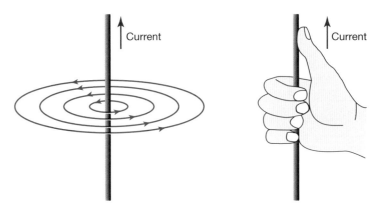

- The **bigger** the current, the **stronger** the magnetic field.

- The **further** the distance from the wire, the **weaker** the magnetic field.

The magnetic field can be made stronger by winding the wire into loops. The more loops there are, the *stronger* the magnetic field. Conducting wire wound into loops is called a **solenoid**.

- Inside a solenoid, the magnetic field is the same everywhere (uniform) and it is strong.

- Outside a solenoid, the magnetic field looks like the magnetic field around a bar magnet.

The magnetic field of the solenoid can be made stronger by wrapping the solenoid around an iron core. A solenoid with an iron core is called an **electromagnet**.

SNAP IT!

Curl the fingers of your left hand as if gripping the solenoid in the above diagram following the direction of the current through the coils of the solenoid pointing your thumb out. Take a photo of your hand. Your thumb should point in the direction of the magnetic field inside the solenoid.

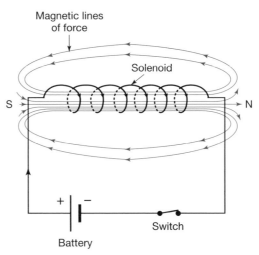

Field lines of the magnetic field through and around a current carrying solenoid

H

Fleming's left-hand rule

When a conductor carrying a current is placed inside the magnetic field of a magnet, the conductor and the magnet exert a magnetic force on each other. This is called the **motor effect**.

We can find the direction of the force generated by the motor effect using **Fleming's left-hand rule**.

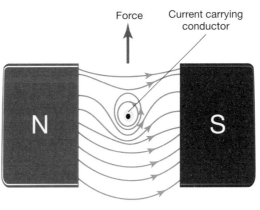

Catapult field

Put the thumb and first finger of your left hand in the shape of an L. Then put your second finger 90° from both your thumb and first finger (all three fingers are 90° from each other). Point each finger in this way:

1 First finger in the direction of the magnetic Field

2 SeCond finger in the direction of the Current

3 The thuMb gives the direction of Movement of the wire (the direction of the force exerted on it).

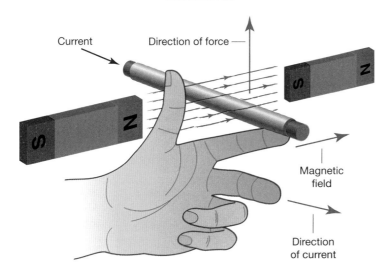

We can calculate the force on a conductor carrying a current at right angles (90°) to a magnetic field using the equation:

$$force = magnetic\ flux\ density \times current \times length$$

$$F = BIL$$

- F = force (unit: **newton**, N)
- B = magnetic flux density (unit: **tesla**, T)
- I = current (unit: **ampere**, or amp, A)
- L = length (unit: **metre**, m)

NAIL IT!

This equation will be given to you, but make sure you know how to use it. As shown by the equation, increasing the magnetic flux density, the current, or the length of the wire will increase the force on the wire.

WORKIT!

Two permanent magnets are put onto a mass scale with unlike poles facing each other. A wire of length 4.0 cm is put between the magnets at 90° from the magnetic field. A current of 0.653 A is applied to the wire. The magnetic flux density is 0.02 T, and the direction of the field and current is such that the force produced on the wire is upward. What mass will the scale read if it was set to zero before the current was turned on? Explain your answer. (4 marks)

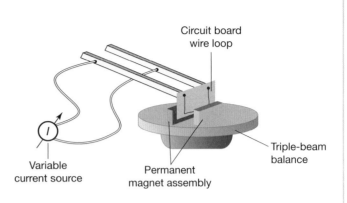

Circuit board wire loop

Variable current source

Permanent magnet assembly

Triple-beam balance

The length of the wire is 0.040 m.

$F = BIl$ ← *Use the equation for the force on a wire carrying a current inside a magnetic field.*

$= 0.02 \times 0.653 \times 0.040$

$= 5.2 \times 10^{-4}$ N (1)

The mass scale measures mass. So, this force will be shown on the scale as the mass of an object of weight 5.2×10^{-4} N. (1) ←

And we can calculate the mass from $W = mg$ and rearrange it to get

$$m = \frac{W}{g} = \frac{5.2 \times 10^{-4}}{9.8} \quad (1)$$

$$= 5.3 \times 10^{-5} \text{ kg}$$

$$= 0.053 \text{ g} \quad (1)$$

Newton's third law tells us that an equal and opposite (downward) force will be applied by the wire on the pair of magnets.

Electric motors

When a coil of wire carries a current in a magnetic field, the coil of wire will rotate. This is how electric motors work:

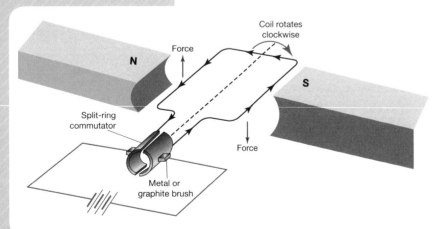

Coil rotates clockwise

Force

N

S

Split-ring commutator

Force

Metal or graphite brush

1 The ends of the coil conductor are connected to the two sides of a 'split-ring commutator'.

2 Each side of the 'split-ring commutator' is connected to the opposite sides of the **dc power supply**.

3 The current in one side of the coil flows in the opposite direction to the current on the other side of the coil.

4 This causes the magnetic field to generate an upward force on one side of the coil and a downward force on the other side, which causes the coil to rotate.

5 When the coil has passed its vertical position, the connections on the 'split-ring commutator' will have swapped polarity, and the forces on the sides of the coil reverse, keeping the coil in rotation.

Loudspeakers

Loudspeakers and headphones use the motor effect to convert changes in the current (signals) in their electrical circuits into changes in the **pressure** of sound waves.

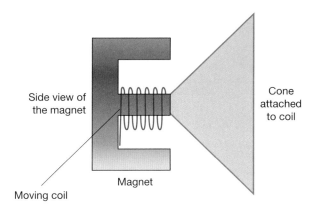

Side view of the magnet

Cone attached to coil

Magnet

Moving coil

1 The music device, for example, an MP3 player, sends an electrical signal with the same **frequency** as the music to the coil conductor.

2 The changes in the electrical current in the coil change the nearby magnetic field.

3 This changing magnetic field interacts with the magnetic field of the permanent magnet, moving the coil backward and forward with the same frequency as the music.

4 The cone attached to the coil is also moved backward and forward with the same frequency.

5 The cone compresses and expands the nearby air with the same frequency, so that sound waves travel through the air particles in the room to your ears.

SNAPIT!

Take a video of a loudspeaker playing some bass music. Use the motor effect to describe what you see happening to the cone of the speaker.

CHECKIT!

1 The current through a wire in a magnetic field is decreased. How does the magnetic force on the wire change?

H2 How can you find the direction of the magnetic force on a wire carrying a current that is placed inside a magnetic field?

H3 A wire carrying 0.210 A experienced a force of 0.046 N when it was immersed perpendicularly in a magnetic field. If the wire is 0.751 m long, what is the magnetic flux density?

1 Draw the magnetic field lines from the two magnets.

2 a Show on the diagram the direction of the current, *I*, the direction of the magnetic field, *B*, and the direction of movement of the wire, *F*.

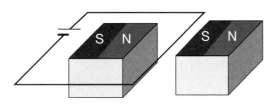

  b Suggest three changes to the system that would increase the force on the wire.

3 Complete the diagrams to show the missing poles, the direction of the magnetic fields and the direction of movement of the wire.

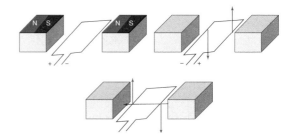

Glossary

Abiotic   An abiotic factor is a non-living condition that can affect where organisms live, e.g. temperature.

Absorb   The process of absorbing substances into cells or across the tissues and organs through diffusion or osmosis.

Abundance   The number of individuals of each species in a sample.

Acceleration   Rate of change of velocity.

Acidic gases   Gases in the atmosphere that can combine with rain water to produce acid rain.

Activation energy   The minimum energy required for a reaction to take place.

Adaptation   Adaptations enable species to survive in the conditions in which they normally live, for example, a cold climate.

Adhesion   The attraction between water molecules and the xylem wall in transpiration.

Aerobic   A process which takes place in the presence of oxygen.

Aerobic respiration   The process of using oxygen to break down glucose to produce energy, making carbon dioxide and water as byproducts.

Air resistance   Frictional force due to air particles hitting a moving object.

Alkali metals   Group 1 in the periodic table, the elements lithium down to caesium. They have similar properties because they all have one electron in their outer shell.

Alkanes   A homologous series of hydrocarbons with the general formula C_nH2_{n+2}.

Alleles   A version of a gene.

Alloys   Mixtures of metals.

Alveoli   Small air sacs in the lungs that are the site of gaseous exchange.

Ammeter   Instrument with virtually no electrical resistance used to measure electric currents.

Ampere   Unit of electric current.

Amplitude   The intensity of a wave, usually measured as the distance between the centre of the oscillation and its peak/trough.

Anaerobic   A type of process which takes place in the absence of oxygen.

Anaerobic respiration   The process of breaking down glucose to produce energy in the absence of oxygen, making carbon dioxide and lactic acid as byproducts.

Angina   Chest pains, often brought on by exercise, as the blood supply to the muscles of the heart is restricted.

Anode   The positive electrode in electrolysis. At this electrode ions lose electrons.

Antibiotic resistance   When bacteria cannot be killed by some or all antibiotics.

Antibiotics   Medicines that kill bacteria, or slow down their growth.

Antibody   A protein that binds to a specific antigen on a pathogen.

Antitoxins   Antibodies that bind to the toxins produced by microorganisms in the body.

Antiviral   Medicines that kill viruses.

Aorta   The artery that carries oxygenated blood away from the heart and the largest artery in the body.

Aqueous solution   A solution in which the solvent is water.

Artificial heart   A mechanical heart that can be used in transplants to aid or replace the heart.

Artificial pacemaker   A small mechanical device that coordinates the resting heartbeat.

Artificial selection   Selective breeding of organisms to produce offspring with the desired characteristics.

Asexual reproduction   A form of reproduction where the offspring are clones of the parent.

Atom   The smallest particle of an element that can take part in a chemical reaction.

Atomic number   The number of protons in the atom of an element.

Automatic control   Processes in the body, controlled by the brain, that are involuntary.

Bacteria   Unicellular, prokaryotic microorganisms.

Balanced symbol equation   An equation where the number and type of r atoms in the reactants are equal to the number and type of atoms in the products. A balanced symbol equation is a consequence of the law of mass conservation.

Bar magnet   A permanent magnet, usually shaped like a bar, or a rod.

Battery   A set of electrical cells connected in series to generate a potential difference.

Behavioural adaptation   Changes to a species' behaviour to help their survival, e.g. penguins huddling together for warmth.

Benign tumour   A growth of abnormal cells, contained in one area, that does not invade other parts of the body.

Bias   A conclusion that may be incorrect.

Bile   A substance produced by the liver that emulsifies fats into smaller droplets.

Bioleaching   Bacteria digest the sulfide in low-grade ores allowing the metal to separate out and be extracted.

Biomass   The total mass of the individuals of a species in a given area.

Biotic   Any living component that affects the population of another organism or the environment.

Bond energy   The energy (in kJ/mol) required to break a covalent bond between two atoms.

Braking distance   The distance a vehicle travels from the moment the brakes are applied until it stops completely.

Bronchi   The two branches from the trachea that lead into the lungs.

Capillary   Small blood vessels that carry blood around the body's tissues.

Capture, release, recapture method   A method of estimating population sizes by capturing organisms, marking them, releasing them and capturing some of them again.

Carbon Capture and Store (CCS)   The storage of carbon dioxide deep under the sea in porous sedimentary rocks especially those near disused oil-wells.

Carbon footprint   The total amount of carbon dioxide and other greenhouse gases emitted during the lifetime of a product. This includes both its production and its disposal.

Carbon monoxide   A toxic gas formed during incomplete combustion of hydrocarbons.

Carbon nanotubes   A hollow tube in which carbon atoms held together by strong covalent bonds.

Carbon off-setting   Measures that reduce carbon dioxide in the atmosphere by planting trees or increasing marine algae in the sea. These organisms carry out photosynthesis which removes carbon dioxide from the atmosphere.

Carbon taxes   Penalties given to companies or organisations which used too much energy or burn excessive amounts of fossil fuels.

Catalyst   A catalyst is a substance that speeds up a chemical reaction and is unchanged chemically and in mass at the end of the reaction. A catalyst speeds up a reaction by helping the reaction proceed via a different pathway which has a lower activation energy.

Cathode   The negative electrode in electrolysis. At this electrode ions gain electrons.

Cell   An electrical component that generates a potential difference.

Cell cycle   The process of division in a cell that duplicates the organelles and the genetic material.

Cellular respiration   The respiration of the cell.

Cellulose cell walls   The cell walls of plant cells.

Central nervous system   The neurones in the brain and spinal cord.

Charge   A physical quantity exerting a force that attracts other unlike charges and repels like charges.

Chemical energy   The energy found in chemicals and biological molecules, e.g. food.

Chemical equilibrium   A dynamic equilibrium in which the rate of the forward reaction equals the rate of the reverse reaction. At equilibrium, properties of the chemical system such as temperature, pressure and concentrations are constant.

Chemical formula   A way of displaying the elements present in a substance and the number of atoms of each element present.

Chemical properties   The properties of a substance that can be seen either during a chemical reaction involving the substance or after the reaction has taken place.

Chemical symbol   Each element has a chemical symbol used to represent it in formulae.

Chemical system   This is the reactants and products of a reversible reaction together inside a closed container.

Chlorophyll   The green pigment in chloroplasts.

Chloroplast   Organelle in plant cells that is the site of photosynthesis.

Chromatography   A separation method used to separate substances in a mixture according to how quickly they rise up paper or another suitable material.

Circuit   A set of electrical components connected by wires to form one or more loops.

Circuit diagrams   An electric circuit represented with drawn symbols and lines.

Coal   A fossil fuel which is a sedimentary rock formed by the compression of dead plants over millions of years.

Cohesion   The attraction between water molecules in the water column in the xylem.

Communities   A group of populations living and interacting with each other in the same area.

Competition   When individuals in a population fight over resources.

Complete combustion   This takes place when a carbon compound burns in enough oxygen to give carbon dioxide and water.

Compound   A substance made of two or more elements are chemically combined.

Compression   Squeezing.

Concentration   The amount in moles or grams of a substance that is dissolved in 1 dm^3 of solution.

Concentration gradient   The difference in the concentration of particles on either side of a partially permeable membrane.

Condensation   When water vapour forms a liquid.

Conduction   The ability of a material to transmit (let through) electric currents. Often used for transmission of energy in the context of thermal conductivity.

Conductor   A material, or component, with high conductivity.

Conjugation   When a bacterium passes on some genetic material to another bacterium.

Conservation of energy   Law of thermodynamics that states energy cannot be created nor destroyed, but it can only be redistributed in different parts of a system, or between systems.

Conservation programmes   Protection of species in their natural habitat or in zoos or botanical gardens to prevent extinction.

Contact forces   Forces between two objects that act when they are touching each other (in contact).

Contract   When the muscle fibres move together to create a high pressure or force.

Coordinator   Usually the brain, controls what will happen in response to the stimulus.

Coronary arteries   Arteries that supply the heart muscle with blood.

Coulomb   Unit of electric charge.

Covalent bond   The bond formed between two atoms by sharing a pair of electrons, one electron coming from each atom.

Crude oil A thick liquid found in the Earth's crust. It is a mixture of hydrocarbons mainly alkanes.

Crystallisation A separation method used to obtain a soluble solid from a solution.

Current Rate of flow of charge.

Deforestation When a large number of trees are chopped down.

Dehydration When the water levels in the body are too low.

Delocalised electrons Electrons that are free to move.

Denature When the hydrogen bonds and other bonds in a protein are broken and the protein no longer functions.

Dendrites A part of a neurone that receives nerve impulses from other neurones.

Density Property of materials that shows the ratio between their mass and the volume occupied by that mass.

Desulfurisation The removal of sulfur from fossil fuels to reduce the amount of sulfur dioxide formed when they burn in air.

Diabetes When either no insulin is produced, or the body stops responding to the insulin in the blood.

Diamond A form of carbon having a giant covalent structure. Each carbon atom is joined to four others by strong covalent bonds.

Diatomic molecules Molecules which contain two atoms. The molecules of hydrogen, oxygen, nitrogen and the molecules of the halogens are all diatomic.

Differentiate When cells change into a specialised cell.

Diffuse reflection Reflection of light from a rough surface that reflects light at many angles.

Diffusion The spreading out of particles resulting in a net movement from an area of higher concentration to an area of lower concentration.

Diode Electrical component that lets current flow in one direction only.

Direct proportionality When the independent variable is doubled the dependent variable also doubles.

Discharged An ion is discharged when it either gains or loses electrons at an electrode during electrolysis.

Disease resistance Organisms that are resistant to disease.

Displacement The shortest way between the initial and the final position. In other words it is the distance with a direction.

Displacement reactions A reaction that takes place when a more reactive element takes the place of a less reactive element in a compound.

Displayed formula This formula shows all the atoms and bonds present in a compound.

Dissipated energy Energy too spread out be used in a useful way

Distribution The way in which a species is arranged in an area.

DNA The molecule that the genetic material is made of.

Dominant The allele that is expressed in the phenotype.

Dose The volume of a medicine that it is safe and effective to take.

Double blind trial When the patient and the doctor do not know which medicine the patient is taking.

Double circulatory system Circulatory system where the blood travels through the heart twice in each cycle.

Double helix The coiled shape of the DNA molecule.

Ductile A substance is ductile if it can be drawn into a wire.

Earth wire Wire connected to the ground as safety for appliances. If there is a surge of current the earth wire is the path of least resistance for the charges to flow through, instead of the user.

Effective collisions An effective collision is one which results in a chemical reaction.

Egg The female sex cell.

Elastic deformation When objects are stretched/compressed and return to their orignial shape when released.

Electrical conductivity The ability of a substance to allow electricity to pass through it.

Electrical impulse A small pulse of electricity that is carried along the neurones.

Electromagnet A magnet that can be switched on and off, as it is magnetic only when current flows through it.

Electromagnetic spectrum The range of electromagnetic waves of different wavelengths and frequencies.

Electromagnetic waves Transversal waves generated by oscillations of electric and magnetic fields that can travel through empty space.

Electron Sub atomic particle carrying negative charge.

Electronic structure or configuration The number of electrons in each electron shell of an atom.

Embryo A developing baby in the uterus.

Embryo screening When embryos are screened for genetic diseases.

Endocrine system The organ system that secretes hormones into the blood.

Endothermic A reaction or change that takes in energy from the surroundings and is accompanied by a decrease in temperature of the surroundings.

Energy The ability to do work e.g. move something, or carry out a reaction.

Energy stores A way to visualise where energy can be stored and measured in a system. They are not physical stores, like containers or boxes, but just a visual representation of a numerical value.

Engulf The ability of phagocytes to take in a pathogen inside itself.

Enzyme Protein that acts as a biological catalyst to speed up reactions.

Epidemiological Data which deals with the incidence, distribution, and other factors relating to health.

Equilibrium When there are the same number of particles on either side of a partially permeable membrane.

Eukaryotic Cells that have genetic material enclosed in a nucleus.

Evaporation The process of changing from a liquid into a gas.

Evolutionary tree   A diagram that shows how closely related two organisms are based on their DNA sequences, physical characteristics and behaviour.

Exothermic   A reaction or change that transfers energy to the surroundings. It is accompanied by an increase in temperature of the surroundings.

Extension   The amount of length an object is extended (or compressed) from its original shape.

Extremophiles   Organisms that live in extreme environments, e.g. volcanoes.

Faeces   The waste from digestion that is removed from the body through the anus.

Fault   A breakage in the circuit that can cause the live wire to be in contact with the outside casing.

Faulty valve   When a valve in the heart no longer functions.

Fermentation   When sugars are broken down to alcohol by microorganisms in anaerobic conditons.

Fertilisation   When a sperm fertilises an egg to produce an offspring.

Fertiliser   Chemical or biological material to improve the growth of plants.

Filtration   A separation method used to separate an insoluble solid from a liquid.

Finite resources   Resources that will run eventually out.

Fleming's left-hand rule   A handy rule to see in with direction a conductor carrying a current will move when it is inside a magnetic field.

Food chain   Feeding relationships within a community.

Food web   Food chains overlap with other food chains to form food webs.

Forces   Pushes or pulls applied between two or more objects. Forces are invisible and unobservable physical quantities, we cannot see forces, but we can measure their effects.

Formulation   A mixture which has components added that are designed to improve the properties or effectiveness of one of the components.

Forward reaction   The reaction taking place when reactants form products.

Fractional distillation   A separation method used to separate two or more miscible liquids with similar boiling points.

Fractionation   The splitting up of crude oil into its constituents or fractions by using fractional distillation.

Fractions   Different parts of crude oil. Each fraction is a mixture of several compounds with similar boiling points.

Frequency (waves)   The number of waves passing through a point in 1 s (the unit of time).

Frequency of collisions   How many collisions that take place per second. It is the frequency of collisions that is a measure of the rate of a reaction rather than how many collisions take place.

Friction   Force between two surfaces rubbing against each other.

Fuel   A substance which burns to give thermal energy or reacts to give electrical energy.

Fullerenes   These can be in the form of spheres or elliptical tubes. An example is the spherical Buckminster fullerene with the molecular formula C_{60}.

Functional adaptation   An adaptation that has occurred through natural selection to overcome a functional problem.

Fungi   A group of organisms that have eukaryotic cells with a chitin cell wall.

Fungicides   A chemical that kills fungi.

Fuse   When two cells or nuclei are joined together.

Fuse (physics)   Small wire with low melting point used as safety precaution in electrical appliances. If the current through the fuse becomes too high, the fuse melts and the circuit is broken.

Gamete   A sex cell.

Gas   A state of matter where a substance has particles freely moving in all directions.

Gas exchange   The exchange of oxygen and carbon dioxide in the lungs.

Gene   A short section of DNA that provides the instructions to make a protein.

Gene therapy   When a faulty allele is replaced with a fuctional allele.

General formula   A general former shows the ratio of carbon to other atoms in a homologous series and represents the composition of the whole series.

Genetic   Variation that comes from your genes.

Genetic counselling   Advice given to people who want to have children and know that they carry a disease allele.

Genetic engineering   The addition of a gene from another organism into an organism's genome.

Genetically modified organisms   Organisms that have genetic material from another organism.

Genome   All of the genes that a person has.

Genotype   The alleles of a gene that a person has in their genome.

Giant covalent structures   A giant structure held together in three dimensions by strong covalent bonds.

Giant ionic structure   A giant three-dimensional lattice of ions held together by strong ionic bonds.

Glucagon   A hormone that causes the liver to breakdown glycogen and fats into glucose.

Glucose   A type of sugar monomer.

Glycogen   An energy storage molecule found in animals.

Graphene   A single layer of carbon atoms, one atom thick from a graphite structure.

Graphite   A form of carbon having a giant covalent structure. The carbon atoms are arranged in layers of hexagonal rings. Each carbon atom is joined to 3 others in the structure by strong covalent bonds.

Gravitational field strength   The value of the gravitational field at a particular place in the field. Near the surface of the earth the gravitational field strength is g = 9.8 N/kg.

Gravitational pull   Gravitational force of attraction between two masses.

Gravity   Force due to the interaction between two masses.

Greenhouse effect   When greenhouse gases build up in the Earth's atmosphere and absorb heat energy from the Sun.

Greenhouse gases   Gases which contribute to the greenhouse effect, e.g. carbon dioxide.

Group   A vertical column of elements with similar properties. The elements have the same number of electrons in their outer shell.

Guard cells   Plant cells on the underside of a leaf that open and close the stomata.

Habitats   The environment in which a species normally lives.

Halide ion   Negative ion with a charge of -1 formed when the atom of a halogen gains one electron.

Halogens   Group 7 of the periodic table– the elements fluorine down to astatine. They have similar properties because they all have seven electrons in their outer shell.

Heart attack   When the blood supply to the muscles of the heart is suddenly blocked, usually by a blood clot.

Heart failure   Muscle weakness in the heart, or a faulty valve, means that the heart does not pump enough blood around the body at the right pressure.

Herd immunity   When many people are vaccinated against a pathogen, to make it less likely that the pathogen will spread through the population.

Hexagonal rings   In graphite and graphene the atoms are arranged in six-membered hexagonal rings of carbon atoms.

Homeostasis   Keeping the internal environment at a set level, despite the external conditions.

Homologous series   A series of compounds with the same general formula, similar chemical properties and same functional group. They show a gradation in physical properties as the number of carbons increases and each member differs from the next one by $-CH_2-$.

Hormone   A protein that travels in the blood to target organs to bring about an affect.

Hydrocarbon   A compound of carbon and hydrogen.

Hyperglycaemia   When the concentration of glucose in the blood is too high.

Hypoglycaemia   When the concentration of glucose in the blood is too low.

In vitro fertilisation   When the egg is fertilised by the sperm in a petri dish.

Inbreeding   When organisms breed and produce offspring with a close family member.

Incomplete combustion   This takes place when a carbon compound burns in an insufficient supply of oxygen resulting in the formation of carbon monoxide or carbon particulates and water.

Inelastic deformation   Stretching/compressing an object that will not return in its original shape after the extension/compression.

Inertia   The resistance of an object (mass) to change its velocity.

Insulators   Objects/materials that are poor conductors of electric currents. Often used for poor conductors of energy in the context of thermal conductivity.

Insulin   A hormone that causes cells to take up glucose from the bloodstream and the liver to store glucose as glycogen.

Interacting   When factors have an effect on each other.

Interaction pairs   Pair of forces caused by the interaction between two objects. When two objects interact they will exert an equal and opposite force on each other.

Interdependence   When species depend on each other for survival.

Intermolecular forces   The weak forces of attraction between molecules.

Internal energy   The total kinetic energy and potential energy of all the particles (atoms and molecules) that make up a system.

Inversely proportional   When one variable increases, the other decreases in proportion.

Ionic bond   The electrostatic attraction between oppositely charged ions in a giant ionic structure.

Ionic equation   An equation describes a chemical reaction by showing the reacting ions only.

Ionised   An atom that has lost or gained one or more electrons in its outer electron shells.

Ionising radiation   Radiation capable of ionising atoms.

Isotopes   Atoms of the same element with the same number of protons and same atomic number but with different numbers of neutrons and therefore different mass numbers.

Joules   Unit of energy.

Kilogram   Unit of mass.

Kinetic   Term referred to moving objects.

Lactic acid   Substance that builds up in the muscles during anaerobic respiration and causes muscle fatigue.

Lamp   electric component that glows when a current flows through it.

Latent heat   The energy needed for a substance to change state.

Law of mass conservation   This states that matter cannot be created or destroyed and it means that a chemical equation must be balanced.

LDR (Light-dependent resistor)   A resistor whose resistance depends on the intensity of light shining on it.

Le Chatelier's principle   This states that when changes are made to a chemical system is at equilibrium then the equilibrium will shift so as to cancel out the effects of the change.

LED (Light-emitting diode)   Electrical component that lets current flow in one direction only and emits light when current flows through it.

Life cycle   The series of changes in the life of an organism including reproduction.

Life cycle assessment (LCA)   An analysis of the environmental impact of a product from its production to its disposal.

Lignin   A waterproof molecule found in the walls of the xylem.

Limit of proportionality   For an elastic material the limit of porportionality is reached when the material has been stretched/compressed so much that it will no longer return to its original shape.

Limiting factor   The factor that is limiting the rate of photosynthesis.

Limiting reactant   The substance that is completely used up when a reaction is complete.

Linear   Arranged in a straight line.

Live wire   Wire in the electric plug that carries the alternating potential difference from the supply to the appliance

Longitudinal waves   Waves whose oscillations are along the same direction (parallel) to the direction of travel of the wave.

Lymphocytes   White blood cells that have a specific immune response.

Magnetic compasses   Small bar magnets suspended and free to rotate, so the compass needle points in the direction of the Earth's magnetic field.

Magnetic field   A force field generated by a magnet.

Magnetic materials   Materials affected by a magnetic force, but that do not generate a magnetic field themselves.

Magnetic poles   The two sides of a magnet where the magnetic forces are strongest.

Magnification   How many times larger an image is compared to the specimen.

Mains electricity   Potential difference and current from our household sockets generated at power stations and carried across the country through the National Grid.

Malignant tumour   A growth of abnormal cells that can spread to different parts of the body and cause secondary tumours.

Malleable   A substance is malleable if it can be hammered into shape.

Mass   Property of all matter. It is the measure of an object's resistance to acceleration when a net force is applied.

Mass number   The total number of neutrons and protons in an atom.

Master gland   A gland that controls all of the other glands.

Medium   Material in which light and other EM waves can travel.

Meiosis   Division of the genetic material to produce four genetically different daughter cells.

Menstrual cycle   The cycle of blood lining in the uterus, ovulation, and blood shedding every 28 days (approximately).

Meristem tissue   Undifferentiated plant tissue that can become different types of specialised cells.

Metal salt   A compound formed when the hydrogen in an acid are replaced by metal. It has a metal part and an acid part.

Metallic bond   The mutual attraction between delocalised electrons and metal cations in a giant metal lattice.

Metals   Elements that form positive ions. Metals tend to be shiny, malleable, ductile and good conductors of heat and electricity.

Metres   Unit of length.

Microvilli   Tiny foldings on the surface of villi in the small intestine to increase the surface area.

Migration   The movement of organisms from one area to another.

Mitochondria   Organelle in the cell that is the site of aerobic respiration.

Mitosis   Division of the genetic material to produce two genetically identical daughter cells.

Mixture   Two or more different substances, elements or compounds mixed but not chemically join together.

Mobile phase   In paper chromatography the mobile phase is the solvent. In gas chromatography it is the carrying gas.

Mole   A mole (abbreviation is mol) is the amount of a substance that contains 6.023×10^{23} (Avogadro's number) particles of that substance.

Momentum   Mass x velocity of an object.

Motor neurone   Neurone that carries nerve impulses from the brain.

MRSA   A strain of *Staphylococcus aureus* that is resistant to many antibiotics.

Muscle cells   The cells that muscles are made of.

Mutations   A change in the order of bases in the DNA sequence.

Natural selection   The process whereby organisms better adapted to their environment tend to survive and produce more offspring.

Negative ions   Ions formed by the gaining of electrons, they move towards the anode in electrolysis.

Nerve impulse   An electrical impulse that travels along the neurones.

Neurones   The cells of the nervous system.

Neutral wire   It completes the circuit in an appliance plugged to the mains.

Neutralisation   A reaction between an acid and an base or alkali to form a neutral solution of pH7.

Neutralises   Makes a neutral pH by adding an acid and an alkali together.

Neutron   A subatomic particle found in the nucleus the atom. A neutron has no charge and has a relative mass equal to that of a proton.

Newton   Unit of force.

Newton's first law   Newton's law describing the tendency of objects to remain in their state of motion.

Newton's second law   Newton's law describing the relationship between the resultant force applied on an object, its mass and acceleration.

Newton's third law   Newton's law describing the relationship between two objects interacting with each other.

Noble gas arrangements   These are the electron arrangements of the noble gases. They are important because they are stable electron arrangements.

Non-communicable disease   Disease that is not infectious.

Non-contact forces   Forces between objects acting at a distance, i.e. without the need for them to touch each other.

Non-linear   Arranged in a non-linear pattern, i.e. not in a straight line.

Non-metals   Apart from the noble gases these elements tend to form negative ions and have properties which are different to metals. For example they are poor conductors of heat and electricity and solid non-metals are brittle.

Non-ohmic A component that doesn't obey Ohm's law, i.e. its resistance will change with the potential difference applied across it.

Non-renewable An energy resource that cannot be replenished.

Non-specific defences General defences by the body to prevent all pathogens from entering the body.

Nucleus (biology) Organelle that contains the genetic material.

Nucleus (chemistry/physics) The centre of the atom containing the protons and neutrons (the nucleons).

Oestrogen A female hormone that causes the development of the sex characterisitcs in women.

Ohm Unit of electrical resistance.

Ohmic conductor A conductor whose resistance remains constant for constant temperature.

Oral contraceptives A pill containing hormones to prevent pregnancy.

Ore A rock that contains enough of a metal to make it worthwhile to use it as a source of that metal.

Organ A collection of several tissues that work together to carry out a particular function.

Organ system A collection of several organs that work together to carry out a particular function.

Oscillation The vibrations of an object, or particles, generating a wave.

Osmosis The movement of water from a dilute solution to a concentrated solution through a partially permeable membrane.

Ovaries Sex organ in women that produces an egg every month for fertilisation.

Ovulation The time in a month when an egg is released from an ovary.

Oxidation The gain of oxygen or the loss of electrons in a reaction.

Oxidised A substance is oxidised when it gains oxygen or loses electrons.

Oxygen debt When the lactic acid is broken down using oxygen after the exercise is finished.

Oxygenated blood Blood that contains high volumes of oxygen.

Parallel Two lines or planes that never touch.

Parasites An organism that lives on anther organism without giving any benefit.

Partially permeable membrane A membrane that allows small molecules to diffuse across.

Particulates Small particles of carbon which cause global dimming (reduction of sunlight) and respiratory problems.

Pathogens A microorganism that causes disease.

Penicillin The first antibiotic that originated in a fungus called *Penicillium*.

Period (periodic table) The horizontal rows of elements in the periodic table. As you go across a period an electron shell is filled up. The period number of an element is the number of its occupied electron shells.

Period (waves) The time taken for a single wave to pass through a point.

Periodic table A table of the elements arranged in order of their atomic number. Elements with similar properties are in vertical columns are called groups.

Peripheral nervous system The neurones that connect the limbs to the spinal cord.

Permanent magnet A magnet that doesn't lose its magnetism and doesn't need a current to become a magnet.

Permeable When water or other small molecules can diffuse across a membrane.

Pesticides A chemical that kills pests.

pH A log scale of hydrogen ions which indicates if a solution is acidic, neutral or alkaline.

pH scale A scale which runs from 0 to 14 and shows the acidity (value less than 7) or alkalinity (value greater than 7). pH 7 is neutral.

Phagocytosis The process of a phagocyte engulfing a pathogen.

Phenotype The expressed characteristics of an organism.

Photosynthesis The light-driven process by which plants and photosynthetic bacteria produce glucose and oxygen from carbon dioxide and water.

Physical properties These properties can be measured or observed without changing the composition of a substance.

Phytomining Plants take up a metal from slag or waste heaps that contain low-grade ores. The metal can then be extracted from the plants.

Pituitary gland The master gland in the brain that controls all of the other glands.

Placebo A fake medicine that looks like the real medicine.

Plasmolysed When a plant cell loses too much water and the cell membrane comes away from the cell wall.

Polarity The property of having poles or being polar.

Pollution Contamination of the natural environment that can cause damage to species.

Polymers Long chains of carbon atoms in a repeating pattern formed from smaller molecules called monomers.

Pooter A device for tapping small flying insects.

Population A group of organisms of the same species living in the same area at the same time.

Positive ions These are ions formed when atoms lose electrons. They migrate towards the cathode in electrolysis.

Potable water Water that is safe to drink.

Potential difference (p.d.) The difference of electrical potential between two points.

Potential stores Energy stores associated with the position of an object (gravity store), or the chemical composition of a substance (chemical store).

Power The rate of transfer of energy, i.e. energy transferred in one second.

Precipitation When water droplets fall as rain or snow.

Predator An organism that eats other organisms.

Predator prey cycle The interaction of prey and predator species.

Pressure The force per unit area, i.e. the force applied on each square metre of surface.

Prey   An organism that gets eaten by another organism.

Products   The substances being formed in a chemical reaction. These are written on the right-hand side of a chemical equation.

Progesterone   A hormone that maintains the uterus lining during pregnancy.

Prokaryotic   Cells that do not have a nucleus.

Protein   A polymer made of many amino acids.

Proton   A subatomic particle found the nucleus the atom. A proton has a positive charge equal to the negative charge of an electron and a relative mass of one equal to the relative mass of a neutron. A hydrogen ion is sometimes called a proton.

Pulmonary artery   Artery that carries deoxygenated blood from the heart to the lungs.

Pulmonary vein   Vein that carries oxygenated blood from the lungs to the heart.

Pulse rate   The number of heart beats per minute.

Punnett square   A genetic diagram used to predict the genotypes of offspring.

Pure   A pure substance can be either a single element on its own or a single compound on its own.

Quadrat   A 1 m by 1 m square made of wire, which may be divided into smaller squares.

Radiation   Emission of energy as electromagnetic waves or as moving subatomic particles.

Random   With no particular pattern, of unpredictable outcome.

Rate of reaction   How quickly a reaction takes place. It is measured by either how quickly reactants disappear or by how quickly products appear.

Ray diagram   Drawing showing the path of light rays.

Reactants   The substances taking part in a chemical reaction. Written on the left-hand side of a chemical equation.

Reaction profile   This shows how the energy changes during the course of reaction when reactants react to form products.

Reactivity series   The order of reactivity of metals. It is usually quoted from potassium as the most reactive down to gold as the least reactive.

Receptor   An area of specialised cells that detect a stimulus.

Recessive   An allele that is not expressed when a dominant allele is present.

Recycling   A way of making new materials from products which have come to the end of their lifetime.

Reduced   A substance is reduced when it either loses oxygen or gains electrons.

Reduction   The loss of oxygen or the gain of electrons in a reaction.

Reflex arc   An involuntary series of nerve impulses that do not involve the brain.

Relative atomic mass   The mass of an atom compared to 1/12 the mass of an atom of carbon-12. The relative atomic mass of an element also takes into account the relative abundances of its naturally-occurring isotopes.

Relative charges and masses   Because the charge on an electron and a proton are so small it is easier to compare them and use relative charges. Similarly is easier to use the relative masses of the neutron and proton.

Relative formula mass   The sum of the relative atomic masses in the formula of a substance.

Relay neurone   A neurone that carries nerve impulses from the sensory neurones to the motor neurones in a reflex arc.

Renewable energy   A way of producing energy that wil not run out in the future.

Renewable resources   Resources, usually from plants, that can be replaced very quickly.

Resistance   Ability to oppose something, e.g. resistance to movement, resistance to a flow of current.

Resistant strain   A strain of bacteria that are resistant to one or more antibiotics.

Resistor   A component designed to control the current through a circuit by applying a resistance to it.

Respiration   The process of using oxygen to break down glucose to produce energy, making carbon dioxide and water as byproducts.

Resultant force   The sum of all the forces acting on an object.

Reversible reaction   A reaction that can proceed in both directions. Reactants can form products in the forward reaction and products can form reactants in the reverse reaction.

Ripple tank   A device used to observe water waves generated in a tank.

Room temperature and pressure (RTP)   20°C and 1 atm. pressure. 1 mol of a gas occupies $24dm^3$ ($24\,000cm^3$) at room temperature and pressure.

Salivary gland   A gland in the mouth that produces saliva.

Saturated hydrocarbon   A hydrocarbon which has the maximum number of hydrogen atoms present in its structure.

Scalar   Physical quantity that does not have a direction.

Secondary sex characteristics   Changes to the human body that occur during puberty that do not involve the sex organs.

Secondary tumours   A tumour formed when a malignant tumour moves from its first location.

Selective breeding   When two organisms are bred together to produce offspring with desired characteristics.

Sensory neurone   A neurone that carries nerve impulses to the brain.

Series   Components connected in the same loop with the source of potential difference.

Sex chromosomes   The chromosomes that control whether an embryo develops into a male or female.

Sexual reproduction   Reproduction that involves male and female gametes.

Sieve plate   The end plate of the cells in the phloem containing small pores for the sugar sap to pass through.

Simple distillation   A separation method used to separate a liquid from a soluble solid.

Smelting   Extraction of a metal by roasting with carbon. The metal is less reactive than carbon.

Solenoid   A long coil of wire used as an electromagnet.

Solute   The solid that dissolves in a liquid to form a solution.

Solvent   The liquid that dissolves a solid to form a solution.

Specialised exchange surfaces   Membranes that are adapted to allow fast diffusion.

Species   A group of similar induviduals that can breed together to priduce fertile offspring.

Specific heat capacity   The amount of energy required to raise the temperature of one kilogram of the substance by one degree Celsius.

Specific latent heat   The amount of energy required to change the state of one kilogram of the substance with no change in temperature.

Specific latent heat of fusion   Specific latent heat in the change of state from solid to liquid.

Specific latent heat of vaporisation   Specific latent heat in the change of state from liquid to gas.

Speed   Distance travelled in the unit of time.

Spinal cord   A collection of neurones that connect the peripheral nervous system and the brain.

Spring constant   Property of elastic materials that determine how stiff and strong they are.

Standard form   Way of writing down very large or very small numbers.

Stationary phase   The stationary phase in paper chromatography is paper. In thin-layer chromatography it is the silica on the plate.

Stem cells   Cells that are not differentiated and can become any type of cell.

Stomach   Organ in the body that uses acid and enzymes to digest food.

Stomata   Small pores on the underside of a leaf that allow gases to enter and exit the leaf.

Store   Referred to energy store, which is a part of the system where energy can be measured and stored.

Stroboscope   A device used to make flashes of light matching the frequency of waves to observe them more easily.

Strong acid   An acid which is completely ionised in water. An example is hydrochloric acid.

Structural adaptation   Adaptations to the body of the organism.

Sub-atomic particles   The particles that make up an atom. These are the electron, neutron and the proton.

Sub-cellular structures   The structures and organelles within cells.

Sugar   Monomers of polysaccharides, e.g. glucose.

Sulfur dioxide   Formed when sulfur burns in air. It causes respiratory problems and when dissolved in water it forms acid rain.

Surface area   The area of a solid that is in contact with other reactants, either in solution or in the gas phase.

Powders have a larger surface area than lumps. The smaller particle the larger the surface area.

Sustainable development   This meets the needs of present development without depleting natural resources.

Switch   A device that can break and close a circuit.

Symbol equation   A way to represent the substances reacting and those being formed in a chemical reaction using chemical formulae. Both sides of the equation are joined by an arrow on which can be written the conditions used.

Synapse   The small gap between two neurones.

System   An object or group of objects.

Tangent   This is a straight line drawn to touch a reaction curve at a certain point. The slope or gradient of the tangent is a measure of the rate of reaction corresponding to that point.

Tension   Force due to the pulling of a rope, string, wire, etc

Tesla   Unit of magnetic flux density.

Testes   Sex organ in men that produces sperm for fertilisation.

Testosterone   A hormone that causes male sexual characteristics to develop.

Thermal   Relating to or caused by temperature.

Thermal conductivity   The ability of a substance to allow heat energy to pass through it.

Thermal decomposition   The breaking up of a compound using heat. Reactions which are thermal decompositions are endothermic processes.

Thermistor   Electrical component whose resistance changes with temperature.

Thinking distance   The distance travelled by a vehicle while the driver reacts to a potential danger they spot before they apply the brakes.

Thyroid gland   A gland in the body that produces thyoxine.

Thyroxine   A hormone that increases the rate of metabolism, generating heat.

Tissue   A collection of similar cells that work together to carry out a particular function.

Toxins   Chemicals released by pathogens which cause disease in the infected organism.

Transect   A long line made with string or a measuring tape, with quadrats placed at intervals along it.

Translocation   The movement of sugar sap up and down the plant.

Transpiration   The loss of water from the top part of the plant.

Transverse waves   Waves whose oscillations are perpendicular to the direction of travel of the wave

Tumour   A group of abnormal cells that divide uncontrollably.

Turgid   When a plant cell swells up with water.

Unsaturated hydrocarbon   A hydrocarbon which has less than the maximum number of hydrogen atoms present in its structure. Examples are the alkenes. Unsaturated compounds decolourise bromine water.

Urea   A less toxic by-product of the breakdown of ammonia, that is lost from the body in urine.

Urine   A liquid lost from the body made of urea, and excess salt and water.

Vaccination   When a small amount of dead virus or antigen is introduced into a body in order to produce an immune repsonse.

Valves   Small flaps in the heart and veins to prevent the backflow of blood.

Variation   Genetic differences between individuals of the same species.

Vector (biology)   An animal that carries a pathogen, or a small amount of DNA that is used to introduce a gene into a new organism.

Vector (physics)   A quantity having direction as well as magnitude.

Velocity   Speed in a direction.

Vena cava   The vein that carries deoxygenated blood to the heart and the largest vein in the body.

Villi   Folding of the inner membrane of the small intestine to increase the surface area.

Visible spectrum   Part of the EM spectrum detectable by the human eye.

Visking osmometer   Visking tubing with a glass tube in the top to measure the amount of osmosis.

Vitamin A   A vitamin that is needed by the eye to make the pigment needed for night vision.

Volt   Unit of potential difference.

Voltmeter   Instrument used to measure the potential difference across two points.

Volume   The space occupied by an object, or gas.

Water cycle   The cycle of water from the sea, to rain, to rivers and then back to the sea.

Watt   Unit of power.

Wave front diagrams   Drawings used to show how waves propagate through media.

Wavelength   Distance between two peaks or two troughs in a wave.

Weak acid   An acid which is only partially ionised in water. An example is ethanoic acid. Ethanoic acid is only 0.4% ionised in aqueous solution.

Work done   Force x distance travelled in the direction of the force.

Yield   The volume of product that is produced.

Zygote   The fertilised egg.

Answers

Cell biology
Review It!

1 No nucleus; no membrane bound organelles; has plasmids.

2 **a** Place the specimen onto the stage; Line up the objective lens and the eyepiece lens; and focus the specimen by moving the stage with the coarse focus.

   b $30\,000 \div 10; = \times 3\,000$

3 **a** Root hair cells have long thin hairs; and do have chloroplasts; The thin hairs increase the surface area to take up water and minerals from the soil.

   b Meristem tissue; in the root differentiates; into root hair cells.

4 **a** Weigh the mass of several cucumber slices; place each cucumber slice in a different concentration sugar solution for 24 hours; weigh the mass of the cucumber slices after being in solution.

   b The independent variable is the concentration of the sugar solutions. The dependent variable is the change in mass of the cucumber slices.

   c 40; 26; 14; −14

   d
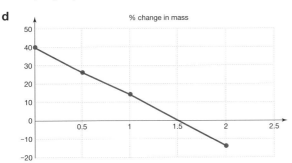

   X axis labelled as concentration of sugar solution (%); y axis labelled as percentage change in mass (%); points plotted correctly; points connected with a straight line.

   e 0%

   f Keep all of the control variables the same; repeat the investigation.

5 **a** Prophase; Metaphase; Anaphase; Telophase.

   b Anaphase

   c Because onion root tips are growing; Root tips contain meristem tissue.

6 **a** Undifferentiated cells; that can become any type of specialised cell.

   b Embryos / umbilical cords; Adult organs contain a few stem cells.

   c No rejection of cells / organs by patient; Transfer of viral infection; No waiting time for transplants; Ethical / religious objections.

7 **a** Diffusion is the movement of particles from an area of high concentration; to an area of low concentration.

   b The salt ions will move towards B.

   c The salt ions are more concentrated on the A side of the membrane; and less concentrated on the B side of the membrane.

8 **a** Facilitated diffusion.

   b In diffusion, particles move from an area of high concentration to an area of low concentration, but in active transport, the particles move from an area of low concentration to an area of high concentration; Diffusion is a passive process, but active transport requires energy.

Tissues, organs and organ systems
Review It!

1 **a** Any two from: Amylase / carbohydrase; pepsin / protease / trypsin; lipase

   b Sugars

   c The enzyme's active site is complementary to the shape of the substrate; The enzyme binds to the substrate; to make a product.

   d The enzyme would become denatured.

2 **a** Add Benedict's reagent; and place into a hot water bath for a few minutes; If sugar is present, the solution will change to green / yellow / orange / brick-red.

   b Add iodine solution to the solution; If starch is present, the solution will turn blue-black.

   c Add Buiret's reagent to the solution; If protein is present, then the solution will turn lilac.

3 **a** **i** Vena cava   **ii** Pulmonary vein

   b The atria contract to move the blood into the ventricles; The valves in the heart prevent the backflow of blood.

   c A small area of specialised cells in the right atrium acts as a pacemaker.

4 **a** Down the trachea; down the bronchi; then the bronchioles; and then into the alveoli.

   b The gaseous exchange surface is one-cell thick to allow for a short diffusion pathway. There are many alveoli to give a large surface area; The blood carries the oxygen away from the alveoli to create a steep concentration gradient for oxygen.

   c $10 \div 0.4 = 25$ mm/s

5 **a** The coronary artery is blocked; and oxygen cannot reach the muscle of the heart.

   b Stents hold the coronary arteries open, and allow blood flow to the heart, and statins reduce blood cholesterol levels by slowing down the rate of fatty material deposit; However, neither of these is a permanent solution. The fatty material will eventually build up if the patient does not eat a healthy diet, and the artery will need to be bypassed.

6 a To allow oxygen and carbon dioxide to diffuse in and out of the leaf; To allow water vapour to diffuse out of the leaf.

b i

Temperature	Number of open stomata			Mean number of open stomata
	1	2	3	
10	14	16	12	14
20	34	36	38	36
30	49	51	44	48
40	87	78	81	82

ii Add °C to column header of first header.

iii The higher the temperature, the higher the number of open stomata.

iv The rate of photosynthesis increases with increased temperature; More carbon dioxide will be needed for photosynthesis, so the stomata are open to allow the carbon dioxide in.

c Increased temperature; increased light; and increased wind speed.

Infection and response

Review It!

1 a A microorganism that causes disease.

b Any two from: Tuberculosis; tetanus; cholera; Salmonella food poisoning; gonorrhoea.

c Any two from: Airborne / coughing and sneezing; direct contact with infected person / animal / sharp object; in food and water.

d Antibiotics

2 a Bacteria are much larger than viruses; bacteria are living cells, whereas viruses are not living / envelopes surrounding some genetic material.

b i Red skin rash; fever.

ii Vaccination; isolating infected people from non-infected people.

c i Tobacco mosaic virus.

ii TMV causes large black patterns to appear on the leaves; that prevents the leaves from photosynthesizing; This means that the plant produces less sugars and has less energy for growth.

3 a Protist / plasmodium.

b i The number of deaths from malaria decreases worldwide from 2000 to 2015; The number of deaths from malaria is highest in Africa.

ii Increased use of mosquito nets / pesticides / antimalarial drugs; The mosquito that carries the protist that causes malaria survives very well in the climate in Africa.

4 a Any two from: Skin is a physical barrier against pathogens; breaks in the skin form scabs; sweat glands in the skin produce sweat that inhibits pathogens; small hairs and mucus in the nose trap airborne particles; mucus in the trachea traps pathogens and is moved up to the throat by small hairs called cilia; hydrochloric acid and protease in the stomach kills pathogens; phagocytes kill any pathogen that invades the body.

b Phagocytes are non-specific and lymphocytes are specific; Phagocytes engulf pathogens by phagocytosis and digest the pathogen; Lymphocytes produce antibodies that attach to pathogens and prevent them from entering cells / target the pathogen for phagocytosis / act as an antitoxin; Lymphocytes have memory cells that remember the specific pathogen and respond more quickly if the same pathogen enters the body again.

5 a i A and B kill approximately the same number of bacteria; C kills more than three times the number of bacteria that A and B kills.

ii The bacteria have become resistant to antibiotics A and B; Antibiotic C is a stronger antibiotic than A and B.

b Bacteria eventually become resistant to antibiotics; The antibiotics become less effective at treating infections.

Bioenergetics

Review It!

1 a $C_6H_{12}O_6$

b Any two from: Used in respiration; converted into glycogen for storage; used to produce fat for storage; combined with nitrogen to make amino acids.

c Animals store glucose as glycogen, whereas plants store glycogen as starch. Animals use glucose to produce fat for storage whereas plants use glucose to make oils for storage.

2 a The breathing rate increases to oxygenate the blood more quickly. The breath volume increases to take in more oxygen with each breath. The heart rate increases so that the blood flows to the cells more quickly.

b Measure the pulse at the wrist or neck for 15 seconds. Multiply this number by 4 to get the pulse rate per minute.

c After anaerobic exercise, there is a build up of lactic acid in the muscles. The lactic acid is transported to the liver where it is converted into glucose using oxygen.

3 a Any example that describes the formation of a complex molecule from one or more simpler ones, or the breaking of a complex molecule into simpler ones.

b i Glucose + oxygen $\rightarrow$ carbon dioxide + water

ii Unbalanced:
$C_6H_{12}O_6 + O_2 \rightarrow CO_2 + H_2O$

Balanced:
$C_6H_{12}O_6 + 6O_2 \rightarrow 6CO_2 + 6H_2O$

4 a Glucose and Oxygen

b Any two from: Increasing light intensity; increasing temperature increasing concentration of carbon dioxide

c When the rate of photosynthesis remains constant as one of the factors that affects the rate of photosynthesis, is limiting photosynthesis.

5 a Award two marks for shape of graph as below.

i

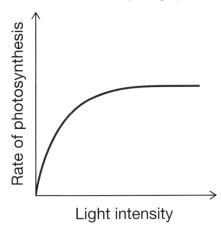

ii X axis labelled as light intensity.

Y axis labelled as rate of photosynthesis.

b As the light intensity increases, the rate of photosynthesis increases, until the rate of photosynthesis remains constant at increasing light levels because the rate of photosynthesis is being limited by a limiting factor / temperature / carbon dioxide concentration.

6 a From top to bottom of rate of photosynthesis column: 40, 30, 10, 0

b 30°C

c i As the temperature increases, there is more kinetic energy and the enzymes and substrates involved in photosynthesis are more likely to collide.

ii After 30°C, the enzymes involved in photosynthesis start to denature. The active site of the enzyme can no longer bind to the substrate.

d Repeat the investigation at least two more times.

Homeostasis and response
Review It!

1 a Any two from: Sensory neurone; motor neurone; relay neurone.

b A voluntary nerve response goes to the brain whereas a reflex does not; Reflexes are faster than voluntary nerve responses; Reflexes involve the relay neurone whereas voluntary nerve impulses do not.

c i No; because there is not enough data to come to this conclusion.

ii Any two from: They should repeat their investigation; They should use a range of volumes of caffeine; They should test more people.

2 a Homeostasis keeps the internal conditions of the body the same; whatever the outside environment may be.

b When blood glucose concentration increases, the pancreas releases insulin; insulin causes cells to take up glucose, reducing the blood glucose concentration; insulin causes glucose to be converted into glycogen in the liver and muscle cells.

H C When blood glucose concentration decreases, the pancreas releases glucagon; glucagon causes glycogen in the liver and muscle cells to be converted into glucose and released into the blood; this increases the blood glucose concentration.

d i *Any two from*: In type 1, no/little insulin is produced; type 1 treated by injecting with insulin; In type 2, the body does not respond properly to insulin. *Accept other correct differences*

ii Blood glucose level would increase after a meal; and remain high for a number of hours.

iii There is little or no insulin [type 1]/body does not respond to insulin [type 2], so cells do not take up glucose from the blood; this causes the blood glucose concentration to remain high after eating.

3 a FSH stimulates the maturation of eggs in the ovary.

b i FSH stimulates many eggs to be released by the ovary; to be fertilised by IVF.

ii Advantages – Baby will be genetic offspring of parents; mother will give birth to her own child; Disadvantages – Success rates are not high; multiple births; emotionally / physically stressful

4 a When a person is under stress, scared or excited.

b Stimulates the metabolic rate.

c If the thyroxine levels in the body are low, the hypothalamus stimulates the thyroid gland to secrete more thyroxine; if the thyroxine levels in the body are high, the hypothalamus stimulates the thyroid gland to stop producing thyroxine.

Inheritance, variation and evolution
Review It!

1 a Reproduction without another parent/no fusion of gametes; using mitosis.

b There are two daughter cells made by mitosis, and four by meiosis; The daughter cells made by mitosis are genetically identical, whereas the daughter cells made by meiosis are genetically different; Daughter cells made by meiosis have half of the genetic information of daughter cells made by mitosis.

c A male and a female gamete join together and the two nuclei fuse; When this happens, the new cell / zygote has a full set of chromosomes; The zygote divides by a type of cell division called mitosis.

2 a A short section of DNA that codes for a particular protein.

b A large structure in the nucleus made of many genes.

c The allele that is expressed in the phenotype.

d Homozygous means that both alleles are the same; heterozygous means that the two alleles present are different.

3 a Genetic; Environmental.

b There is variation within a species. Individuals with characteristics that are more suitable for the environment are more likely to survive; and have offspring with these same characteristics.

c Disease resistance; size.

d Insert a gene for insect resistance into the plant which makes the plant produce an insect toxin; the insect pests die when they eat the plant containing the insect toxin.

4 a Each species has a two part name; describing their genus and species.

b Kingdom – Animal; Genus – Canis.

c Bones/shells have not decayed; parts of the dead organism are replaced by minerals as they decay; so traces of an organism e.g. bones/footprints/leaf prints are preserved.

d Some bacteria are resistant to an antibiotic; these bacteria survive and produce offspring; the offspring share the antibiotic resistance.

5 a i

	B	b
B	BB	Bb
b	Bb	bb

ii 1:2:1

b 50% probability.

H6 a The introduction of a gene from one organism; into another organism.

b The desired gene is cut out the original genome using enzymes called restriction enzymes; The same enzymes are used to cut open a plasmid; The gene is inserted into the plasmid; The plasmid is inserted into the bacterial cells.

c Could produce medicines more quickly / cheaply; Could insert a functional gene into a cell with a faulty gene.

Ecology
Review It!

1 a A group of organisms of the same species living in the same area at the same time.

b A group of populations living and interacting with each other in the same area.

c Any two from: temperature; light intensity; moisture levels in the soil / air; pH of the soil; wind intensity / direction; carbon dioxide levels; oxygen levels of the water.

2 a Used in photosynthesis; to make sugars.

b Aerobic respiration; and combustion.

c water in the oceans is warmed by the Sun and evaporates; in the atmosphere, the water condenses into clouds; the clouds are transported inland by the wind, and the clouds rain/precipitate onto the land; the water moves into streams/rivers back to the oceans.

3 a i Global temperatures are increasing; due to the increased amounts of greenhouse gases / carbon dioxide / methane in the atmosphere.

ii Water availability is decreasing; as higher temperatures causes desertification of some areas.

iii Carbon dioxide in the air increases; as photosynthesis decreases due to fewer trees / plants.

b Global weather patterns will change, causing decreased availability of food and water; sea levels will rise, decreasing habitats; increased extinction of species as some species will not be able to adapt quickly to the changing climate.

4 a Biodiversity is the variety of living organisms on the earth or in an ecosystem.

b The human population is increasing; more pollution/ sewage/toxic chemicals; land is cleared to make room for houses/quarries/farming/dumping waste; fewer habitats and food sources for species.

c Conservation/breeding programmes; using renewable energy/using less energy; reducing waste/recycling waste; *any other sensible answer*

5 a Intraspecies competition.

b As the number of hares increases, the number of lynx increases after a short delay; after the number of lynx increases, the number of hares decrease; this decreases the number of lynx, which in turn increases the number of lynx.

6 a i Total number of buttercups – 34

ii $34 \div 10 = 3.4$;

Area = 10m × 15m = 150m^2 ;

Estimated number of buttercups = 3.4 × 150 = 510 buttercups

b $\dfrac{25 \times 18}{10} = 45$

Atomic structure and the periodic table
Review It!

1 a $2Na(s) + Cl_2(g) \rightarrow 2NaCl(s)$

b i Both have only 1 chemical symbol and therefore only 1 type of atom.

ii Na is a shiny silver white solid; Cl is a pale green gas.

iii Ions

c Because the sodium chloride is not chemically combined with the water and they can be easily separated by physical means.

2 The group is a vertical column of elements and a period is a horizontal row

3 Each element can only have one atomic number and that number is unique to that element. If it had an atomic number of 12 it would not be sodium.

4 Group 6, period 3.

5 a X is found in the middle of the periodic table in the transition elements.

b Shiny; good electrical conductor; good thermal conductor; malleable; ductile; denser than the group 1 elements.

6 a -1

b When they react the group 7 elements gain one electron to form a stable outer shell, their reactivity depends on their ability to gain this extra electron. As the group descends, the outer shell is further from the positively charged nucleus and is shielded from the nucleus by an increasing number of electrons. This means that as you go down the group the attractive force on an electron being gained gets less and it gets harder to capture the extra electron.

7 a It has 12 protons and 12 electrons.

b These are isotopes. Each isotope has the same number of protons but a different number of neutrons.

c Let there be 100 atoms of gallium. 60 atoms have a mass number 69 with a total mass of 4140 amu. 40 atoms have a mass number of 71 the total mass of 2840 amu.

Therefore 100 atoms have a total mass of 4140 + 2840 amu = 6980 amu. The relative atomic mass is the average mass of each atom = $\frac{6980}{100}$ = 69.8 amu

8 Elements with similar properties were placed in vertical columns and ordered by their relative atomic masses. Where the known elements did not fit the pattern he left spaces for elements which had not yet been discovered.

9 Group 1 elements lose their outer electron when they react. As the group is descended, this outer electron is further from the attractive force of the nucleus and there are more shielding electrons between the nucleus and the outer electron. This means that the outer electron feels less of an attractive force and is more easily lost therefore making the lower elements more reactive.

10 The noble gases have stable full outer electron shells. This means they do not have to gain or lose electrons to become stable.

11 a Mendeleev's periodic table was organised in groups of elements with similar properties. If argon had very distinct properties then it had to fit into its own group and therefore they had to be a group of elements with similar properties.

b Because it did not react with any other elements.

12 a As the group is descended the elements get darker in appearance. Iodine is a dark grey solid; astatine is below iodine and would be darker in colour which suggests that it is black.

b At_2

c NaAt

d -1. The ion is At^-.

Bonding, structure and the properties of matter
Review It!

1 The lithium atom loses its outer electron to form the Li^+ ion. The ion has a stable full outer electron shell.

2 a $\left[Mg \right]^{2+}$ $\left[\overset{\bullet\bullet}{\underset{\bullet\bullet}{\overset{x}{Cl}}} \right]^-$

b $MgCl_2$

c The ionic bonds between the magnesium and chloride ions are very strong and because it is a giant structure all the bonds have to be broken. This requires lots of energy and a high melting point.

d In solids the ions are not free to move, therefore they cannot carry the current and do not conduct electricity.

3 a

b Methane is a neutral molecule and the intermolecular forces between methane molecules are weak and require a small amount of energy to break them. Therefore methane has low melting and boiling points making it a gas at room temperature.

4 a Giant covalent structure

b The bonds between the carbon atoms in both diamond and graphite are strong covalent bonds, all these bonds have to be broken. This requires lots of energy, so the melting point is high.

c The bonds in the layers of graphite are strong covalent bonds but between the layers the intermolecular forces are weak and easily broken allowing the layers to slide over each other easily.

d In graphite each carbon is bonded to 3 other carbons leaving a spare electron. These spare electrons are delocalised in the layer and can carry an electric current making graphite a good electrical conductor. In diamond there are no spare electrons or charged particles making it a poor conductor.

5 A – Simple molecular

B – Giant ionic

C – Giant metallic

D – Giant covalent

6 a Methane has a simple molecular structure with weak intermolecular forces so it has low melting and boiling points. Potassium chloride has a giant ionic structure with strong ionic bonds between the ions. All these bonds need lots of energy to break them and therefore it has high melting and boiling points.

b The ions in magnesium oxide are Mg^{2+} and O^{2-}. In potassium chloride they are K^+ and Cl^-. The larger charges on the Mg^{2+} and O^{2-} means that their ionic bonds are stronger than those between the K^+ and Cl^- ions, these need more energy to break and therefore magnesium oxide has a higher melting point.

c Zinc ions are larger than copper ions in the giant metallic lattice, this means that the layers of ions cannot slide over each other as easily, making the alloy a harder material.

d Sodium has a giant metallic structure in which there are delocalised electrons in both the solid and liquid states and these delocalised electrons can carry an electric current. This means that sodium is a good electrical conductor in both the solid and liquid states. Sodium chloride has a giant ionic structure. In the solid state the ions are not free to move and cannot carry an electric current so as a solid sodium chloride is a poor conductor. In the liquid state they can move and carry the current making sodium chloride a good electrical conductor.

Quantitative chemistry
Review It!

1 a i 8.33×10^{-2} ii 2.23×10^5
      iii 8.561×10^2 iv 4.53×10^{-5}

  b i 4.00 ii 6.57×10^{-2}
      iii 4.55×10^{-2} iv 4.39×10^{-4}

2 a **i** $H_2(g) + Cl_2(g) \rightarrow 2HCl(g)$

    ii $2Na(s) + Br_2(l) \rightarrow 2NaBr(s)$

    iii $6K(s) + N_2(g) \rightarrow 2K_3N(s)$

    iv $Mg(s) + 2AgNO_3(aq) \rightarrow Mg(NO_3)_2(aq) + 2Ag(s)$

    v $4Na(s) + O_2(g) \rightarrow 2Na_2O(s)$

b The law of conservation of mass states that the mass of the reactants = mass of products; this means that the number and type of atoms on left-hand side of the equation must be the same as those on the right-hand side.

3 a 96   **b** 74
   c 148   **d** 61
   e 174   **f** 134.5   **g** 60

H 4 **a** 44

   b 0.1

   c $0.1 \times 6.02 \times 10^{23} = 6.02 \times 10^{22}$

   d $0.1 \times 24\ dm^3 = 2.4\ dm^3$

5 a Atom economy method I $= \frac{44}{173} \times 100\% = 25.4\%$

   Atom economy method II $= \frac{44}{44} \times 10\% = 100\%$

   b Reduces waste

H 6 $HCl(aq) + NaOH(aq) \rightarrow NaCl(aq) + H_2O(l)$

Volume of NaOH $= \frac{30}{1000} = 0.03\ dm^3$ Vol of HCl $= 0.02\ dm^3$

No. of moles of NaOH $= C \times V = 1 \times 0.03\ mol = 0.03\ mol$

No. of mol of HCl = no. of mol of NaOH $= 0.03\ mol$

Concentration of HCl $= \frac{n}{V} = \frac{0.03}{0.02} = 1.5\ mol/dm^3$

H 7 **a** $\frac{6}{24} = 0.25\ mol$

   b No. of mol of HCl $= C \times V = 1 \times \frac{200}{1000} = 0.2\ mol$

   c From the equation 1 mol of magnesium reacts with 2 mol of HCl. Therefore 0.25 mol of magnesium react with 0.5 mol of HCl. There are only 0.2 mol of HCl and this is the limiting reactant.

Chemical changes

Review It!

1 a copper, iron, zinc, aluminium, magnesium

   b **i** $\rightarrow$ zinc sulfate(aq) + copper(s)

     ii $\rightarrow$ NO REACTION

     iii $\rightarrow$ aluminium oxide(s) + iron

   c **i** magnesium(s) + carbon dioxide(g) $\rightarrow$ magnesium oxide(s) + carbon(s)

     ii $2Mg(s) + CO_2(g) \rightarrow 2MgO(s) + C(s)$

     iii The magnesium gains oxygen – oxidation and the carbon dioxide loses oxygen – reduction

     iv **I** magnesium oxide

       II carbon

2 a The gas, hydrogen, is produced which can be tested for using a lighted splint. The gas pops.

   b The copper is less reactive than hydrogen and will not displace it from the acid.

H c **i** $Zn(s) + 2H^+(aq) \rightarrow H_2(g) + Zn^{2+}(aq)$

    ii The zinc loses electrons when going from the neutral Zn to the positive Zn^{2+} ion. This is oxidation. The H^+ ions gain electrons when forming H_2 and this is reduction.

3 a Hydrogen at cathode; bromine at anode.

   b A solution of potassium hydroxide

4 a Its solution is a weak alkali

   b Phenol solution is a very weak acid

Energy changes

Review It!

1 It decreases

2 A is an endothermic reaction because it is a thermal decomposition reaction.

B is an exothermic reaction because it gives out heat to warm up the food.

3 a **i** 23; 41; 7

    ii The units for temperature /°C.

   b Use the same amount of acid (same volume/same concentration), the metals should have the same surface area (e.g. all 3 are powders), use either the same calorimeter/reaction vessel or identical ones.

   c Thermometer or temperature datalogger, top-pan balance, well insulated calorimeter, spatula.

   d Least reactive X, Z, Y Most reactive. X gives the lowest temperature rise then Z than Y which gives the greatest temperature rise.

4 a A Heat of reaction or energy change for reaction

    B reactants   C Activation energy

    D Products   E Energy

    F Course of reaction

   b Exothermic

   c The energy required for the reaction to take place.

H 5 **a** $4C-H, 1C == C, 1\ Cl-Cl \rightarrow 4C-H, 1C-C, 2C-Cl$

The 4 C−H bonds are unchanged and can be omitted from the calculation.

   1×610 1×245
   1×350 2×345

The energy taken in to break bonds $= 610 + 245 = 855\ kJ$

   b The energy given out when forming bonds $= 350 + 690 = 1040\ kJ$

   c **i** 185 kJ/mol

    ii The energy given out is greater than the energy taken in, therefore the reaction is exothermic.

Rates of reaction and equilibrium

Review It!

1 Measure the volume of carbon dioxide produced with time. Measure the loss in mass as time progresses.

2 a Measure the gradient of the tangent to the graph at any point.

   b Rate $= \frac{25}{15}\ cm^3/s = 1.67\ cm^3/s$

3 a C because it has the steeper gradient at the beginning.

   b C because increasing the temperature increases the rate.

c As the temperature increases the particles collide more frequently and with greater force, so the frequency of effective collisions increases and so does the rate.

4 a i A catalyst speeds up a chemical reaction and is unchanged chemically at the end of the reaction.

ii 0.10 g

b

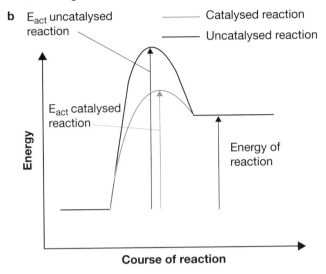

c A catalyst lowers the activation energy and this means that more particles have enough energy to react so that when they collide the collision is more likely to produce a reaction.

5 Temperature and surface area of a solid if one is involved. If two or more solutions are involved, then one concentration can be varied (the independent variable) whilst the others are kept constant.

6 a The $\rightleftharpoons$ sign shows that the reaction can proceed both ways

b It is exothermic

c It turns blue and there is heat given out

17 a The equilibrium will shift to the right.

b An increase in pressure favours the side with fewer gas molecules and this is the left-hand side.

Organic chemistry
Review It!

1 a Crude oil is composed of a mixture of miscible liquids with similar boiling points.

b i The fractions become less viscous

ii The boiling points decrease

iii The fractions become easier to light

c The fractions contain smaller molecules and this means that there are weaker intermolecular forces, therefore the molecules are more easily separated to become gases.

2 a C_nH_{2n+2}

b i $CH_4(g) + 2O_2(g) \rightarrow CO_2(g) + 2H_2O(l)$

ii $C_2H_6(g) + 3\frac{1}{2}O_2(g) \rightarrow 2CO_2(g) + 3H_2O(l)$

c i A = cobalt chloride paper;
B = limewater; C = to pump

ii The cobalt chloride paper changes form blue to pink. This shows that water is formed.

iii The limewater goes cloudy showing that carbon dioxide is formed.

3 a The amounts produced of the fractions with large molecules are more than required. At the same time the amounts of the fractions with smaller molecules are less than required. Cracking converts the larger molecules into smaller ones.

b i C_6H_{14} **ii** C_9H_{20}

Chemical analysis
Review It!

1 a Effervescence/fizzing

b i Yellow precipitate

ii White precipitate

iii Cream precipitate

iv White precipitate

c i White precipitate

ii No change/no reaction

Chemistry of the atmosphere
Review It!

1 Carbon dioxide, water vapour, methane, nitrogen and ammonia

2 Dissolving in the water forming the oceans. Uptake by plankton in the sea which then form their shells; they are compressed by sediments and form limestone. Photosynthesis by plants. Plankton covered by sediments in the absence of oxygen form oil. Plants covered by sediment then compressed form coal

3 78% nitrogen; 21% oxygen; the remaining 1% consists of noble gases and approximately 0.04% of carbon dioxide

4 Carbon dioxide and methane absorb infrared radiation and then are re-radiated back to Earth, warming up the atmosphere.

5 The recent rises in carbon dioxide levels are mirrored by increased temperature in the atmosphere.

6 The increased global temperature causes the ice caps to melt, thus increasing the water in the oceans and the water levels rise given rise to floods. Also severe storms lead to greater rainfall and flooding.

7 The total amount of carbon dioxide emitted over the lifetime of an activity or product.

8 • Energy conservation will reduce the amount of carbon dioxide produced by burning fossil fuels.

• The use of alternative energy resources will also reduce the amount of carbon dioxide produced by burning fossil fuels.

• In carbon capture and storage carbon dioxide produced in power stations is pumped into deep underground porous rocks at the sites of exhausted oil wells.

• Carbon taxes penalise people/companies/organisations that use too much energy and this inhibits people from overuse of energy.

• Carbon offsetting is when plants are planted, taking carbon dioxide through photosynthesis thus reducing the amount of carbon dioxide in the atmosphere.

9 • People are reluctant to change their lifestyle. For example, they still use large cars which consume more energy.

   • Countries do not cooperate with each other.

   • Some countries still believe that global warming is a natural phenomenon and is not caused by humans.

   • People are still unsure of the facts and the consequences of global warming.

   • Countries still find it economical to use fossil fuels

10 An atmospheric pollutant is something that is introduced into the atmosphere and has undesired or unwanted effects.

11 a i CO

   ii By the incomplete combustion of carbon-containing fuels.

   iii It is toxic because it reduces the amount of oxygen getting to the brain.

   b i Oxides of nitrogen are formed by the reaction between nitrogen and oxygen at high temperatures.

   ii For example, in car engines and exhausts and in thunderstorms.

12 Sulfur dioxide is formed by the reaction of sulfur and oxygen. It dissolves in water to form acid rain and it causes respiratory problems.

Using resources

Review It!

1 a A finite resource will run out but a renewable one can be replaced.

   b The fermentation of sugar is the sustainable process because the sugar can be regrown again whilst the ethene comes from the cracking of crude oil which is a finite resource.

2 a i Large objects are screened out of the water

   ii Aluminium sulfate is added to make small particles clump together and settle to the bottom of the tank

   iii It contains bacteria

   iv Chlorine is added OR the water is treated with UV light to kill bacteria

   b Distillation or reverse osmosis

   c i Evaporate off the water

   ii The boiling point

   d i In the absence of oxygen

   ii Methane and fertilisers

3 a i Roasting with carbon

   ii Plants which absorb copper are planted on sites where there are low-grade copper ores. After they have grown they are harvested and burned to leave copper deposits in the ashes.

   iii Bacteria use low grade copper sulfide ore in heaps for an energy source and they oxidise the ores. The liquids leaching from the heaps contain copper ions.

   b Electrolysis and displacement of the copper by adding a more reactive metal such as iron.

4 a A life cycle assessment is an analysis of the environmental impact of a product at each stage of its lifetime from its production all the way to its disposal.

   b The extraction/production of raw materials. The production process – making the product, including packaging and labelling. How the product is used and how many times it is used. The end of the life of the product – how is it disposed of at the end of its lifetime. Is it recycled?

Energy

Review It!

1 Renewable: water waves, bio-fuel, hydroelectricity, the Sun, geothermal, the tides.

   Non-renewable: nuclear fuels, coal, oil, natural gas.

2 a *kinetic energy = 0.5 × mass × (speed)²*

   $E_k = \frac{1}{2}mv^2 \rightarrow m = 2\frac{E_k}{v^2}$

   b *elastic potential energy = 0.5 × spring constant × (extension)²*

   $E_e = \frac{1}{2}ke^2 \rightarrow e = \sqrt{2\frac{E_e}{k}}$

   c *g.p.e = mass × gravitational field strength × height*

   $E_p = mgh \rightarrow h = \frac{E_p}{mg}$

3 $\Delta\theta$ is a temperature difference, and degrees Celsius and kelvin have the same magnitude; so using degrees Celsius or kelvin will affect the result of the calculation because the taking difference gives the same result; For example, the temperature increase between 2°C and 20°C is $\Delta\theta = 20 - 2 = 18°C$; which expressed in kelvin becomes $\Delta\theta = 293.15 - 275.15 = 18$ K.

4 a Lift B has greater power than lift A; because it can lift the same load the same height in less time; The rate of energy transferred is higher for lift B.

   b Energy transferred is

   $E_p = mgh$ and $P = \frac{E}{t} = \frac{mgh}{t}$

   $h = \frac{Pt}{mg} = \frac{343 \times 15}{70 \times 9.8} = 7.5\ m$

   So, each floor is $\frac{7.5}{3} = 2.5$ m high.

5 Aluminium has a higher thermal conductivity than plastic; so it will transfer energy by conduction to the ice cube at a higher rate than the plastic block; so the ice cube on the aluminium block will melt faster than the cube on the plastic block.

6 First we need to calculate the energy in the kinetic store when the car is at top speed.

   $E_k = \frac{1}{2}mv^2 = \frac{1}{2}0.520 \times 2.2^2 = 1.26$ J

   So, at top speed the toy car transfers 1.26 J to the kinetic store each second; This means a useful power output of 1.26 W; So, to find the efficiency of the toy car:

   Efficiency $= \frac{Useful\ power\ output}{Total\ power\ input}$

   $= \frac{1.26}{1.5} = 0.84$

   We can multiply by 100 to find the percentage efficiency, i.e. 84%.

Electricity

Review It!

1

Light Dependent Resistor		It is a resistor that changes its resistance depending on how much light is shone on it.
Light Emitting Diode		It lets current flow through it in one direction only, but when current flows through it emits light.
Lamp		It has a very thin filament (wire), usually made of tungsten, which gets really hot when a current flows through it, so it glows and emits visible light.
Variable Resistor		It opposes the flow of current with a certain resistance and it can be set to different values of resistance within a specific range.
Battery		It is a collection of cells connected in series with each other.
Thermistor		It is a resistor that changes its resistance depending on its temperature.

2 a Electric current is the rate of flow of electrical charge, i.e. the amount of charge passing a point of a circuit in a second.

 b If the ammeter had a resistance larger than zero, the current in the circuit would decrease; so the ammeter would not read the real value of current through the circuit.

3 $t = \frac{Q}{I}$

4 a An I-V graph would have I on the y-axis and V on the x-axis; so we would need to rearrange the equation to:

   $I = \frac{1}{R}V \iff y = mx + 0$

   The gradient of this line graph

   would, therefore, be $m = \frac{1}{R}$.

 b $I = \frac{V}{R} = \frac{9}{12000} = 7.5 \times 10^{-4}\ A$

 c Between 140–150 Lux of light intensity.

 d First we find the current using

   $I = \frac{V}{R} = \frac{4.5}{2500} = 0.0018\ A$

   The charge flowing through a point in a certain time can be calculated using:

   $Q = It = 0.0018 \times 90 = 0.162\ C$

Particle model

Review It!

1 a kg/m³

 b i Ethanol is less dense than olive oil; so it will float on top of it.

   ii The ratio $\frac{m_e}{m_o} = 0.97$

 c $r = \frac{m}{V} = \frac{0.437}{\pi r^2 h} = \frac{0.437}{\pi \times 0.012^2 \times 0.120}$

   $= 8050\ kg/m^3$

2 a Latent heat of fusion is between solid and liquid; latent heat of vaporisation is between liquid and gas.

 b $E = mL = 0.185 \times 59000$

   $= 10915\ J$

3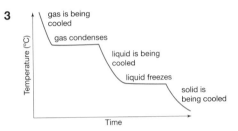

4 Increasing the temperature of the gas increases the average energy in the kinetic store; As the volume of the gas does not change, the increase in the kinetic store causes more collisions of particles per second and the force from each collision will also be greater due to the increased speed; and both effects produce an increase in the pressure of the gas.

5 The pressure on the piston is the same as the atmospheric pressure outside the syringe; so the pressure of the air inside the syringe must also be the same.

   When the temperature of the gas is higher (in hot water) the pressure of the air inside the syringe will increase; so the piston will rise until the volume of air increases enough to allow the gas pressure to rise to atmospheric pressure.

   When the temperature of the gas is low (icy water) the pressure of the air inside the syringe will decrease; and the piston will drop until the volume of air reduces sufficiently to allow the gas pressure to rise to atmospheric pressure.

6 a $p_1V_1 = p_2V_2 \iff p_2 = \frac{p_1V_1}{V_2}$

   $= \frac{101325 \times 20}{55} = 36845\ Pa$

 b We had to assume that the temperature remained constant.

Atomic structure

Review It!

1 a $^{23}_{11}Na \rightarrow$ 11 electrons, 11 protons and 12 neutrons

 b $^{14}_{7}N \rightarrow$ 7 electrons, 7 protons and 7 neutrons

 c $^{235}_{92}U \rightarrow$ 92 electrons, 92 protons and 143 neutrons

 d $^{208}_{84}Po \rightarrow$ 84 electrons, 84 protons and 124 neutrons

 e $^{9}_{4}Be \rightarrow$ 4 electrons, 4 protons and 5 neutrons

 f $^{14}_{6}C \rightarrow$ 6 electrons, 6 protons and 8 neutrons

2 Bohr suggested that electrons orbited around the nucleus of atoms at specific distances, while in Rutherford's model electrons could orbit at any distance from the nucleus.

3 **a** Half-life of element A = 2 days; and half-life of element B = 3.5 days.

 b The count rate per minute drops to 6 after approximately 7 days, which is 3.5 half-lives.

 c After two half-lives ¼ is left; The half-life of B is 3.5 days, so it takes two half-lives = 7 days to drop to ¼ the original value.

4 Radioactive contamination is the presence of radioactive atoms within other materials and irradiation is the exposure of an object to nuclear radiation.

Forces

Review It!

1 **a**

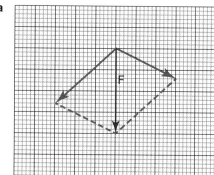

 b

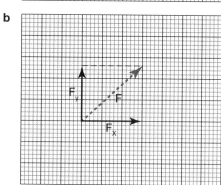

2 Due to Newton's 3rd law, the Earth and the Moon exert forces on each other equal in magnitude and opposite direction:

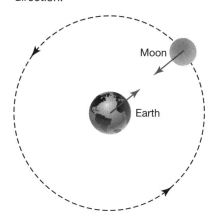

3 The yellow arrow represents the resultant force in this situation; Resultant force = 1784 N to the right.

4 Work done is W = Fs, so the distance travelled will be:

$$s = \frac{W}{F} = \frac{378652}{2 \times 523} = 362 \text{ m}$$

5 s = vt, so:

$$v = \frac{s}{t} = \frac{30}{3.78} = 7.9 \text{ m/s from}$$

stationary $v = \frac{30}{3.95} = 7.6$ m/s

from the starting gun.
There is a difference of 0.17 s because of Usain Bolt's reaction time.

6 We need to calculate the resultant force first 950 N – 320 N = 630 N; Newton's 2nd law states that the resultant force is F = ma, so:

$$a = \frac{F}{m} = \frac{630}{81.5} = 7.7 \text{ m/s}^2$$

Waves

Review It!

1 **a**

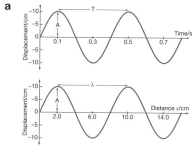

 b $v = f\lambda = \dfrac{0.080 \text{ m}}{0.4 \text{ s}} = 0.2$ m/s

2

Long wavelength						→ Short wavelength
Radio waves	Microwaves	Infrared	Visible light	Ultraviolet	X-rays	Gamma rays
Low frequency						→ High frequency

3

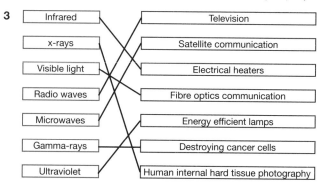

4

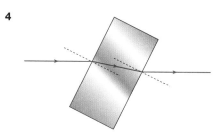

Electromagnetism

Review It!

1

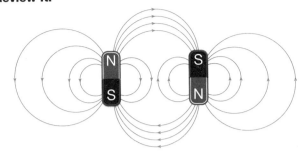

2   a

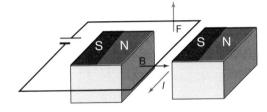

b   Increase the current through the wire by adding
more cells in series; increase the magnetic field
by using stronger magnets or moving them closer
together; increase the number of loops in the wire.

3

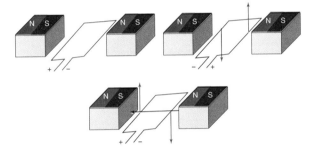

REVIEW IT!

Index